Public Space

In both the UK and the US there is a sense of dissatisfaction and pessimism about the state of urban environments and particularly with the quality of everyday public spaces. Explanations for this have emphasised the poor quality of design that characterises many new public spaces; spaces that are dominated by parking, roads infrastructure, introspective buildings, a poor sense of place, and which in different ways for different groups are too often exclusionary.

Yet many well designed public spaces have also experienced decline and neglect, as the services and activities upon which the continuing quality of those spaces depends have been subject to the same cuts and constraints as public services in general. These issues touch upon the daily management of public space, that is, the coordination of the many different activities that constantly define and redefine the characteristics and quality of public space.

This book draws on four empirical research projects to examine the questions of public space management on an international stage. They are set within a context of theoretical debates about public space, its history, contemporary patterns of use, and the changing nature of Western society; and about the new management approaches that are increasingly being adopted.

Matthew Carmona is Professor of Planning and Urban Design and Head of UCL's Bartlett School of Planning. He is an architect-planner with research interests that span the policy context for delivering better quality built environments. His books are: *The Design Dimension of Planning* (1997), *Housing Design Quality* (2001), *Delivering New Homes* (2003), and *Measuring Quality in Planning* (2004), all published by Routledge, and *Public Places Urban Spaces* (2003), and *Urban Design Reader* (2007).

Claudio de Magalhães is Senior Lecturer at UCL's Bartlett School of Planning, and Director of the MSc in Urban Regeneration. He is an architect-planner, with research interests focusing on the relationship between urban governance, urban design, public space and an increasingly globalised property market. He has published widely on these topics including his book *Urban Governance, Institutional Capacity and Social Milieux* (2002).

Leo Hammond took his MPhil Town Planning at UCL's Bartlett School of Planning, following which he was appointed as Research Fellow. Subsequently he has worked as an urban designer in private practice and in the public sector.

Public Space

The management dimension

Matthew Carmona

Claudio de Magalhães

Leo Hammond

Routledge
Taylor & Francis Group

LONDON AND NEW YORK

First published 2008
by Routledge
2 Park Square, Milton Park, Abingdon, Oxon OX14 4RN

Simultaneously published in the USA and Canada
by Routledge
270 Madison Avenue, New York, NY 10016, USA

Routledge is an imprint of the Taylor & Francis Group, an informa business

© 2008 Matthew Carmona, Claudio de Magalhães and Leo Hammond

Typeset in Optima by HWA Text and Data Management Ltd, London
Printed and bound in Great Britain by TJ International Ltd, Padstow, Cornwall

British Library Cataloguing in Publication Data
A catalogue record for this book is available from the British Library

Library of Congress Cataloging in Publication Data
Carmona, Matthew.
  Public space : the management dimension / Matthew Carmona, Claudio de Magalhães, Leo Hammond.
    p. cm.
  Includes bibliographical references and index.
  1. City planning – Great Britain. 2. Public spaces – Great Britain. 3. Sociology, Urban – Great Britain.
  4. City planning – New York – New York. 5. Public spaces – New York – New York. 6. Sociology, Urban.
  I. Magalhães, Claudio de. II. Hammond, Leo.   III. Title.
  HT169.G7C347 1980
  307.1'2160941--dc22   2007046451

ISBN10: 0-415-39108-3 (hbk)
ISBN10: 0-415-39649-2 (pbk)
ISBN10: 0-203-92722-2 (ebk)

ISBN13: 978-0-415-39108-5 (hbk)
ISBN13: 978-0-415-39649-3 (pbk)
ISBN13: 978-0-203-92722-9 (ebk)

Contents

Illustrations

Acknowledgements

Profound thanks are due to all those who have engaged in different capacities with the research projects on which this book is based. This amounts to many hundreds of individuals, and far too many to thank individually here. Nevertheless thank you for your time, input and wisdom. Without you the book would not have been possible.

Particular thanks go to Simon Pinnegar and Rachel Conner at ODPM (now DCLG), Rachael Eaton and Edward Hobson from CABE, to all our international partners, to John Hopkins, Jeremy Caulton, Hayley Fichett, Doshik Yang, to colleagues at Ipsos MORI, and most of all to Ruth Blum.

The authors wish to thank the following individuals and organisations for permisstion to reproduce copyright illustrations which appear in the book:

Adam Clevenger – Creative Commons Attribution, Figure 2.1
Adrian Pingstone – Creative Commons GFDL, Figure 10.2
Aromano – Creative Commons, Figure 10.13
ArtkraftStrauss LCC, Figures 2.18 and 2.19
British Museum Images, Figure 2.9
Bureau of Urban Development Tokyo Metropolitan Government, Box 7.1
CABE, Boxes 5.1 and 5.2
Creative Commons GFDL, Box 8.8
Dillif – Creative Commons GFDL, Figures 2.8 and 2.12
Doshik Yang, Figure 3.15

The Duke of Bedford and the Trustees of the Bedford Estates, Figure 2.10
FOLP.free.fr, Box 8.3
Georges Jansoone – Creative Commons GFDL, Figure 2.4
Guildhall Library, City of London, Figure 2.6
Historical Publications Ltd, Figure 2.11
Hsuyo – Creative Commons License, Figure 2.16
Ipsos MORI, Figure 1.1
Karen Atwell, Box 7.3
Kirsten Mingelers STAD BV, Box 8.4
Leonardo Cassarani – Creative Commons GFDL, Figure 10.8
Lou Stoumen Trust, Figure 2.20
Louis Hellman, Figure 3.5
Medienserver Hannover & Region, Box 8.2
Minneapolis Park and Recreation Board, Box 7.2
Museum of London, Figure 2.7
News International Limited, Figures 2.13 and 2.14
Peter Braig, Parks Victoria, Box 8.7
Steven Tiesdell, Figures 2.21, 2.24 and 3.21
Surprise Truck – Creative Commons, Figure 10.7
The New-York Historical Society, Figure 2.17
Tim Delshammar, Box 8.6
Times Square Alliance, Figure 9.1
Wellington City Council, Box 8.5
Xenones – Creative Commons License, Figure 2.2

Part ONE

Conceptualising public space and its management

Most need improving in this local area

*Q Thinking about this local area, which of these things, if any, do you
think most need improving? Again you may choose up to 5.*

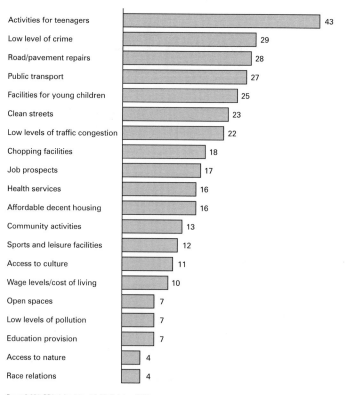

Activities for teenagers	43
Low level of crime	29
Road/pavement repairs	28
Public transport	27
Facilities for young children	25
Clean streets	23
Low levels of traffic congestion	22
Chopping facilities	18
Job prospects	17
Health services	16
Affordable decent housing	16
Community activities	13
Sports and leisure facilities	12
Access to culture	11
Wage levels/cost of living	10
Open spaces	7
Low levels of pollution	7
Education provision	7
Access to nature	4
Race relations	4

Base: 2,031 GB adults, 15+, 18–22 October 2001
Source: MORI

1.1 Liveability: a top priority

Chapter 1

The use and nature of public space

This first chapter introduces the concept of public space and seeks to explore the complexity of both public space as a concept, its use and users, and the management of public space as an aspiration and set of activities. The chapter is in three parts. In the first section, the inspirations and objectives underpinning the writing of the book are presented in order to establish the purpose of the book, and equally its limitations. A brief overview of how the book is structured is included here. This is followed by a second section in which public space is deconstructed. This is done in order to draw out and understand the physical and human components of urban public space, in other words, the subjects of management. The third section draws out and discusses the welter of roles and responsibilities for actually managing public space.

The chapter begins the process of unpacking (at least conceptually) the issues that provide the focus for the rest of the book.

The book

Inspirations and objectives

In recent years there has been considerable and growing interest amongst academics worldwide concerning the role of public spaces in urban life. Works emanating from disciplines such as geography, cultural studies, politics, criminology, planning and architecture have tried to define and explore that role, and understand current changes and their consequences. In part, it would seem, this interest was sparked by the almost complete absence of interest in the subject amongst the policy community in many parts of the world in the last decades of the twentieth century, and the impact this disinterest has had.

But recent research has demonstrated that people place the quality of their local environment high on the agenda of issues that concern them and most need improving, and often higher than the 'headline' public services such as education and health (MORI 2002 – Figure 1.1).

This reflects the fact that people use the street outside their front door, their local neighbourhood and the environment around their workplace on a daily basis, and as a result, the quality of streets, parks and other public spaces affects everyone's daily life, and directly contributes to their sense of wellbeing.

Yet, in many parts of the world, considerable evidence has been gathered to demonstrate a shared sense of dissatisfaction and pessimism about the state of urban environments, particularly with the quality of everyday public spaces. Explanations for this dissatisfaction have emphasised the poor quality of design that characterises many new public spaces; spaces that are typically dominated by parking, roads infrastructure, introspective buildings, a poor sense of place, and which in different ways, for different groups in society, are often exclusionary.

However, the research upon which this book draws suggests that this is not the whole story. Many contemporary and historic spaces are well designed but have nevertheless experienced decline and neglect. In part this is because the services and investment upon which the continuing quality of those spaces depends have been subject to the same constraints and pressures as public services in general. Changes in the roles of the state and civil society, of government and the governed, shifts in modes of provision of public services, and so forth, have all played a part. These issues touch upon the management of public space, and reflect the impact (positive or negative) of the many different activities that constantly define and redefine the characteristics and quality of that space.

The basis of the book

The book draws upon four empirical research projects as well as a wide body of literature to examine questions of public space management on an international stage. The first project examined the management of everyday urban public spaces in England, the second, the management of green parks and open spaces in eleven cities around the world, the third, three iconic public spaces in New York and London, and the fourth, real users' perceptions and aspirations for public space in England. The empirical research is set within a context of theoretical debates about public space, its history, contemporary patterns of use and its changing nature in Western society, and about new management approaches that are increasingly being adopted as a response to public space problems in an evolving urban governance scenario.

In undertaking the research over a period of five years, the authors have become increasingly aware that despite the many critiques of public space, its generation and evolution, and despite the voluminous tomes on how to design new public space, relatively little academic literature exists on the subject of its long-term management. In a very real sense, public space management has been a forgotten dimension of the policy discourse, perhaps because so many of the solutions are, on the face of it, quite prosaic: designing with maintenance in mind; regular street cleaning; coordinating management responsibilities; and so forth. Yet, proper management, or the absence of it, can impact in a profound way on the key urban qualities that other policy areas increasingly espouse: connection; free movement; provision of social space; health and safety; public realm vitality; and the economic viability of urban areas.

The four projects were an attempt to understand these issues. In reporting on them, the book addresses one of the big cross-disciplinary debates: how to deal more effectively with the quality of public spaces? In the process it aims to forward a range of practical and sometimes more fundamental solutions to better manage public space.

Defining public space ... and the research limitations

Unfortunately, debates about public space are situated within a literature characterised by a host of overlapping and poorly defined terms: liveability, quality of place, quality of life, environmental exclusion/equity, local environmental quality, physical capital, well-being, and even urban design and sustainability. These are all concepts that overlap and which are often used as synonyms, but equally are frequently contrasted, or used as repositories in which almost anything fits (van Kamp *et al.* 2003: 6; Brook Lyndhurst 2004a).

Broadly, the different concepts owe their origins to different policy-making traditions, each being multi-dimensional and multi-objective. Thus Rybczynski (1986, cited in Moore 2000) describes them as being like an onion: 'It appears simple on the outside, but it's deceptive, for it has many layers. If it is cut apart there are just onion-skins left and the original form has disappeared. If each layer is described separately, we lose sight of the whole'. To add to the complexity, some aspects are clearly subjective, related to the way places are perceived and to how individual memories and meanings attach to and inform perception of particular places. Others are objective, and concerned with the physical and indisputable realities of place (Massam 2002:145; Myers 1987: 109).

Van Kamp *et al.* (2003: 11) usefully distinguish between the various concepts by arguing that some are primarily related to the environment, whilst others are primarily related to the person (liveability and quality of place being in the former camp, and quality of life and well-being in the latter). Moreover, some concepts are clearly future-oriented (i.e. sustainability), whilst others are about the here and now (i.e. liveability and environmental equity).

What is clear is that the quality of the physical environment, and therefore physical public space and space as a social milieu, relates centrally to each of these, yet each is also much broader than a concern for public space management. In this regard, defining public space too widely may result in a nebulous concept that is difficult for those charged with its management to address. Conversely, defining the concept too narrowly may exclude important areas for action which, once omitted from policy, may undermine the overall objective of delivering better managed public space.

Debates about the nature and limits of public space will be discussed in some depth later in the book (see in particular Chapters 2 and 3), but for the purposes of defining the limits of this book it is worth presenting, up front, the definition adopted in the various research projects on which Part Two of the book is based. Two definitions are offered. First, an all-encompassing definition of public space that defines the absolute limits of the subject area, and second, the narrower definition, that was adopted as the focus of the empirical research.

A broad definition of public space could be constructed as follows:

Public space (broadly defined) relates to all those parts of the built and natural environment, public and private, internal and external, urban and rural, where the public have free, although not necessarily unrestricted, access. It encompasses: all the streets, squares and other rights of way, whether predominantly in residential, commercial or community/civic uses; the open spaces and parks; the open countryside; the 'public/private' spaces both internal and external where public access is welcomed – if controlled – such as private shopping centres or rail and bus stations; and the interiors of key public and civic buildings such as libraries, churches, or town halls.

This wide definition, encompasses a broad range of contexts that can be considered 'public', from the everyday street, to covered shopping centres, to the open countryside. Inevitably the management of these different types of context will vary greatly; not least because:

- the latter two examples are likely to be privately owned and managed, and therefore subject to private property rights, including the right to exclude;
- the shopping centre is internal rather than external and likely to be closed at certain times of the day and night;
- the intensity of activity in the open countryside is likely to be vastly less (at least by people) than in the other two contexts.

For these reasons, a narrower definition of public space would exclude private and internal space, as well as the open countryside. This definition provides the basis for the work:

Public space (narrowly defined) relates to all those parts of the built and natural environment where the public has free access. It encompasses: all the streets, squares and other rights of way, whether predominantly in residential, commercial or community/ civic uses; the open spaces and parks; and the 'public/private' spaces where public access is unrestricted (at least during daylight hours). It includes the interfaces with key internal and external and private spaces to which the public normally has free access.

This second definition does not imply that the wider definition is invalid; merely that it is possible to interpret a term such as public space in many different ways. For the purposes of this book, the narrower definition helps to focus attention on the areas where many have argued the real challenge for enhancing public space lies, in the publicly managed, external, urban space. It sets the limits and limitations of this book, which are further limited by a focus on public space in the context of (predominantly) Western, developed countries.

How the book is structured

The structure of the book aims to gradually unpack the range of issues discussed so far, initially by focusing in greater detail on the nature and evolution of public space, and then on its management. To do this, the book is structured in two parts. *Part One: Conceptualising public space and its management*, constitutes the first four chapters of the book and sets the scene for the empirical research that follows in Part Two. It airs a range of theoretical and practical debates around public space and its management.

In Part One, this first chapter introduces the concept of public space and explores issues surrounding its inherent complexity and the complexity of its management. Chapter 2 then provides a historic context for the discussions that follow by tracing the evolution of public space through history from antiquity to the modern era. Examples of public spaces in London – the historic market place, Georgian residential square and the grand civic square – are contrasted with spaces from New York – the town square, downtown space and the corporate plaza. The historical discussion draws out the changing balance between public and private in the production, use and management of urban space, and key issues for the contemporary management of public space.

The historical review is followed in Chapter 3 by a discussion of contemporary debates and theories concerning public space. The intention here is to draw from a range of literature from different scholarly traditions – cultural geography, urban design, property investment, urban sociology, etc. – to establish the key tensions at the heart of public space discourse. Conflicting definitions of public space will be discussed, and evidence presented about the use and changing nature of contemporary public spaces. The chapter concludes with a new classification of urban space types.

The final chapter in Part One focuses on the management literature, aiming to draw out discussions about the nature of public sector management as an activity and a policy field, and how it relates to public space. A typology of approaches is presented encompassing the paternalistic management of public space by the state, privatised models of public space management, and devolved community-based models. Drawing from the literature, the pros and cons of the different models are articulated, as well as the implications of each for some of the debates discussed in Chapter 3.

Part Two: Investigating public space management presents the four empirical research projects in turn, projects that have systematically addressed the different challenges for public space management identified through the literature discussed in Part One. Together, the projects extend across the national and international stages, and from strategic to local dimensions of public space and its management.

In Chapter 5, a first research project examining the management of everyday urban public spaces in England is introduced. This, the first of two chapters dealing with the project, examines typical practice through interrogating the results of a national survey of local authorities in England and findings from interviews with a range of key stakeholder groups. The intention is to understand the multiple drivers and barriers confronting public space decision-makers in their attempts to improve the quality of public space. Chapter 6 is a second linked chapter which examines a range of innovative practice via case studies identified through the national survey and interviews. Each case study featured one or more initiatives intended to address the perceived decline in public space quality. Lessons with wider application to the barriers identified in Chapter 5 are drawn out from the experiences.

In Chapter 7 discussion moves on to the international stage but focuses on a research project that examined the management of a particular type of public space – urban public open spaces. In this chapter, the stories of eleven cities from around the world with a reputation for the high quality of their open space environments are begun. The particular focus here is the context within which open space management occurs. Chapter 8 is the second chapter in this pair which re-focuses the discussions of the eleven cities onto the day-to-day practice of open space management as a means to extract common lessons with wider application elsewhere. In both chapters a common structure is used to aid comparison and to enable key lessons to be extracted.

A third project is examined in Chapters 9 and 10, focusing in some depth on three internationally iconic public spaces. In these chapters, discussion moves from strategic management concerns to a focus on particular spaces and their place-specific requirements. Chapter 9 focuses on Times Square in New York and also includes an overview of the research methodology for both chapters. Chapter 10 focuses on Leicester Square and Piccadilly Circus in London. In both, an in-depth analysis of the spaces based on detailed on-site observation and related interviews is presented. The chapters discuss how new management vehicles are challenging the status quo, but also raising profound questions about exclusion, ownership and the future of public space.

Chapter 11, the final chapter in the book, revisits the previous discussion and attempts to link in a systematic manner the theoretical discussions in Part One with the empirical findings presented in Part Two. The use and nature of public space is discussed, and the argument is made that too often academic discourse has seen public space in black-and-white terms, whereas public space management is in practice far more complex and nuanced. As a postscript to the book, the results of a fourth and final empirical research study are used to illustrate this. The project addressed the issue of what the users of public space actually want, as opposed to what academics, public space managers, politicians, or other interested parties think is good for them.

Understanding public space

Why is public space and its management important?

Most writers on public space issues recognise a general decline in this realm, although the causes and the cures prescribed are often very different. Broadly, the literature demonstrates a dichotomy amongst critics.

Many of the best-known critics choose to focus on what they view as the over-management of some types of external (and internal) public spaces that manifests itself in what they see as the commodification and homogenisation of space (for example, Sorkin 1992; Boyer 1994; Zukin 1995; Loukaitou-Sideris and Banerjee, 1998 – see Figure 1.2). Others focus on what they view as the under-management of external public spaces and paint a picture of a rubbish-strewn, poorly designed and insecure public realm (Figure 1.3). Many of the former set of concerns revolve around formal, high profile public space types that, through a wide variety of development and policy processes, have become increasingly privatised and therefore more or less exclusionary. These are very real concerns which are dealt with in some depth in Chapter 3, and which underpin critiques of some of the recent trends in public space management that are discussed in Chapter 4.

Critics of the latter type are not new. Classic urban design texts such as Jane Jacobs (1961) and Oscar Newman (1973) have long since bemoaned the tendency to design environments that encourage uncivil behaviour and a heightened fear of crime. In this tradition, Alice Coleman's (1985) work examined how the design of the built environment could support activities such as littering, graffiti, vandalism and other anti-social behaviour, leading all too quickly to a degraded environment and a disadvantaged community. A huge literature has spawned from these pioneering studies, much of which challenges the details, although perhaps not the fundamentals, of the early work.

THE VALUE OF PUBLIC SPACE

The existence of literature from both sides of the Atlantic making essentially the same observations about the deterioration of public space illustrates the portability of such concerns. In fact, as shall be demonstrated in Chapters 7 and 8, these concerns about public space quality and its better management are shared across the developed world; and in many parts

1.2 Privatised public space: Euston, London

1.3 Deteriorating public space: The Bund, Shanghai

of the developing world (Zetter and Butina-Watson 2006). Arguably they are underpinned by a growing awareness of the value of public space that now reaches to the highest political levels.

In the UK, for example, in his Croydon speech of April 2001, former Prime Minister Tony Blair marked a decisive shift in national policy by calling for cleaner and safer streets where communities are given the opportunity to thrive and not just survive. This interest from the very top reflects an increasing perception about the importance of public space issues as a political concern (see Chapter 5), but also an awareness of a growing body of evidence that public space is able to deliver a range of benefits across economic, social and environmental spheres (see Woolley *et al.*, n.d.). Empirical evidence now strongly suggests that public space:

Economically,

- can have a positive impact on property prices – research suggests variously by between 5 per cent (Colin Buchanan and Partners 2007) 8 per cent (Luttik 2000) and 15 per cent (Peiser and Schwann 1993) or even up to 34 per cent in some circumstances (CABE 2005a);
- is good for business – boosting commercial trading by 40 per cent in one case (DoE and ATCM 1997);
- raises land value and levels of investment (Luther and Gruehin 2001; Phillips 2000);
- helps boost regional economic performance (Frontier Economics 2004).

For human health,

- can encourage exercise with associated health benefits – for example reducing the risk of heart attack, diabetes, colon cancer and bone fractures (Hakim 1999; Diabetes Prevention Group 2002; Slattery, Potter and Caan 1997; Grisso, Kelsey and Stom 1991);
- can influence a longer life (Takano *et al.* 2002);

- provides a space for formal and informal sports and games (Woolley 2003; Woolley and Johns 2001);
- reduces stress and enhances mental health (Hartig *et al.* 2003; Halpern 1995);
- enhances child health – for example helping parents manage children with attention deficit disorder (Taylor *et al.* 2001).

Socially,

- delivers learning benefits to children, creative play, and reduces absenteeism (Fjortoft 2001; Taylor *et al.* 1998);
- nurtures social and cognitive skills (Pellegrini and Blatchford 1993);
- can help to reduce incidents of crime and anti-social behaviour (McKay 1998; Conolly 2002; Painter 1996; Loukaitou-Sideris *et al.* 2001; CABE 2005b);
- promotes neighbourliness and social cohesion (Baulkwill 2002; Massey 2002; Quayle and Driessen van der Lieck 1997; Kuo *et al.* 1998; Appleyard 1981);
- provides a venue for social events (Schuster 1995);
- reduces child mortality – by avoiding car-dominated environments (Living Streets 2001; Maconachie and Elliston 2002);
- provides a venue for social interchange and for supporting the social life of communities (Mean and Tims 2005; Dines and Cattell 2006; Jones *et al.* 2007; Watson 2006).

Environmentally,

- can encourage the use of sustainable modes of transport (Gehl and Gemzøe 1996; 2000);
- improves air quality, reduces heat island effects, pollution and water run-off (Littlefair *et al.* 2000; Whitford *et al.* 2001; Shashua-Bar and Hoffman 2000; Upmanis 2000);
- creates opportunities for urban wildlife to flourish (Shoard 2003).

Public space therefore has the potential to influence a wide range of benefits: as a stage to encourage social cohesion and interaction and

1.4 Inadvertent impacts: the humble wheelie bin

1.5 A standards-based approach to public space design

build social capital; as a venue for economic exchange and element in determining economic competitiveness and investment decisions; as an environmental resource and direct influence on energy use; and as an important contributor to the liveability or urban places and influence on the health and well-being of local populations.

The nature of public space

Of course not all public space is deteriorating and much is well-designed and managed. Nevertheless, if a general perception exists that the 'quality' of public space is deteriorating, then it can be argued that it is beholden on those responsible for its up-keep to understand why this is so, and what can be done about it. It may be, for example, that a lack of understanding of the nature of public space is a root cause behind the deterioration, perhaps because the delivery of space quality does not feature as a significant objective of many key stakeholders (see below).

It seems that in order to manage public space more efficiently, there has been a tendency to carve up the field into smaller units of responsibility, sometimes contracted out to a multitude of private contractors. This has replaced multi-tasking and holistic approaches to public space management that were epitomised in the guise of, for example, the park keeper or estate caretaker. A consequence seems to be the loss of key individuals who take an overview across all the elements of public space and its management, and a culture of delivering only what is specifically contracted or specified. This issue of the disaggregation of responsibilities for public space and its management will be a key theme, supported through empirical evidence, that is returned to throughout the book.

For now, the failure to understand the connections between different public space management objectives can be illustrated by way of a simple example effecting residential streets throughout the UK. Efficient refuse collection is a vitally important component in managing the urban environment by keeping streets sanitary and clean. In order to more efficiently (and cheaply) manage this process, many local authorities have given their residents wheelie bins that not only securely hold significant quantities of rubbish (so avoiding the problem of rubbish spilling onto streets), but also allow operatives to clear rubbish with less chance of injuries to themselves. Despite these benefits, in some environments

where houses open directly onto the street, the inadvertent side effect has been a negative impact on the urban environment as wheelie bins come to visually dominate the street scene, as pavement space for play is reduced, and as accessibility, particularly for those with disabilities, is compromised (see Figure 1.4).

The illustration demonstrates the need to carefully consider the impact of one policy decision upon others, to consider their impact in different contexts, and to be able to predict where conflicts might occur. In other words, to make the connections.

The illustration also demonstrates the need for a deep awareness of outcomes, the optimisation of which might be seen as the first and over-riding public space management objective, but which needs managers who understand the interlinkages between different policy responses. Unfortunately, it seems that rather than skilling-up to meet the challenges, coping methods have often been found to simply avoid the worst effects of contemporary public space pressures, whilst still maintaining functionality. The inevitable result is the crude application of standards-based approaches to service delivery: planning and highways standards, road adoption specifications, police 'designing out crime' principles, accessibility regulations, road safety markings and signage, corporate street furniture, public transport infrastructure, and so forth, with little real understanding of the overall impact (Figure 1.5). This question of skills will also be returned to throughout the book.

A DESIGN-LED MANAGEMENT PROCESS?

Some have argued that what is required is a design-led approach to public space management in order that the complexities are full understood. In England, the government-convened Urban Task Force (1999) contended that 'More than 90 per cent of our urban fabric will be with us in 30 years time' and that as a consequence this is where the real 'urban quality' challenge lies, rather than with the much smaller proportion of newly designed spaces created each year. They argued, however, that the way spaces look and feel today and the ease with which they can be managed relates fundamentally to how they were designed in the first place. Moreover, because every subsequent intervention in space (following its initial development) has an impact upon its overall quality, the importance of design skills remains fundamental.

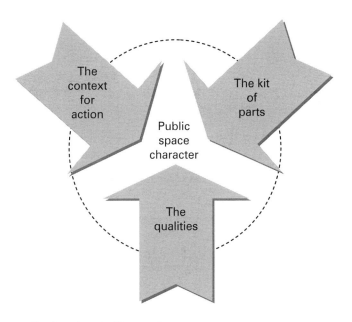

1.6 The dimensions of public space character

This does not imply that all those involved in the management of public space need to be designers in an artistic sense, and some have argued that the over-design of spaces to the detriment of other factors can be problematic when much everyday space is often (and quite appropriately) banal or untidy in order to be functional and versatile, for example, street markets (Worpole and Knox 2007: 3). It does imply, however, that interventions (no matter how small) should be considered creatively and sensitively, involving weighing-up and balancing options and impacts in order to find the 'optimum' given solution within the constraints set by context and resources. As the wheelie bin example indicates (alongside countless other more significant public space management decisions taken every day), this frequently does not happen.

Focusing on the issue, the Urban Design Skills Working Group (2001) argued that rectification of the problem must begin with four things:

- on the demand side, reawakening the public's interest in the quality of public space through adequate community participation and the stimulation of grassroots involvement;
- on the supply side, increasing the skills base available to design and produce better places;
- reaching a position where local authorities make use of those skills in administering their functions;
- bridging the divide between the different disciplines concerned with the built environment by focusing on the common ground – the public realm.

However, given the range and diversity of activities required to successfully manage public space (see below), it may be that for the majority of those involved, all that is required is an 'awareness' of their role in, and responsibilities to, the overall and ongoing design process. For others, a more complete understanding of the total urban environment and all the contributions to its upkeep is necessary in order to establish a vision, define the roles and responsibilities of constituent services, and reconcile possible conflicts.

This is likely to require a good understanding of the nature and complexity of public space, which, for the purposes of this book, is conceptualised in terms of three key dimensions that together define its character (Figure 1.6):

- the key elements that constitute public space – in other words, the 'kit of parts';
- the particular characteristics of public spaces – the 'qualities' that different spaces possess;
- the range of socio/economic and physical/spatial contexts – or the 'context for action'.

A similar division was used by Bell (2000: 21) in her work developing Urban Amenity Indicators for New Zealand in which she usefully distinguishes between 'amenity attributes', representing the tangible and measurable elements, and 'amenity values', or the less tangible perceptions people have about these. In each case, she argues, context it vital: 'We all know what amenity means to us, but it means different things to different people depending on where we live work and play'. In England, government guidance on design also adopts a similar division (DETR and CABE 2000). As well as defining seven 'Objectives for Urban Design', the guidance distinguishes between eight 'Aspects of Development Form' to which the objectives relate, and argues that the patterning together of the two in different places can help in understanding the local context and therefore in drawing up appropriately responsive policy and guidance frameworks for different areas.

The kit of parts

Starting therefore with the 'kit of parts', this first element of public space character is on the face of it the most basic, representing the constituent components of public space. Taking a pseudo-morphological approach to the character of public space (see Carmona *et al.* 2003: 61–6), it is possible to envisage a kit of parts that disaggregates space into four key elements (Box 1.1):

1 buildings
2 landscape (hard and soft)
3 infrastructure
4 uses.

BOX 1.1 PUBLIC SPACE, THE KIT OF PARTS

Buildings	**Infrastructure**	**Landscape**	**Uses**
Walls	Roads and cycle lanes	Trees	Events
Structure	Bus stops/shelters	Planting beds and areas	Gatherings
Windows	Tram/bus lanes	Lawns and verges	Street entertainment
Entrances/exists	Traffic lights/road signage	Planters/hanging baskets	Street trading
Balconies/projections	Telegraph polls	Paving	Markets
Shopfronts	Telecommunications	Road surfaces	External eating/drinking
Signage	equipment	Traffic calming	Kiosks
Building lighting	Street lighting	Steps	Play grounds
Floodlighting	Telematics	Boundary walls/fences/railings	Parks
Artwork	Parking bays/meters/car parks	Fountains/water features	Sports facilities
Decoration	Public toilets	Public art	Retail uses
Canopies	Waste and recycling bins	Signage	Leisure uses (active/passive)
Colonnades	CCTV polls and cameras	Advertising	Community uses
Skyline/roofscape	Telephone/post boxes	Street furniture	Homes
Corners	Gutters/drainage	Bollards	Workplaces
Flags and banners	Utilities boxes	Shelters/band stands	Industrial uses
Monuments/landmarks	Underground services	Festive decorations	Tourism
	Servicing bays/turning heads		

The first three categories are entirely physical in nature, whilst the last encompasses a set of human activities and is therefore perhaps the most challenging to manage, and also – arguably – the most significant in giving public space its character. The first three also delineate the physical urban form (the streets, spaces, urban blocks, and key routes and connections) that define the limits of external public space, and which between them create the venues for human activity.

When considered by management responsibility, buildings and uses tend to be privately owned, with responsibility for their upkeep largely in the hands of companies, institutions and individuals. Motivations for managing these assets will therefore be influenced by an assessment of their economic value and the costs and benefits of maintaining them. Conversely, most of the landscape between buildings in urban areas, and much (although not all) of the infrastructure will be owned and managed by the public sector, whose motivations for its management will be determined by competing local and national priorities and available resources. The distinction reinforces the fact that despite perceptions to the contrary, in almost all environments effective management will be a direct result of a formal or informal partnership between public and private interests.

The issue of time also distinguishes the different elements of the kit of parts, as the buildings and much of the infrastructure will tend to change only very slowly over long periods of time, emphasising, with regard to its long-term management, the need to get the design right in the first place. By contrast, elements of the landscape, and in some environments the uses in and surrounding external public space, will tend to change more quickly (Buchanan 1988: 33). It is these elements that can have the most decisive short-term impact on the way public space is perceived by its users. Therefore, although at any one time most of the physical environment already exists and changes only very slowly, the way the different elements are cared for, and the impact of those elements that change most frequently – the paving, street furniture, shop-fronts, signage, soft landscaping, building uses, and public space activities, etc. – are likely to be decisive in determining users' perceptions of quality.

Moreover, in an evermore complex built environment, the 'kit of parts' that contemporary public spaces need to accommodate have increased dramatically, whilst the intensity with which many spaces are used and the hours in the day over which activities happen have also multiplied. The result is inevitable conflicts that are difficult to resolve and which can undermine quality (Audit Commission 2002a: 3–5). This is hardly surprising when one considers the range of functions that many streets and spaces accommodate:

1   pedestrian thoroughfares
2   traffic arteries
3   retail destinations
4   market venues
5   venues for civic functions
6   places of relaxation
7   places to congregate
8   venues for public and political meetings
9   places for cultural exchange
10 opportunities for car parking
11 gateways to the private realm
12 places for social interaction
13 servicing arteries (gas, water, electric, cable, telephone)
14 play spaces
15 venues for eating and drinking
16 public transport arteries (bus, tram, taxis)
17 containers for landscaping
18 sources of information and communication (signs, advertisements, public phones)
19 opportunities for building servicing
20 breaks for light, sun and air penetration..

Examples of conflict include: between the needs of drivers and public transport versus the needs of pedestrians; the needs of utility providers to supply and maintain underground infrastructure versus the space required for street trees to grow and flourish; or the needs of commercial and entertainment premises versus the needs of local residents for peace and quiet. When the functions that spaces accommodate conflict, the overall quality of the space is often the first casualty. The challenge is therefore to manage the conflicts whilst enhancing quality and maintaining functionality. This question of managing conflict within public space represents another overarching theme of the book.

Public space qualities

Awareness of the kit of parts is by itself of little value without an awareness of how the parts are patterned together to optimise the 'qualities' of public space that make it conducive to human activity. The influential Copenhagen-based architect Jan Gehl (credited with the transformation of much of his own city) has argued that public space activities are particularly important in perceptions of public space. They are also particularly sensitive to the physical quality of environments. Gehl (1996) has characterised outdoor activities into three categories:

- necessary activities that we have to engage in – walking to work or school, waiting for a bus, shopping for food, etc.;
- optional activities that we choose to do if the time and place is conducive – walking for the sake of it, watching the world go by, sunbathing, window shopping, sitting at a pavement cafe, etc.;
- resultant (social) activities which are dependent on the presence of others in public space – children playing, casual greetings, conversations, communal activities, etc.; social activities are resultant because they occur spontaneously as a direct result of the other two forms of activity.

Based on extensive research across the world, Gehl has concluded that necessary activities are influenced only slightly by the physical quality of the environment because they are necessary for life to continue. Optional activities, by contrast, only take place when conditions are optimal, and are therefore a direct barometer of the quality of public space. They also effect users' perception of space because if people are choosing to stay in spaces rather than hurrying through, the space itself seems more 'liveable'. Finally, social activities happen whatever the physical context, although their quality and intensity will be affected by both the numbers of people in a space, and by the extent to which the quality of space encourages users to linger.

It is therefore a mistake to think of better quality public space as purely a visual concern, of interest only to a minority of aesthetes. Instead, these are fundamental issues that impact directly on the way all users perceive, function, and socialise in public space, and by implication on the viability of public space for different economic activities.

TANGIBLE QUALITIES

A wide range of publications focus on the design of urban space, setting out key aspirational principles for designing new and enhancing existing public spaces. Some of these are summarised in Table 1.1, which indicates that most converge on a set of widely accepted urban design principles. However, managing rather than designing public space is a broader concern that encompasses, but extends beyond, design objectives. It is also constrained by the fact that in most environments, the 'kit of parts' is already in place and unlikely to substantially change over the short or medium term.

Successive polls from MORI have focused on what residents perceive will most improve their areas, work which repeatedly throws up a consistent range of factors (MORI 2002), including:

- crime reduction
- activities for young people
- removal of rubbish/litter
- reduction in noise/disturbance
- better lighting
- reduced traffic
- better parks and open space
- less dog mess
- better street cleaning
- better maintenance i.e. of pavements.

The Association of Town Centre Managers have also attempted to gauge public perception of factors that make for a 'good' local environment through assessment of local authority enhancement initiatives. As well as basic 'Objectives of Urban Design', they cite cleanliness, a lack of graffiti, low transport emissions, safety and security, access for all, and quietness as preferred qualities, as well as a desire for basic amenities, including: good pedestrian routes and car parks, cycle routes, benches, places to meet and shelter, toilets, and clear signage. Indeed these represent reoccurring issues across a range of research projects (Williams and Green 2001: 4).

MORI (2000), for example, found that in the case of parks, people expect safety, cleanliness, tidiness, access for all, and provision for dogs; the University of Sheffield (1994) found that when looking specifically at children's requirements for good public space, they wanted clean streets, less litter, graffiti and traffic, places to meet, better street furniture, and a reduction of anti-social behaviour, especially alcoholics in city centres. Pan-European research, discovered that factors that make public spaces popular include, places for sitting and relaxing, something to watch (preferably other people), sufficient pedestrian through-flow, and 'ambience', whilst low levels of vehicular traffic was not viewed as a problem (Hass-Klau et al. 1999).

Llewelyn Davies (2000: 99–105) confirms the importance of a good ambience, arguing that a comfortable and stimulating public realm requires activity, with uses related to public spaces in such a way that animation, diversity and versatility results. They call for public space that stimulates the senses, visually, but also by sound, touch and smell; places that are distinctive and interesting, building on local character; places free of clutter, but which nevertheless exploit the power of public art; and places with are legible through good lighting and signage.

The Audit Commission (2002a: 3–6) define this as the 'liveability agenda' which to them aims to strengthen local communities, to make streets safer, cleaner and better managed and to provide high quality public spaces. Their analysis shows that people want streets that are:

- pleasant
- attractive
- well designed
- free from danger pollution and noise
- functional
- litter free
- not repeatedly dug up
- diverse, to cater for all needs – peaceful and lively, business and play.

By contrast, the Project for Public Space (2000), based on their analysis of hundreds of public spaces around the world, conclude that four key qualities are required for a high-quality environment:

- access and linkage – convenient to use, visible, easy to get to and move within;
- uses and activities – providing a reason to be there, vital and unique;
- comfort and image – safe, clean, green, full of character and attractive;
- sociability – fostering neighbourliness, friendship, interaction, diversity, pride.

Table 1.1 Conceptualisations of urban space design

Lynch (1981)	Jacobs and Appleyard (1987)	Bentley et al. (1985)	Tibbalds (1988)	Congress for New Urbanism (1993)	Urban Task Force (1999)	DETR and CABE (2000)	Llewelyn Davies (2000)	Carmona et al. (2002)
Control	Identity and control community and public life	Personalisation	Cater for all sections of the community and consult them	Cities and towns shaped by community institutions				
	Urban self-reliance	Robustness	Build to last and adapt	Design for pedestrians and transit	Building to last, sustainable buildings, environmental responsibility	Adaptability	Manage the investment / Design for change	Sustainable urban design
	Imagination and joy	Richness	Human scale, intricacy, joy and visual delight		Context, scale and character	Character	Work with the landscape	Townscape
Fit			Avoid change on too great a scale	Physically defined public spaces	Optimising land use and density	Continuity and enclosure	Mix forms	Urban form
Sense	Liveability		Places before buildings		Public realm	Quality of the public realm	Places for people	Public realm
	Authenticity and meaning	Legibility	Legible environments			Legibility		
Vitality	Access to opportunities	Variety	Mixing of uses	Diversity in use and population	Mixing activities, mixing tenures	Diversity	Mix uses	Mixed use and tenure
Access	An environment for all	Permeability	Freedom to walk about	Universal accessibility	Access and permeability	Ease of movement	Make connections	Connection and movement
		Visual appropriateness	Learn from the past and respect context	Celebrate local history, climate, ecology and building practice	Site and setting	(Application through eight aspects of urban form)	Enrich the existing	Application to context

For them, places without these characteristics are likely to be alienating, uncomfortable or simply unusable, indicating that something is wrong with the design, management or both. Smith *et al.* (1997), based on an extensive analysis of place-based physical visions, developed a similar list of qualities that urban environments should fulfil: liveability, character, connection, mobility, personal freedom and diversity; whilst Carr *et al.* (1992: 87–136) conclude that five types of reason account for people's needs in public spaces: comfort, relaxation, passive engagement with the environment, active engagement, and discovery (the desire for stimulation), and that any one encounter with a place may satisfy more than one purpose. They argue,

> it is important to examine needs, not only because they explain the use of places, but also because use is important to success. Places that do not meet people's needs or that serve no important functions for people will be underused and unsuccessful.
>
> (Carr *et al.* 1992: 91–2)

Numerous physical prescriptions have also been established for what makes a good space. William Whyte (1980), for example, concluded his observations of public squares in New York with the following requirements, that:

- public spaces should be in a good location (preferably on a busy route and both physically and visually accessible);
- streets should be part of the 'social' space (cutting off a space from the street with railings or walls will isolate it and reduce its use);
- the space should be level or almost level with the pavement (spaces raised significantly above or below the pavement were less used);
- there should be places to sit – both integral (e.g., steps, low walls, etc.) and explicit (e.g., benches, seats, etc.);
- moveable seats facilitated choice and the opportunity to communicate character and personality.

Less important factors included sun penetration, the aesthetics of the space, and the shapes and sizes of spaces. By contrast Amos Rapoport (1990: 288) identified 36 supportive characteristics of successful street spaces that are almost all to do with their size and shape. These he grouped into six categories, successful streets are likely to: have high levels of enclosure; be narrow; have complex profiles (i.e. variation in width, turns and twists, subspaces, projections, etc.); have short blocked views; have highly articulated surfaces and enclosing elements; and be part of a complex pattern of routes and sequences of space.

Other writers, Bill Hillier (1996) for instance, have focused on the interconnectivity (visually and physically) of spaces as the key determinant of their functional success, whilst Jan Gehl (1996: 135), amongst others, has argued that all these factors – size, shape, connections, the disposition of elements within space, and their detailed design – are important in determining the quality of public space and therefore the types of human activities they will sustain. For him, moreover, all are both measurable and tangible.

INTANGIBLE QUALITIES

Despite the level of agreement across the literature, research undertaken by DEMOS (2005) has shown that many of the needs that determine how the public environment is perceived are often intangible, reflecting the diverse motivations, needs and resources available to different groups and users. Moreover, they argue the core ideal of public space being free and open to all is increasingly being undermined by a focus on safety, creating bland places with no real ability to draw or retain people. Elsewhere, environments are becoming 'specialised' in order to cater for diverse lifestyles, incomes, ages, ethnicities and tastes. The findings are particularly valuable in highlighting the dangers of over-emphasising particular qualities to the detriment of others, or of taking a narrow view of what constitutes the 'public environment'. Solutions include:

- spaces that enable users to participate in the space, by creating activities of their own;
- environments that encourage a diversity of user groups, and avoid domination by one group or use;
- creating spaces that were available 'on tap', at any time.

The research supports the historically important role of public space for social exchange, and suggests that non-traditional public spaces – the car-boot sale or skate park, for example – have an increasingly important role in encouraging socialisation, although the environmental qualities sought by users of such spaces may be very different from traditional public space.

Lloyd and Auld (2003) confirm the central importance of social space as a dimension of quality. For them, the extent to which environments encourage socialisation impacts directly on the quality of life of those who use them. In this regard, trends of commercialisation, privatisation and commodification in public spaces and facilities (see Chapter 3) can act to undermine this vital role by making the use of many spaces transitory, linked solely to commercial rather than social exchange. Their answer to the problem is the need, as they see it, to create or refurbish local environments, to make them conducive to social interactions that extend across successive visits. They argue that 'research must go beyond counting heads and observing

Table 1.2 Universal positive qualities for public space

Clean and tidy	Well cared for	Clear of litter, fly tipping, fly posting, abandoned cars, bad smells, detritus and grime; adequate waste-collection facilities; provision for dogs
Accessible	Easy to get to and move around	Ease of movement, walkability; barrier-free pavements; accessible by foot, bike, and public transport at all times; good quality parking; continuity of space; lack of congestion
Attractive	Visually pleasing	Aesthetic quality; visually stimulating; uncluttered; well-maintained paving, street furniture, landscaping, grass/verges, front gardens; clear of vandalism and graffiti; use of public art; coordinated street furniture
Comfortable	Comfortable to spend time in	Free of heavy traffic, rail/aircraft noise, intrusive industry; provision of street furniture, incidental sitting surfaces, public toilets, shelter; legible; clear signage; space enclosure
Inclusive	Welcoming to all, free, open and tolerant	Access and equity for all by gender, age, race, disability; encouraging engagement in public life; activities for young people; unrestricted
Vital and viable	Well-used and thriving	Absence of vacant/derelict sites, vacant/boarded-up buildings; encouraging a diversity of uses, meeting places, animation; availability of play facilities; fostering interaction with space
Functional	Functions without conflict	Houses compatible uses, activities, vehicle/pedestrian relationships; provides ease of maintenance, servicing; absence of street parking nuisance
Distinctive	A positive, identifiable character	Sense of place and character; positive ambience; stimulating sound, touch and smell; reinforcing existing character/history; authentic; individual
Safe and secure	Feels and is safe and secure	Reduced vehicle speeds, pedestrian, cyclist safety; low street crime, anti-social behaviour; well lit and good surveillance, availability of authority figures; perception of security
Robust	Stands up to the pressures of everyday use	High-quality public realm, not repeatedly dug up; resilient street furniture, paving materials, boundaries, soft landscaping, street furniture; well-maintained buildings; adaptable, versatile space
Green and unpolluted	Healthy and natural	Better parks and open space; greening buildings and spaces; biodiversity; unpolluted water, air and soil; access to nature; absence of vehicle emissions
Fulfilling	A sense of ownership and belonging	Giving people a stake (individually or collectively); fostering pride, citizenship and neighbourliness; allowing personal freedom; opportunities for self-sufficiency

behaviour. It must illuminate the lived experience of individuals and groups in relation to public leisure spaces' (Lloyd and Auld 2003: 354).

The trends raised by Lloyd and Auld (2003) also reflect the dangers of the social exclusion of key groups (i.e. the young or economically inactive) from some types of contemporary public space such as shopping centres, reinforcing for the researchers the key principles of equity, citizenship and access as qualities to be natured in the local environment. Related research examining the use of public space in the East End of London confirmed the importance of these social roles (Dines and Cattell 2006: xii). The study concluded that 'people need a variety of public open spaces within a local area to meet a range of everyday needs: spaces to linger as well as spaces of transit; spaces that bring people together as well as spaces of retreat'. Queens Market, for example, a long-established street market has evolved to reflect the different needs of the populations arriving in the area. As such it has provided (Dines and Cattell 2006: 32–3):

- a strong and enduring element in the area's identity and peoples' attachment to it;
- an important local social arena and venue for unexpected encounters;
- a local place where people felt comfortable, safe and able to linger;
- a multi-ethnic and multi-lingual place of interaction between different communities;
- a familiar and uplifting place that contributed directly to a sense of well-being in users.

Although these perceptions were not shared equally by all groups in the area (younger people and children were far more negative about the market as a social space), they nevertheless demonstrate the importance of seeing public spaces as social venues and as an important resource for individuals and communities; not just as physical containers. These qualities were considered fragile, raising concerns that they could easily be damaged by otherwise well-meaning processes of 'regeneration' or management that are often unaware and unconcerned about this important social role (Dines and Cattell 2006: 17–18).

DESIRABLE QUALITIES

The discussion above presents just the tip of the iceberg of literature dealing with the desirable qualities of public space. Combined with the range of urban design objectives drawn from various sources (see Table 1.1), it is possible to identify a set of – arguably – 'universal positive qualities' for public space that reflect the complex and overlapping social, economic, and environmental characteristics of local places (see Table 1.2).

Inevitably, as writers such as Kevin Lynch and many others have long since argued, relative judgements about the importance of various qualities are matters of individual perception, and different users will value different qualities more or less highly. Consequently, the emphasis placed on different qualities by local public space services will be matters for local judgement. But, just as Lynch (1960: 48–9) argued that the component images of place pattern together to create one overall image of place in users minds, so will the qualities pattern together to form an overall experience of public space. Therefore, concentrating on some qualities to the detriment of others may simply undermine attempts to improve the overall quality of space.

The context for action

The final conceptualised dimension of public space character adds yet further complexity to the management of public space by introducing the notion of a range of physical/spatial 'contexts for action' to which public space management processes need to respond. The contexts are initially generated by the patterning together of the different elements from the 'kit of parts' to create the networks, densities, mixes, urban typologies (urban, suburban, rural) and urban forms that constitute particular places.

For example, perceptions will vary considerably depending on whether the area being described is rural or urban. Rural areas are – perhaps unsurprisingly – considered to be more friendly, safer and greener by their residents (by a factor of two, three and three respectively). They are also much less likely to be characterised as shabby, dangerous or run down (MORI 2005: 23). Perceptions that higher density or mixed-use environments offer lower environmental quality are also well established in the literature (Carmona 2001a: 201–5).

The socio-economic context also dictates a separate set of factors that are likely to impact on local environmental quality. Such factors include:

- choice and opportunity open to residents
- levels of owner occupation
- child density levels
- levels of economic activity and employment
- levels of community engagement.

A range of research provides powerful evidence to back up these relationships. For example, evidence gathered together to test the concept of environmental exclusion (Brook Lyndhurst 2004b) indicated a particularly strong relationship between levels of deprivation in an area and the quality of the immediate local environment. Drawing on the English Housing Condition Survey, the report suggested that twice as many dwellings in areas characterised by multiple deprivation are effected by worse air quality than other districts; with litter, rubbish, graffiti and dumping experienced fourfold in deprived areas. A sister report (Brook Lyndhurst 2004a) suggested that two fundamental factors underpin perceptions of local environmental quality in deprived areas: public safety and public health. Parks and play areas, for example were only seen as

benefits if residents could also be confident that such spaces were secure from crime (the overriding concern), clean (from litter, dog fouling, broken glass, and drug needles), and safe from road traffic.

Other research has demonstrated how the socio-economic context can impact on the ability to deliver neighbourhood environmental services. Hastings et al. (2005), for example, have found that there is a gap between the environmental amenity of deprived and non-deprived neighbourhoods. They show that poor neighbourhoods have more environmental problems than affluent neighbourhoods, and that these include a greater range of problems, and problems that are more severe, particularly graffiti, litter, fly-tipping, and generally the poor maintenance of public and open spaces. They identify a complex range of reasons (Hastings et al. 2005: viii):

- greater use of the neighbourhood environment with associated rubbish and wear and tear, due to higher rates of economic inactivity and higher population densities, particularly child densities;
- built forms that are more difficult to manage, including large open spaces, undefined front gardens and high housing densities and a predominance of flats;
- the presence of a higher proportion of vulnerable households, less able to manage their neighbourhood environment;
- diminishing social responsibility within the community, and less motivation amongst residents to tackle the up-keep of their neighbourhood, leading to less effort amongst residents to control their local environment;
- reduced concern amongst frontline workers for deprived neighbourhoods because of the scale of problems and the difficulties in working in some places – fear, threats, violence, etc.

By contrast, the research recoded the increased motivation amongst operatives when working in affluent areas, driven as much by the fear of complaints following shoddy work as by the knowledge that they could work effectively in such areas (Hastings et al. 2005: ix). The result was further polarisation between poor and wealthy neighbourhoods.

MORI's work on physical capital (2005: 23) supports these findings. Their polling reveals residents of deprived areas are three times more likely to consider their area noisy and four times more likely to describe their area as shabby, whilst residents of affluent areas are significantly more likely to describe their areas as friendly, safe and green.

Other contextual factors are also important. The argument has already been made that policy approaches that are both effective and efficient in one circumstance may have unintended consequences in others, and therefore that sensitivity to context is required. Streets in predominantly

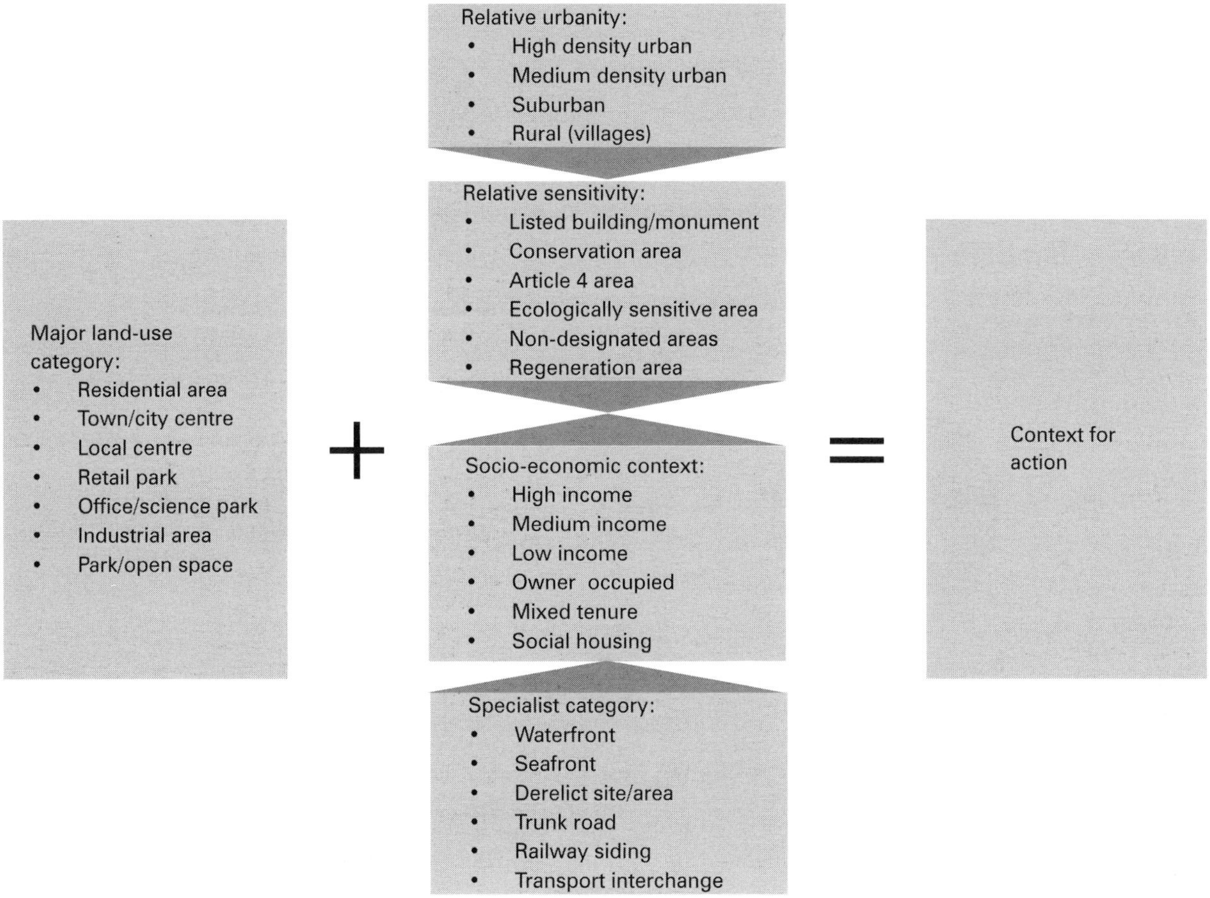

Relative urbanity:
- High density urban
- Medium density urban
- Suburban
- Rural (villages)

Relative sensitivity:
- Listed building/monument
- Conservation area
- Article 4 area
- Ecologically sensitive area
- Non-designated areas
- Regeneration area

Major land-use category:
- Residential area
- Town/city centre
- Local centre
- Retail park
- Office/science park
- Industrial area
- Park/open space

+

Socio-economic context:
- High income
- Medium income
- Low income
- Owner occupied
- Mixed tenure
- Social housing

=

Context for action

Specialist category:
- Waterfront
- Seafront
- Derelict site/area
- Trunk road
- Railway siding
- Transport interchange

1.7 The context for action

residential use, for example, will have different physical characteristics to commercial streets and are also likely to be subject to different pressures and statutory/management regimes. Waste collection, street cleansing, and on-street parking, for example, tend to be handled very differently in residential areas to commercial high streets. The degree of urbanity will also affect the relative emphasis on the natural environment and on soft (green) landscape as opposed to hard landscape.

Moreover, some areas are classified as particularly sensitive contexts though conservation (and other) designations, while others are not. The result is that some particularly sensitive contexts will require a further layer of regulatory processes (and sometimes resources) with a consequent influence on the quality of public space. To a lesser degree, the same often applies to areas receiving funding under different area-based regeneration and renewal programmes. Finally, a set of 'special' contexts can be identified with particular management requirements by virtue of the intensity (or lack) of their use (e.g. city centre parks); their particular patterns of ownership (e.g. private shopping centres); or their relationship to natural features or infrastructure (e.g. waterfronts or roadside verges).

CATEGORISING CONTEXT

Figure 1.7 summarises the range of major land-use categories and refinements thereof that form the context for action, and the final element

in determining the character of public space, and the management response required. Although it might be argued that the pursuit of high-quality public space should remain the same as contexts change, it is likely that the relative emphasis on different aspects of management will vary. In very-high-density areas, for example, the emphasis will be on designing accessible, robust space that can cope with the demands. In suburban areas, the emphasis may be on making a more attractive environment through greening streets and spaces and through the seasonal maintenance that this dictates.

This question of managing within and in response to context, and that aspirations should be shaped by an understanding of both the limitations and opportunities that different contexts present, represents a further overarching theme pursued throughout the book.

Responsibilities for action

Whose responsibility?

The final theme to be introduced in this chapter relates to the notion of a multiplicity of stakeholders and their complex interrelationships when managing public space. Interviews conducted with 150 everyday users

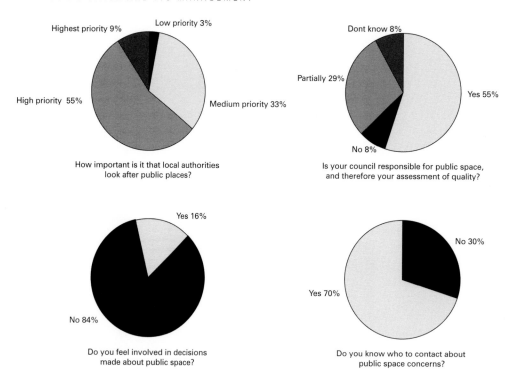

Highest priority 9%

Low priority 3%

High priority 55%

Medium priority 33%

How important is it that local authorities
look after public places?

Dont know 8%

Partially 29%

Yes 55%

No 8%

Is your council responsible for public space,
and therefore your assessment of quality?

Yes 16%

No 84%

Do you feel involved in decisions
made about public space?

No 30%

Yes 70%

Do you know who to contact about
public space concerns?

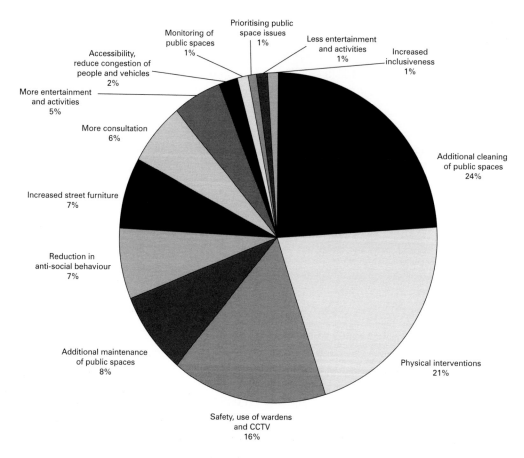

Monitoring of
public spaces
1%

Prioritising public
space issues
1%

Less entertainment
and activities
1%

Increased
inclusiveness
1%

Accessibility,
reduce congestion of
people and vehicles
2%

More entertainment
and activities
5%

More consultation
6%

Increased street furniture
7%

Reduction in
anti-social behaviour
7%

Additional maintenance
of public spaces
8%

Additional cleaning
of public spaces
24%

Physical interventions
21%

Safety, use of wardens
and CCTV
16%

How can your council improve public space quality?

1.8 External public space quality audit by users

18

of the public space (residents, visitors and local businesses) in the English case-study areas discussed in Chapter 6 were illuminating (Figure 1.8). They revealed that:

- users place a high priority on how local authorities look after public space;
- almost nobody considers public space to be a low priority;
- users do not feel involved in decision making;
- users are not knowledgeable about who makes decisions, although they knew how to make a complaint when necessary.

Most revealing was the finding that over half (56 per cent) of users felt that the management of public space was the sole responsibility of their local council, and that whether they perceived that their council was doing a good job, or not, it was the council that was responsible for the state of public space. In this regard, a clear misapprehension was evident concerning the responsibility for, and extent of influence of, local authorities for public space.

A smaller percentage (29 per cent) regarded space management as a joint civic responsibility in which individuals, businesses and other organisations also had a role to play.

A clear incentive for local politicians was therefore revealed. Namely, if authorities are going to get the blame whenever things go wrong (and the credit when things go right), there should be a direct political incentive for councillors to prioritise public space quality through the actions of their local authority, whilst doing all they can to encourage other stakeholders to do the same. Users of public space most frequently considered that councils could improve public space through additional cleaning, a range of place-specific physical interventions and through safety initiatives, including CCTV and street wardens.

Polls conducted for CABE (2002) reinforce the finding. They reveal that almost half of those who say they vote in council elections said they would be more inclined to support a different party if there was a significant deterioration in the quality of the local environment. A similar proportion of non-voters said that the issue alone would make them more inclined to vote. Clearly this is – or has the potential to be – a live political issue.

The multiplicity of stakeholders

Many current public space management regimes are still largely based on the 'traditional' local government model (see Chapter 5). This presents a range of challenges and restrictions, principal among which are the uncertainties inherent in local political contexts for what are typically discretionary services. In reality, however, much of the management of public space lies outside the direct control of local authorities. Instead, responsibilities lie across a wide range of stakeholders, both public and private. Therefore, although public perceptions may be that urban space is the sole responsibility of the public sector, in most contexts, the delivery of high-quality public space will be dependent on a diversity of interests working together (or not).

Moreover, the increasing complexity of public spaces as physical entities is mirrored by the increasing complexity of the stakeholders engaged – either positively or negatively – in public space management. In part this is a result of the impact of the increasing numbers of private stakeholders with a part to play, including contractors engaged by, and working for, local authorities. It also reflects the diversity of public and semi-public agencies involved in managing public space, and the complex range of public space types (see Chapter 3).

Broadly, stakeholders might be split into four groups:

1. Private, including private property owners and developers, but also utility providers
2. Public/private, including the range of arms-length pseudo-government agencies and operators (e.g. of public transport)
3. Local government, including a wide range of services across one or more tiers of local government
4. Community, including residents and special interest societies and local groups.

Significantly, each of these stakeholders is likely to have a very different set of motivations informing their approach to public space, and few will have the overall quality of space as a primary motivation. Based on a range of recent research undertaken in England (Audit Commission 2002a; CABE and ODPM 2002; Stewart 2001; ICE 2002; ODPM 2002), Table 1.3 postulates on what these might be.

Finding means to ensure that 'outcome quality' is factored into the decision-making processes of key players is therefore likely to be an important prerequisite for enhancing public spaces; and to achieve this, it may be that as the dominant player, the public sector will need to take the lead role through the processes discussed in Part 2 of this book. Sometimes this will require direct action, and sometimes the guidance, incentive or control of others, but if authorities are not there to lead, then it is hypothesised that widespread public space quality is unlikely to be secured outside of private enclaves. This issue of the balance between different stakeholders, and particularly between the public and private sectors, is another theme of the book.

Table 1.3 Stakeholders and their anticipated aspirations (England)

PRIVATE

Developers/contractors
- housing developers
- commercial developers
- contractors (both in construction and public space management)

Typical motivations
- Motivations vary, but generally developers are concerned with developments that are buildable, marketable and profitable. Because marketability is affected by the quality of the environment, developers are concerned with these issues, but only to the extent that they do not impact negatively on profitability. This will be a commercial judgement based on the requirements of likely purchasers.
- Contractors will rarely be concerned with the quality of the end product beyond delivering that which is specified in their contract with either the public or private client. They will generally do the minimum to meet the terms of the contract.

Property owners
- residents
- businesses
- investors
- landlord/registered social landlords

Typical motivations
- Property owners will generally be deeply concerned with the quality of he environment, not least because it will negatively or positively impact on the value of their investment, and on the quality of life of themselves (in the case of residents) or their employees (in the case of businesses) and tenants (in the case of investors, registered social landlords (RSLs) and other private landlords.

Property occupiers
- residents
- businesses

Typical motivations
- Occupiers will be less concerned about the knock-on property value consequences of public space quality. They will nevertheless be concerned about quality of life issues, and in the case of businesses, about employee productivity and the image their business environment suggests to clients.

Licensed operators
- billboard/street furniture
- fly posting
- gas/electricity
- cable/telephone
- water/sewerage/drainage
- pay phones
- Post Office

Typical motivations
- Advertising in public space functions through legitimate and non-legitimate operators, the former mainly on permitted billboards/hoardings and on a wide range of street furniture (bus shelters, benches, telephone kiosks, etc.), and the latter through fly-posting or non-permitted billboards. Both have as their primary objective to maximise coverage and visual impact for their advertising.
- Utility providers are concerned with the establishing and maintaining a high quality infrastructure network at lowest possible cost. They will generally not be concerned with the visual impact of their infrastructure on the street scene (whether above or below ground) or with the impact of street works.
- Public payphone providers (and to a lesser degree the Post Office) will, within limits, be concerned about the visual impact of their equipment in order to encourage customers. They will also wish their equipment to make a positive statement about their company.

PUBLIC/PRIVATE

Public transport operators
- network rail
- rail operators
- bus operators
- tram operators

Typical motivations
- Public transport operators will also wish to make a positive statement about their companies to customers and to thereby increase custom, through the quality of their stations/stops, but will also wish to control expenditure on non-essential maintenance to enhance profitability. They will generally not be concerned with the visual impact of infrastructure that is not directly at the customer interface.

Conservation agencies
- English Heritage
- Environment Agency
- British Waterways

Typical motivations
- Conservation agencies will regard the quality of public space as a top priority and will from time to time offer grant aid to improve its quality. They will be particularly concerned that schemes are distinctive (not standardised) and sensitive to the historic context. As owners of public space themselves they will also be faced with many of the same management challenges as local authorities i.e. the cleanliness of canal towpaths.

Partnerships
- regeneration partnerships
- community safety partnership
- local strategic partnerships (LSP)
- Local Agenda 21

Typical motivations
- Regeneration partnerships (initiatives) will often aim to improve the quality of the environment as a key objective and the subject of direct investment. Occasionally, investments in the social and economic infrastructure will be undermined if comparable investments in the physical infrastructure are not made.
- Community safety partnerships are focused on reducing crime and the fear of crime at the local community level.
- LSPs will be concerned with a wide range of cross-cutting and sometimes conflicting economic, social and environmental objectives and with enhancing the basic wellbeing of the communities they serve. Within this complex field of responsibility, local priorities will inevitably differ, and will be shaped by the representation in the partnership. Improving the management of public space is therefore frequently not identified as a priority in the resulting community strategies, although invariably different elements of the agenda are i.e. reducing crime, conservation, greening.

LOCAL GOVERNMENT

Local planning

- forward planning
- urban design
- economic development
- development control
- conservation
- enforcement
- building control

Typical motivations
- Local planning encompasses a range of services that have a decisive impact on public space across policy, implementation and regulatory roles. At the policy level planning is motivated by a wide range of complex economic, social and environmental objectives, only part of which concerns the quality of public space. At the control level, much that impacts on public space quality is outside of their control i.e. permitted development. Going forward, planning is increasingly motivated by space quality, and by the impact of development activity (large and small) in creating new and modifying existing public space. A lack of skills (particularly in design) and resources has held back both the potentially positive, creative and proactive role that planning can play and authorities willingness to enforce planning control.

- Building control impacts on public space through implementation of accessibility and fire regulations, and through policing on-site building works. Its motivation is purely technical, the delivery of the building regulations.

Highways and transport
- highways engineering
- traffic management
- street lighting
- roads/pavement maintenance
- street furniture
- car parks
- parking control
- public transport coordination

Typical motivations
- Highways and transport is the responsibility of county councils in two tier areas, and of unitary authorities elsewhere (in partnership with central government – the Highways Agency is responsible for trunk roads). Motivations have invariably been driven by three key concerns: rights of way (as opposed to qualities of place); a heavy emphasis on planning for vehicles as opposed to pedestrians and cyclists; and on vehicle flow speed and efficiency. Practice has been driven by an emphasis on engineering solutions and standard approaches to highway design as opposed to the qualities of particular places, and by a 'play it safe', rather than evidence-driven approach to pedestrian safety.
- The 'engineering'-driven approach has often been extended to pavements, roads and street furniture maintenance, with cheap (in the short-term), standardised approaches favoured, usually reflecting the 'corporate' livery and colours of the local authority whatever the context.
- Car parking policy has often been driven on the basis of the line of least resistance rather than any clearly defined vision of balancing need with impact on the local environment.
- Public transport (particularly local bus services and facilities) has rarely been a high priority in local government, and municipal bus stations, like many municipal car parks have suffered a lack of investment and vision. Within limited resources authorities will wish to deliver high quality, reliable public transport, but in order to do so, will tend to invest in services, rather than facilities.

Parks and recreation
- sports and leisure
- parks

Typical motivations
- Like many other aspects of non-statutory external public space, parks and external sports facilities have suffered a historic decline in resources and quality. Authorities tended to see parks and external sports facilities as a lower priority than other formal recreational facilities, and have had to reduce levels of management and maintenance in order to make a thinner slice of the resources cake go further.

Environmental (street scene) services
- waste collection/recycling
- environmental health
- trader licensing
- public toilet provision
- street cleaning (fly-sweeping, poster, graffiti, abandoned cars, dumping)
- landscape maintenance (trees, verges, hanging baskets, planters, public art, fountains, decorations)
- town centre management (TCM)
- events management
- alcohol licensing

Typical motivations
- The environmental category covers a wide range of local authority services concerned with managing public space. Collectively they have tended to be seen as routine local authority services that lack glamour and therefore political attention. Some services such as waste collection or environmental health have a significant impact on public space quality, yet the motivation driving them is generally the efficiency and cost of delivery, the meeting of targets, or technical health concerns, rather than their impact on space quality. The unintended impact is often negative.

- Other services such as street cleaning or landscape maintenance have environmental quality more directly as an aim, although the tendency is to pursue minimum standards, rather than to enhance space over time. Like the first group, the tendency has been to contract out many of these services, and in so doing, to narrowly define each for the purposes of contracts. The local authority role is then reduced to a monitoring role.
- By contrast the remit of TCM (which is often a separate semi-independent local authority service) is wide-ranging, and motivated by urban space quality, as well as town centre vitality and viability. Town centre managers focus on the delivery of many of the local authority services better in urban areas, in partnership with their retail partners. Town centre managers are sometimes also responsible for arranging and managing special events. Local authorities are also responsible for licensing premises for the sale of alcohol and are increasingly under pressure to grant extended licensing hours.

Housing (and estates)
- estate management
- grounds maintenance

Typical motivations
- Housing services are primarily concerned with housing their clientele, often to a minimum standard. Lack of resources and the expense associated with maintaining the poor quality post-war housing stock has reduced standards of grounds maintenance. As many spaces within social housing developments are effectively 'public', this has sometimes resulted in a poor perception of the quality of social housing. Other local authority estates services have suffered from the paring back of budgets.

COMMUNITY
Policing
- crime detection
- crime prevention
- traffic control
- CCTV
- street wardens

Typical motivations
- Police authorities are the combined responsibility of local and national government. Their focus is largely on reducing crime and the impact of crime on communities, but extends to the management of traffic. Their motivations are therefore focused on only a part of the space management agenda but have an important influence over both the design of new public space (through architectural liaison officers) and to maintaining day to day civility.
- New space management approaches coordinated variously through the police and other local authority departments include CCTV and street/neighbourhood wardens. Each are concerned with deterring criminal behaviour, maintaining civility, and with maintaining a visible presence on the street.

Residents' societies
- residents societies
- tenants groups

Typical motivations
- Research indicates that the primary motivations of residents groups and societies and tenants groups focuses on the quality of public space, including issues of cleanliness and safety and security.

Special interest groups
- amenity societies
- campaigning groups

Typical motivations
- The motivations of special interest groups varies according to their remits. Many are concerned with maintaining the distinctive qualities of their environment and with resisting proposals seen as detrimental to that quality

Community policing
- neighbourhood watch

Typical motivations
- Neighbourhood watch is focused on reducing crime and the fear of crime in residential neighbourhoods. This includes environmental crimes such as graffiti and vandalism.

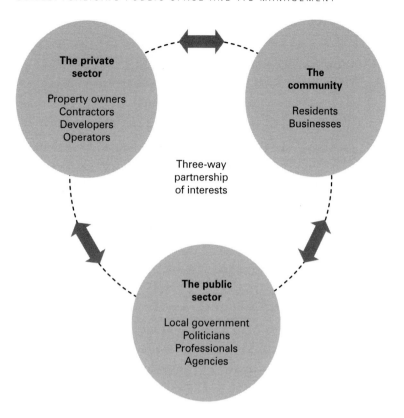

1.9 The idealised three-way partnership

In reality, rarely will a single agency have overall responsibility to coordinate the actions of all the others. Even local government, which typically has a diverse range of powers, frequently fails to take a joined-up approach on such matters, even, as Chapter 5 will show, internally between different departments and services within the same authority. This is because the activities of different local government services are themselves driven by very different motivations; many of which do not relate to delivering a better quality environment.

Authorities responsible for highways, for example, are often primarily concerned with the efficient flow of traffic and with the safety of highways users (including pedestrians). Their approach is invariably to give vehicles precedence while pedestrians are kept behind a surfeit of signs, barriers and lights to keep them safe. Recent reviews of street quality (CABE and ODPM 2002; ICE 2002) have focused on these concerns, and argue that there is need for a clearer line of responsibility for public space, centring on local authorities.

But if local authorities take on a more central and coordinating role, then the contemporary context of multiple stakeholder roles and responsibilities suggests that this may only be successful if it also recognises the important role of the private sector and the community, alongside local government. A core three-way partnership of interests would seem to be critical (Figure 1.9).

Conclusions

This chapter has begun to unpack some of the complexity inherent in public space management by, first, exploring the nature and different dimensions of public space itself, and second, the network of those responsible, in theory, for its management. Although, to some, questions of how we manage public space may at first seem rather prosaic, this fails to grasp either the true nature of public space and how it profoundly impacts on peoples' daily lives across the world, or the web of inter-connecting roles and responsibilities that play a part in ensuring that this impact is a positive one.

Examination of actual public space management practice in Part 2 of the book will assess to what extent any partnership does exist, and in different contexts, where the balance of power lies. First, however, a historical perspective on public space is taken.

Chapter 2

Public space through history

This chapter traces the evolution of Western urban public space through history from antiquity to modern times. Predominantly, the discussion is on the public square (and its variants) as a type of space, but the issues relate equally to the other forms of urban public space discussed elsewhere in the book. The first section outlines the complex evolution of public space in Europe, from the ancients through to the Renaissance and Baroque, and identifies the main functions of pre-modern European public space. The second and third sections focus on the 'modern' age, examining, respectively, a series of public spaces in London and then New York. A fourth section looks briefly at modernist space, and the recent return from there to designing more 'positive' urban space today. From the discussion, the complex and shifting relationships between spaces and their functions is identified, whilst the changing balance between public and private in the production, use and management of public space is drawn out. This is not a history book, but this brief look back helps to demonstrate how many of our contemporary preoccupations with the nature of public space in fact have very deep roots.

European public space from antiquity to renaissance and baroque

The form and function of Western urban public space today has its origin in the ancient civilisations. The Greek polis started to flourish around the fifth century BC while the Roman city began to flourish around the third century BC, the two civilisations overlapping for some centuries. Both the Greeks and the Romans valued urban public space greatly as places for social interaction, and this was epitomised in the aesthetic qualities that these spaces came to possess.

In modern times the result has been a tendency for many planners, architects and historians to eulogise the approach to public life these civilisations took. Most famously, Camillo Sitte, an Austrian architect disillusioned with the public space in his native Vienna around the turn of the nineteenth century, became a particular advocate of the ancient approach. Sitte (1889: 4) paid particular tribute to the public space of the ancient Greeks and Romans, arguing 'public squares, or plazas, were then of prime necessity, for they were the theatres for the principal scenes of public life'. Any history of Western public space should therefore begin with the ancients.

The ancients

To understand how public space functioned in ancient Greek society, it is important to understand the Greek 'polis', or 'city-state'. LeGates and Stout (2000: 31) emphasise the importance of public space to life in the polis, arguing 'Public life was essentially communistic. The polis as a social institution defined the very nature of being human for its citizens', whilst the physical form of the polis stressed public space.

Public space in ancient Greece therefore had a crucial role in the politics of the polis, particularly as many were self-governing. As such, public space in the ancient Greek poli is often described as democratic space, a much cited example being the agora in ancient Athens where citizens could vote on issues of government and justice. However, governmental systems varied between poli, and over the course of the Greek civilisation, with some having monarchs or oligarchies. Moreover,

2.1 The acropolis of Athens

as Mumford (1961: 138–9) notes, during the height of Greek civilisation, only one-seventh of the population were citizens; women, foreigners, and slaves could not be citizens, and therefore did not have access to many public spaces. Therefore, whether one was citizen, foreigner, woman, or slave not only dictated one's place in society but also what public space one had access to. Issues of exclusion, a theme returned to time and time again in this book, are therefore nothing new.

The focal point of the early Greek polis was the acropolis (Figure 2.1). Starting as a hilltop fortification, the acropolis evolved into a public space for religious and secular assemblies and commerce (Kitto 2000: 33). As Greek civilisation developed, public space shifted towards the agora (Figure 2.2), with less emphasis on fortification and religion. Mumford describes the agora as a place where citizens could meet for 'daily communications and formal and informal assembly' (quoted in Carr et al. 1992: 52), while Hall (1998: 38) describes it as 'no mere public space, but the living heart of the city'. The agora, like the acropolis, had multiple functions, but evolved principally into a marketplace, and in this function was open to all, not just 'citizens', despite the calls of some – Aristotle and Plato, for example – for greater restrictions (Hall 1998: 39).

The earlier Greek polis developed in an organic fashion containing few planned public spaces. Yet as Greek cities began to be developed in a more formal and organised way – often around a gridiron structure – public space took on a greater prominence. As organic growth gradually gave way to planned urban form, the conscious design of public space increasingly reflected the notion that its aesthetic quality would impart an experience to the soul of its users. For example, applying a non-axial design to public space in order to emphasise the three-dimensional qualities of the space (see Figure 2.2; Goldsteen and Elliott 1994: 74–7).

Discussion of the ancient Greek polis therefore identifies several key themes that still have resonance in debates about public space today:

- the notion of public space having multiple functions
- public space being democratic space, where citizens can interact and discuss issues pertaining to the city

- public space being used for commercial purposes
- public space as an informal meeting place and community space
- the aesthetic qualities of public space giving rise to pleasure
- notions of restricting access to public space, with some people having greater rights than others.

From Greece to Rome

Roman cities were far larger than the Greeks polis, Ancient Rome itself reaching a population of over one million. The Roman urban fabric was therefore richer and more varied.

The nucleus of the Roman city was the forum. Carr et al. (1992: 53) describe the Roman forum as a combination of the Greek acropolis and agora. The larger forums contained open, semi-enclosed, and enclosed spaces, while their functions embraced markets, religious meetings, political events, athletics, and informal meetings. They contained piazzas, important civic buildings such as basilicas, and temples to the different Roman gods. The basilica was an indoor space that could be used for judicial or commercial purposes. Temples fulfilled a double role, being used as a meeting place (e.g. for the senate in the case of the Temple of Concord), as well as for religious purposes.

Despite these early spaces being used for formal and commercial purposes, the former always took priority. Therefore, as Roman cities grew, single function forums were established, with forums being cleared of the clutter of statues, arches, monuments and so forth that had built up over time (Mumford 1961: 221–3). Hall (1998: 625) notes that by 113AD Rome had 'vast spaces for walking, business and pleasure'. The cities of the Roman Empire had pushed forward urban civilisation well beyond that of the polis by this time, not least because of the high level of public works.

Roman cities had therefore introduced a more planned approach to the production of urban public space. It was carefully integrated into the fabric of the city, creating a downtown area with social spaces, cultural

2.2 The Agora in Athens

2.3 The Imperial Forum in Rome

2.4 Medieval winding street in Assisi

spaces, shopping spaces and spiritual spaces, very similar to Western cities today. The Romans also understood the semiotic qualities of public space. Examples of this are the strong symbolism of state and religion in Roman piazzas, where surrounding buildings contained the senate and temple, accompanied by monuments and statues. This is a tradition that has continued in towns and cities through to today.

Imperial Rome used this method to impress an image on its population (Figure 2.3). While the Greeks recognised that the aesthetic qualities of space could beautify the soul and exalt the mind, Imperial Rome recognised that the design of space could have controlling influences on the population, and imperial and totalitarian regimes throughout history have used this principle.

The middle ages onwards

After the fall of the Roman Empire in the fifth century, city life declined in Western Europe. The church became powerful with the decline of the state, and expanded its influence within the walled cities the Romans had left behind. The walls of the settlements that protected against marauding tribes constricted the development of the settlements (Pirenne 2000: 39–41), and as the settlements declined, so consequently did urban public space. However, Mumford (1961: 255) stresses that these small settlements continued to contain commercial activity through the dark ages, and when international trade routes reopened, urban growth was accelerated.

Medieval public space was framed and often controlled by the church. Often the only planned public space was in front of the church to accommodate the congregation entering and leaving, with markets often sharing the same space, and operating on a weekly cycle. The church was the centre of the settlement and public life, with religious festivals, pilgrimages and processions used to bring the community together. Growth became organic and ad hoc with an emphasis on defence. Webb (1990: 40–1), for example, describes medieval streets as utilitarian, and it was only latterly as towns prospered that streets and spaces could be beautified. This also created public spaces that were independent of the church but still within the narrow confines of the town wall.

Despite the lack of formal design, the results often had their own innate qualities. Alberti and Sitte both admired the medieval winding street as an aesthetic, producing unpredictability and excitement for users walking through the city (Figure 2.4), as opposed to the rigidity of the gridiron (Mumford 1961: 261–314). Furthermore the medieval city was a more egalitarian place than its ancient predecessors. 'The medieval town had succeeded as no previous urban culture had done. For the first time the majority of inhabitants were free men… city dweller and citizen were synonymous terms' (Mumford 1961: 316), and this was reflected in the unrestricted use of public space. The street systems that developed were organic, commercial, and vibrant public spaces.

2.5 Michelangelo's Renaissance Piazza del Campidoglio in Rome

2.6 Cheapside, showing a medieval market cross and multifunctional space

Renaissance and baroque

As the power of secular rulers and interests increased, from the middle of the fourteenth century new piazzas began to emerge in many Italian cities. As Girouard (quoted in Carr *et al*. 1961: 55) notes, 'the idea of a piazza expressing civic dignity and therefore unsuitable for commercial activities had clearly crystallised'. The grand piazzas of the renaissance sought to emulate the classical world, aesthetically and politically.

These ruling interests sought to 'regenerate' the medieval cities by employing artists and architects to beautify them, as well as their own grand palaces, and in the process to assert and display their own status and wealth (Figure 2.5). As Webb (1990: 68) observes, 'the link between art and power is as old as civilisation'. With the flourishing of the arts in fifteenth-century Italy, aesthetic principles, particularly scale and proportion, became essential in the design of urban space. Italian piazzas were beautified piecemeal from their medieval structure, or created afresh by the demolition of part of the town. Commercial traffic and markets were often banned from the centre, while architecture and sculpture reflected the monumental.

Royal patronage lead to similar developments in baroque Spain, and the Spanish in the New World used principles of renaissance city planning drawn from the Laws of the Indies (Broadbent 1990: 42–8). Baroque Paris, again via royal patronage, built its first planned square in 1605, the Place Royal (now the Place des Vosges). Increasingly, therefore, the spaces that resulted were designed to display as publicly as possible, the status and wealth of the ruling classes. Again, the parallels are clear to see in the design of many contemporary public spaces, designed to show off the power and wealth of the corporate/business sector. The balance between public and private interests in the provision and management of public space represents an issue with very deep roots indeed.

Into the modern era 1: space types in London

This and the next section of the chapter outline three studies each from London and New York respectively, in the evolution of urban public space. The studies continue chronologically the history of urban public space and introduce contemporary case study material for London and New York that is picked up again later in the book (see Chapters 9 and 10).

The English marketplace: commerce and community

In most historic English towns today, the chief public space is the marketplace. Girouard (1990: 10) notes 'Many markets have been held in the same place for eight hundred years, and a few for over one thousand. The only centres of resort to rival them in age and importance are the churches'.

In 1600 there were approximately 800 market towns in England. The commercial success of marketplaces in medieval England ironically meant that the open space of the market began to be reduced (and privatised) as stalls evolved into the frontage of built shops (Girouard 1990: 11). Despite the loss of public space, the fact that markets were only held once or twice a week gave the marketplace the opportunity to host other functions. These were invariably other formal or informal public occasions when local people could interact in public life, often around the focus of the market cross. The marketplace was the centre for news and gossip in the town, as well as for buying and selling, and the market cross was the focal point of the market and therefore the town.

The market and the market cross were a crucial part of life for the English urban dweller in a similar way to the agora for the Greek citizen; it offered the urban dweller a chance to partake in public life: religious, political, commercial and informal. The market cross continued its civic development by becoming the market hall or town hall in many market towns as corporations, created by royal charter, replaced the authority

26

2.7 Covent Garden piazza with church and market, 1751

of the lord of the manor, the church, or merchant guild (Girouard 1990: 9–30).

The English marketplace served numerous functions, with commerce being of prime importance. This also demonstrates the reliance of community functions on essentially commercial space; with political, religious and social functions occurring on space created for commercial reasons. However, the space was not always in 'public' ownership, as often the owner was the church or the lord of the manor. Indeed, today, some traditional marketplaces are still owned by the church, although the majority are now owned by the state in the shape of local councils.

For its part, London was a city that had grown rich from trade, and consequently had numerous markets early on in its history. However, unlike the rest of England, London never had market squares, but rather street markets or covered markets. The public space of these markets was almost entirely commercial, while other functions of the space, such as civic life, were negligible (Clout 1991: 148–9). The exception was Cheapside, the centre of London's retail trade, which was often the site of royal and civic pageantry and popular celebration (Figure 2.6). As such, public space in London tended to be more differentiated than in the rest of the country.

The centre of London's government, for example, was at the Guildhall, the site not of commerce but of civic life since Roman times. Social urban public space in London was to be found developing around St Paul's. Mitchell and Leys (1958: 142) describe old St Paul's Church and courtyard as a 'thoroughfare for citizens', and note that by the late sixteenth century the central aisle of the nave of old St Paul's – Paul's Walk – had become the greatest promenade in London. Here the news of the day was whispered, or spoken aloud, here assignations were made and kept, and it was said that more business deals were carried out in Paul's Walk than in the whole of the Royal Exchange.

The middle ages also witnessed London markets specialising in certain goods, such as Billingsgate for fish, and Smithfield for meat. However, many of the oldest London markets were destroyed in the Great Fire of 1666, and were rebuilt as covered markets under royal charter. New markets were also beginning to be built on what was then suburban land by aristocratic landlords, including Covent Garden.

EVOLVING PRIVATE/PUBLIC SPACE

Covent Garden provides a valuable example of how the character of public space can change over time, raising different challenges for those responsible for its management. Covent Garden was originally a residential square begun in the 1630s (see below). Designed by Inigo Jones it was London's first planned formal open space, originally intended for use by courtiers. However, during the Civil War a local produce market sprang up on one side of the piazza, usurping its original function. In 1671 the Duke of Bedford applied to Charles II to make official this daily fruit and vegetable market (Figure 2.7). The market vastly altered the genteel activities and aesthetic of the residential square, which quickly became more insalubrious and public, despite being on private land. The square took on a shabby appearance as stalls became permanent shops and spread across the space, reflecting the general experience of English marketplaces elsewhere. Crime also increased as there were no police patrolling the area until the nineteenth century.

In 1830 a design response was made in the form of the New Market, created with avenues, colonnades, and conservatories in three parallel ranges (Ackroyd 2000: 332). While the Covent Garden area might still have been raw and dangerous, it contained a certain 'social realism' associated with Victorian public urban space. This was despite the contradiction of ownership still being with an aristocratic landlord (Rasmussen 1934: 153–7). Although the Covent Garden fruit and vegetable market was moved out to Nine Elms in 1974, a 'festival marketplace' re-opened in the conserved and renovated structure in 1980. The covered market and a portion of the surrounding open space is now owned by a private insurance company, Scottish Widows, who are solely responsible for managing what was a new type of public space.

With this, the character dramatically changed once again, whilst the company employed private security and CCTV in an effort to keep the

2.8 Covent Garden Market

2.9 Leicester Square with private garden for residents, 1721

market clear of any 'undesirables'. Users can be removed, with force if necessary, for drinking alcohol, playing music, leaflet distribution, or preaching. Furthermore all street entertainers and buskers must have a permit to perform in the area, and to obtain this they are required to undergo an audition to ensure they meet certain standards. A timetable is given to each permit performer who performs at certain times and on certain days of the week. Design and aesthetic changes also seek to keep out those who are not wanted. This is most noticeable through changes in floorscape and street furniture, the placement of which demarcates legally which land is owned by Scottish Widows and which is public, owned by Westminster Council.

For some, Covent Garden, is now facilitating a homogenised commerce aimed at an international clientèle rather than locals. Image and history are used to create a consumable vision of urban public space for tourists (Figure 2.8). Critics argue that the multifunctional market or the social realism of the Victorian market are now largely reduced to pure commercial exchange, where the fostering of civility and community are consequently diminished (Franks 1995). For tourists and many Londoners however, Covent Garden represents one of the great destinations of the capital, and a success story in how to re-invent public space.

The evolution of the London residential square: access and control

The evolution of the London square demonstrates the changing attitudes in England towards public and private urban space, particularly when compared to post-renaissance continental Europe. Webb (1990: 91) observes that with the exception of Trafalgar Square and Sloane Square, all London's planned squares were intended as the private domains of residents in the surrounding properties. As such, the residential square contrasts sharply with the public marketplace as regards access, activity, and therefore design and setting.

The London residential square was first developed in the seventeenth century as commercial speculation by aristocratic landlords who had obtained tracts of land from the church via Henry VIII a century earlier. The first London square to be laid out was Covent Garden in 1631. When

2.10 Gates with gatekeeper on the Taviton Street entrance of Gordon Square, Bloomsbury, shortly before their removal in 1893

the Earl of Bedford applied to Charles I to develop the land, he intended to create a residential area for aristocratic families. These new London residents were country house dwellers who increasingly wanted a winter residence in London for business and for socialising (Girouard 1990: 156).

If Covent Garden did not fully realise its potential for residential development (see above) for the new society Londoners, the squares of Bloomsbury that followed certainly did. The land in Bloomsbury was also owned by the Bedfords, who, with the help of speculative builders, built many of the residential squares in Bloomsbury: from Bedford Square in 1776 to Gordon Square in 1860. The development process employed created new typologies of space and ownership.

The central squares in Bloomsbury were gravelled and fenced off with wooden rails so the affluent residents of the square could promenade in semi-privacy (Girouard 1990: 158). Railing off the central space of the London square to residents became the norm after Covent Garden, preventing stalls, hawkers, carts and so forth from entering, an example being Leicester Square (Figure 2.9).

2.11 Victorian buskers in Leicester Square

Harwood and Saint (1991: 95–7) note that many London squares did not legally achieve full privatisation until the eighteenth century. St. James's Square was one of the first to achieve this by Act of Parliament in 1726. Today, a few – for example Bedford, Fitzroy, Kensington, and Belgrave Squares – are still restricted to key-holding residents. However, the privacy of Bloomsbury went further than this 'The Bedfords maintained an unwavering course, insisting on the finest materials and the largest houses the market would bear, excluding such undesirables as tradesmen and taxis by gates and by strictly enforced regulations' (Webb 1990: 95). Bloomsbury effectively became a gated community until the end of the nineteenth century (Rasmussen 1934: 166); privatised for the exclusive use of the gentry, residents and their servants (Figure 2.10).

Later the Reptonian garden revolution and the park movement of the early nineteenth century had a profound influence on the design of the central squares, chiefly through landscaping and the addition of monuments and statues. Early examples included Grosvenor Square in the 1770s and Soho Square some years later. Still, however, they remained private gardens, for the use of residents only.

THE OPENING OF THE PARK SQUARES

Public promenades did exist at old St Paul's, but formal landscaped public spaces, such as those within the private realm of the residential squares, were not available to most Londoners. Formal 'public' walks were created in London's royal parks, namely Hyde Park, St James's Park, Green Park, and Kensington Gardens, which were all originally royal hunting grounds. The Mall in St James's Park was the first example of this, planned and planted in 1660. Despite these parks being opened to the public in the sixteenth century (apart from Kensington Gardens), official public access was not granted until the early nineteenth century when pressure for urban public space was heightened with the rapid urbanisation of the industrial revolution (Girouard 1990: 269).

Golby and Purdue (1984: 90) describe related attempts made by the middle and upper classes to 'civilise' the new working-class urbanite, attributing the philanthropy of the nineteenth century to the guilt of the middle and upper classes:

By the 1820s and 1830s there was a growing feeling, especially among reforming and Evangelical groups, that although the lower orders seemed to have an inbuilt disposition towards spending any free time they had in sexual excesses, gambling and drinking, the middle and upper classes were not entirely free from blame or responsibility for this state of affairs.

These events brought about the public park movement and access to many of the Georgian residential squares, creating in the process new 'park squares'.

Leicester Square, by contrast, became public through a story of private neglect and public rescue. In 1630 the land was granted to the Earl of Leicester who built Leicester House and laid out public walks as a condition of the grant in what became known as Leicester Fields. After 1660 the Earl undertook the development of a residential square (Kingsford 1925: 53–6). The first formal garden was established in 1727 and later an equestrian statue of George I was added.

However, by the end of the eighteenth century, Leicester Square had ceased to be a fashionable residential quarter. Rather it was becoming a place of popular resort and entertainment (Figure 2.11) which ranged from theatres, bagnios, buskers, and gaming rooms, to collections of curiosities and spectacular exhibitions, such as the Royal Panopticon of Science and Art and the Great Globe, which was built on the (by then) derelict gardens of the square. These and other ambitious ventures were short-lived and the decline of the square accelerated until the vandalisation of George I's statue prompted an outcry which led to an act of parliament enabling the recently formed Metropolitan Board of Works to acquire the gardens. In the event the land was bought by the MP Albert Grant who commissioned a redesign of the gardens to a typical Victorian layout. To raise the tone further there was a statue of Shakespeare and busts of Reynolds, Hogarth, Hunter, and Sir Isaac Newton who had lived nearby.

In 1874 Grant transferred ownership to the Metropolitan Board of Works (Tames 1994: 115). By this time the square was dominated by several major theatres specialising in light entertainment which attracted respectable as well as raffish pleasure seekers. In 1894 the Purity Campaign agitated against the Empire Theatre, part of which served as a promenade for prostitutes, leading to the intervention of the new London County Council to clean things up (Tames 1994: 132).

In the twentieth century, cinemas replaced the theatres but while these flourished the square declined as a public space, dominated by traffic which impeded access to the gardens. Only in the 1990s did Westminster City Council undertake a redesign of the square and the gardens. Leicester Square was pedestrianised and reinvented with an American flavour, including bright lights, glitzy movie premieres, funfair rides, and celebrity concrete handprints in the pavement. A statue of Charlie Chaplin was placed in the gardens to invoke the entertainment tradition of the square.

Today the area has a reputation for danger, excitement, and debauchery, as well as the attraction of its major cinemas. This is in keeping with a history where the respectable and dissolute have inhabited the same space. Westminster Council approved an action plan in 2002 aimed at regenerating the square and surrounding area, a repetition of the cycle of 'plans' and redevelopments through the square's long history. The action plan implemented changes in management to control who can use the square, and what activities are allowed, and followed a 'zero tolerance' campaign by the local police and Westminster Council. It has now been supplemented by the designation of a 'business improvement district' (BID) in the surrounding area, formed in 2005 to support business interests by tackling the square's complex management problems (see Chapter 10 for a detailed discussion and analysis).

PRIVATE – AND PUBLIC – EXCLUSION

Despite demonstrating that on occasions, the private sector can be the root cause of neglect, the history of Leicester Square remains distinct from London's other residential squares. The history of these squares generally demonstrate the deep-seated desire of some sections of society to restrict access to certain types of public space; extending in the case of Bloomsbury to whole neighbourhoods (echoing the debates over gated communities today). Planned public space was for the privileged few, and, initially at least, there was no recognition that the design and aesthetics of urban public space could foster civility and health among the masses, as was widely accepted in continental Europe.

In one respect Leicester Square does typify many of London's other residential squares, where the public sector (as opposed to the private) is now increasingly behind attempts to restrict user freedom in the broader 'public interest'. While Bloomsbury's streets and many of its squares are now in the public domain, being owned and managed by the local authority, restrictions on behaviour through a host of restrictive bye-laws still remain. Increasingly council-owned squares such as Russell, Bloomsbury and Gordon Squares have had their design and management

altered to deter the homeless, beggars, street vendors and homosexuals who, until recently, used to cruise there.

So, in one form or another, restrictions remain, effectively deterring certain cultural and social groups. The story illustrates how another type of public space has been gradually transformed, first by way of a transition from an elite space type to a shared space, but latterly, through restrictions on use, designed to curb some of the perceived excesses of the users to whom the space has been opened up.

Civic space: display and public gathering

The final type of space represents the various forms of space that exist primarily for gathering and display. Classic examples include two of London's most famous spaces, Trafalgar Square and Piccadilly Circus.

GATHERING AND CIVIC DISPLAY

When designed and built, Trafalgar Square was the only purpose-built public square in central London. A space that is framed by the cultural institution of the National Gallery, it contains the symbolism of a bygone empire, and has a history of public gatherings of demonstration and celebration. The square was conceived by the architect John Nash as part of plans for the beautification of the vistas around Charing Cross.

Mace (1976: 31–42) notes that Nash first proposed the project in his report of 1812, but the project was formalised by an act of parliament in 1826 to enable the public purchase of land for the creation of a 'large splendid quadrangle … to embellish and adorn the metropolis'. The new space was to have strict rules to prevent commercialisation with a fine of 20 shillings a day for 'all signs or other Emblems, used to denote Trade, Occupation, or Calling of any Person or Persons'. The National Gallery, established in 1824, was in need of a new building as it was growing out of its premises at Pall Mall. It was John Nash in his original plan who suggested that an institution could be placed on the north side of the new space, in so doing helping to frame it. The National Gallery was to turn the square into a cultural space, and was eventually completed by William Wilkins in 1840.

There had been much discussion of a monument to commemorate the death of Nelson and the British victory at Trafalgar, but this was independent of the newly named square. Nelson's Column was seen as a fitting tribute, and the new public Trafalgar Square had the name to fit. Charles Barry, who became the chief architect of the scheme protested that it would be out of scale with surrounding buildings, particularly the new National Gallery, and would block the vista. Nevertheless Nelson's Column was erected in 1842.

2.12 St Patrick's Day in Trafalgar Square

For his part, Barry thought that the first public square in London should be more artistic in nature:

> Giving scope and encouragement to sculptural art of a high class, and … giving that distinctive and artistic character to the square, which is so needed in public areas and squares of London, to excite among the classes that respect and admiration for art, so essentially necessary to the formation of a pure and well grounded national taste.
>
> (quoted in Mace 1976: 77)

In this regard, the debate reflected notions that aesthetic properties could give rise to public pleasure, and reflected the Ancient Greek view that at least one role of public space was to meet the higher spiritual needs of onlookers. Barry suggested that the sculpture be grouped on pedestals in the square in a regular axial form, an arrangement believed to be in keeping with the idea of a cultural space. The designs also included two fountains, a design feature considered helpful for a baser function, the policing of a possible 'urban mob' or riotous public assembly.

Over the following hundred years, Trafalgar Square turned into a home for statues of military and naval war heroes, who occupied the plinths Barry had intended for great art. Trafalgar Square also soon became a meeting area for gatherings and demonstrations; its central location and public nature making it a natural arena for Londoners to assemble, particularly since there was nowhere else in central London. Significantly, assemblies were banned in 1848 soon after completion of the square, but this was later relaxed until Bloody Sunday in 1887, a demonstration against unemployment. This event led parliament to debate the nature of the square in 1895. If the space was public, a Liberal MP claimed, then the

right of assembly at the square should be permitted. The Tories claimed that the land belonged to the queen and was therefore private (Mace 1976: 155–200). This debate echoes public versus private arguments seen in connection with public space to this day.

In recent times Trafalgar Square has been the scene of many assemblies, whether demonstrations – such as the suffragettes at the start of the century, nuclear disarmament in the 1960s, trade unions in the 1970s, anti-apartheid rallies in the 1980s, and anti-war rallies in the 2000s – or regular celebrations, such as New Year's Eve and sporting victories (Figure 2.12). However, it also functions as a cultural space, representing London to tourists, and high culture via the presence of the National Gallery. Until recently it has also been a largely barren place for much of the year, encircled by traffic and containing only tourists and pigeons.

Richard Rogers and Mark Fisher (1992: 105–6) were amongst the first to suggest that traffic on the northern side of the square could be re-routed, connecting the National Gallery to the central space. Later, under the 'World Squares for All' project, Norman Foster suggested better access to the square's monuments, a redevelopment that commenced in December 2001 and was completed in 2003, and that includes a new chain café on the square itself.

The Greater London Authority (GLA) now manage the space and the square has its own bye-laws and 24-hour 'heritage wardens' preventing spontaneous activities such as music and non-planned demonstrations. In addition, London's first mayor, Ken Livingstone, managed to clear the square of pigeons by banning the sale of pigeon feed, until then the only legal commercial activity allowed in the space.

So, like many of London's public spaces, Trafalgar Square has been transformed, but in this case bringing its function more into line with that originally envisaged, as a space for gathering and civic display, rather than what it had become, a traffic roundabout. Trafalgar Square has always been heavily managed in one form or another since its original construction, but in common with many other spaces around the world, this management did not extend to perhaps the most pervasive and character-changing element of the street scene, the growth and growth of the private car. The appropriate balance between people and vehicles in public space marks a source of considerable management conflict in many cities throughout the world. In Trafalgar square, this relationship has been shifted somewhat back in favour of the space, and away from the car, although three sides of the space are still dominated by heavy traffic.

GATHERING AND COMMERCIAL DISPLAY

While Trafalgar Square is an example of a high-profile planned urban public space for public gathering and civic display, an example of high-profile organic urban public space for informal public gathering, and latterly commercial display is Piccadilly Circus. Piccadilly Circus in its original form of 1819 was part of Nash's grand scheme that linked Charing Cross and Regents Park, its function being no more than a road junction. The public space was created in the 1880s when Shaftesbury Avenue was cut through from the north-east, enlarging the circus area. A focal point was added in 1893 with the erection of the Shaftesbury Memorial Fountain (Figure 2.13), better known as Eros (GLC 1980: 7).

Even before Eros, the circus was widely known as a centre of entertainment and popular pleasures (Tames 1994: 119). This role rapidly expanded in the late Victorian period with the building of theatres, music halls, shops, and restaurants around and near the circus. Meanwhile Shaftesbury Avenue added to the acute traffic congestion that contended with the movement and congregation of pedestrians (GLC 1980: 7). Yet by popular choice reflected in and reinforced by the new picture postcards, this constricted and misshapen space became the 'hub of the Empire', a magnet for Londoners and visitors (Oxford 1995: 7). Piccadilly Circus, like Leicester Square, but more emblematically, became the focus for a rejuvenated London life.

It was commerce that gave Piccadilly Circus the glamour and significance the public expected of it. In the 1890s electric advertising signs began to appear on buildings on the north-east side, which the leases previously granted by the Metropolitan Board of Works were unable to prevent, although the rest of the circus which was in the ownership of the crown remained clear of signs (GLC 1980: 11).

Refined critics deplored 'those many-coloured electric illuminated advertisements' as 'blatant, vulgar and useless' and 'a hideous eyesore which no civilised community ought to tolerate' (Ditchfield 1925: 102).

2.13 Piccadilly Circus in 1897

2.14 Victory celebrations and electric advertising in Piccadilly Circus, 1945

32

But the lights quickly became an essential part of the circus's ambience and perhaps its most publicised aspect through picture postcards featuring night time views (Oxford 1995: 56). The lights were even used to display election results on an electrical zipper in the 1920s (Ditchfield 1925: 102). In this setting, crowds gathered and on occasions of celebration or major events filled the entire space (Figure 2.14).

In contrast to Trafalgar Square, Piccadilly Circus organically became an important space for civility and community, for celebrations and for democratic exchange, with the steps of the Eros fountain becoming a major meeting and hanging out space for Londoners and visitors. The steps of Eros, and the space as a whole contain the same mixture of the respectable, in the form of the shopper or tourist, and dissolute, in the form of the drinkers and loiterers for whom the space has long been a magnet. In contrast also, the space demonstrates, like no other in London, the dominance of commercial interests, encouraging users to consume, and the results of fragmented management responsibilities resulting in a light touch management framework. Piccadilly Circus is now, however, part of the same 'business improvement district' as Leicester Square which is marketing the space for the large private-sector landlords who own the advertising space and buildings around Eros (see Chapter 10).

Into the modern era 2: space types in New York

Continuing the historical evolution of public space, three further studies on the other side of the Atlantic are explored in central New York. Like the London studies, these were chosen for their historical importance and for the contemporary trends they help to illuminate.

Town squares and parks: the Americanisation of European culture

Early settlers in North America from Spain, France and England each brought with them their own traditions of public space. The spaces created in the New World by each group of settlers had their provenance in parts of Europe, but soon evolved into a new and distinct typology.

The Spanish influence is to be found in the southwest of the United States, particularly in California. Public space in Spain is typically organised around a central plaza, usually in the form of a paved square. Early plazas in America hosted a marketplace in the centre, often containing a corral for animals, while also being used for formal public events such as celebrations and bullfights. The French influence can be found in the southern states, principally in Louisiana. Loukaitou-Sideris and Banerjee (1998: 37–9) note that the main public space for the French settlers was the place d'armes, a more formal space than the plaza, that was initially intended for military parades and training, but came to host civic events and public celebrations.

The English influence was most obvious through the Puritan communities of New England, and brought from home the tradition of common land that was originally used for the communal grazing of cattle and horses, and for the training and parading of local militias. Building plots around the common became built up in time, first with meeting houses, then courthouses, churches, shops, and schools (Webb 1990: 116–118). The common developed into the village green in the early nineteenth century after Puritan religious dominance was broken and the militias disbanded. Boston Common in Massachusetts is an example and is today the oldest public park in the United States, dating from 1634.

The plans for settlements were often laid down by respective colonial governments, such as through the Spanish colonial Laws of the Indies. 'Colonial towns represented integrated wholes; their public was more or less a homogeneous, uniform entity. The town centre, represented by the square, was conceived as a setting for collective action. People went there to participate in public activities that were often political and carried communal meanings' (Loukaitou-Sideris and Banerjee 1998: 37–8). Urban public space was therefore developed to serve social and democratic needs. But the square was often viewed as a central modular part of an otherwise organic settlement.

Some colonial public squares still exist in central locations, but many were lost as American cities expanded massively in the nineteenth century. Rapid immigration and urbanisation meant that despite the abundance of land elsewhere, urban space was in short supply. Moreover gridiron plans lent themselves to quick and easy speculative development.

PARKS IN NEW YORK

In New York, as in the United Kingdom, a park movement was formed to lobby for citizens to have greater access to open space. Heckscher (1977: 161–70) suggests this was to 'moralise' New York's citizens, particularly immigrants, a comparable movement to the attempts to moralise the 'working classes' in nineteenth-century Britain. Yet it was also touted as a method to increase real estate values. Central Park, the first landscaped park in the US, was made possible by demolishing many central Manhattan blocks – between 59th and 110th Street, and 5th Avenue and 8th Avenue – creating over 800 acres of space in the late 1850s. Frederick Law Olmstead and Calvert Vaux created a park in the English romantic

2.15 Bryant Park with New York Public Library in the background

tradition, with many design features conducive to health, morality and civility, as they saw it. Various other smaller scale parks and squares were designed and built throughout the rest of the nineteenth century.

One of these, Bryant Park, is a park square in midtown Manhattan, on 42nd Street (Figure 2.15). The park site originally contained a reservoir surrounded by public space, and became the site for the 1853 World's Fair. After troops were stationed there in the American Civil War the area became a public park, named after the leading advocate of the creation of Central Park, William Bryant. In 1911 New York Public Library was opened adjacent to the park (www.bryantpark.org).

The recent history of Bryant Park typifies the approach to public space in New York. Zukin (1995) suggests that New York is marketing itself through culture and a commercial economy based on cultural symbols and links these ideas with the increasing privatisation of urban public spaces, claiming that cultural symbols and design can be used to include or exclude certain social, cultural and racial groups. As post-industrial New York cut much of the funding for the parks department in the 1960s, amid general social decay Bryant Park became a haven for 'undesirables' such as drug dealers and the homeless. Latterly the park has been taken over by the Bryant Park Restoration Corporation, a private company that manages and decides on commercial and design issues related to the park (www. bryantpark.org).

By reclaiming the park for office workers through 'pacification by cappuccino', Zukin (1995: 31) notes that 'The cultural strategies that have been chosen to revitalise Bryant Park carry with them the implication of controlling diversity while re-creating a consumable vision of civility'

That the square is utilising cultural symbolism through design to attract and deter specific social, cultural and racial groups is, for some, evident not just in the cappuccinos but in the expensive restaurant, bar and grill, an open-air cinema screen sponsored by HBO, that shows old Hollywood movies in the summer, and in the Google-sponsored wireless network. To add to the deterrents, the park features benches that prevent lying down and sprinklers that prevent lying down on the grass. Private security regulates activity and behaviour and keeps out 'undesirables'. Bryant Park certainly illustrates the notion that public space in New York is increasingly consumed, something very different from the diverse traditions and cultures that arrived in the United States from Europe.

Downtown space 1: skyscrapers and corporate space

Rem Koolhaas (1978: 18) tells a story of a city without a manifesto, which the author then retroactively constructs and analyses. The construction of the gridiron system on Manhattan Island is described by Koolhaas as 'the most courageous act of predication in Western civilisation'. The gridiron divided Manhattan into real estate blocks suitable for speculation, essentially privatising the whole island, in much the same way as the aristocratic landlords did in London.

Girouard (1990), in describing the rise of New York, noted that the only substantial urban public space in the plan was a 'parade ground' between 23rd and 24th Streets (Girouard 1990: 314). New York emerged into Loukaitou-Sideris and Banerjee's (1998: 5) 'walking downtown phase' of development as mercantile economies industrialised, creating the first specialised central business district (CBD) in the Wall Street area. At this time the squares and markets of the old colonial settlement in the south of Manhattan were now replaced by businesses – banks, insurance companies, trust companies – located around Wall Street and the new stock exchange.

Merchants and outdoor markets felt the pressure to move from Manhattan's harbours to establish stores on Broadway, New York's Main Street. With the loss of public space under the 1811 gridiron, avenues and streets became increasingly valuable urban public spaces in New York. Retailers were aware of this and store windows and advertising began to adorn the streets. Stores grew in size quickly, and department stores began to appear, such as Macy's in 1857. Loukaitou-Sideris and Banerjee (1998: 6) observe that it was department stores that brought women into a downtown area that previously had been a largely male domain. These interconnected themes produced a city where public life had begun to commercialise.

> In the eighteenth century people gathered at the town centre to participate in civic functions or public events. One century later, people came to the CBD to conduct their own business. Buying, selling, trading, and window shopping became the primary activities conducted in American city centres.
>
> (Loukaitou-Sideris and Banerjee 1998: 7)

The commercialisation of urban public space produced two distinct types of downtown urban public spaces. First, there is the entertainment downtown: the theatre and shopping districts of 5th Avenue, 42nd Street, and Times Square. Today in New York, and many other cities, this is the symbolic downtown. Second, the business district, creating skyscrapers that gave way, via zoning regulations, to public space requirements.

2.16 The Sony Plaza, public/private space

THE RISE OF PUBLIC/PRIVATE SPACE

Taking the latter first, as transport improved and residents of New York moved out, the headquarters of the large corporations downtown expanded in the only direction they could, up. During the first fifteen years of the twentieth century the downtown city blocks of New York were transformed into a vertical Gotham skyline. But while New York's skyline was becoming increasingly dramatic when viewed from afar, the streets below were becoming dark, airless spaces, generating the need for some form of intervention.

Kayden (2000) charts the evolution of semi-public corporate space as a reaction to changes in city planning zoning ordinances in Manhattan, Brooklyn, and Queens, and argues 'The history of privately owned public space is inextricably linked to the history of zoning in New York City'. He describes the history of planning ordinances in the city, starting with the New York City Building Ordinance, adopted by the Commission on Building Districts and Restrictions in 1916: the first comprehensive zoning ordinance in the US. The 1916 ordinance sought to protect the interests of wealthy influential businessman as well as the health of the public. Loukaitou-Sideris and Banerjee (1998: 48) observe that the former was clearly the strongest incentive, as a tactic to help stabilise or even enhance land values. Health issues were nevertheless a salient topic during the early years of the nineteenth century, and Ken Worpole (2000: 10) describes how reformers on both sides of the Atlantic 'strongly predicated the benefits of clean water, sunlight, and fresh air'. This was therefore also a major factor.

The 1916 ordinance did not, however, stop the podiums of skyscrapers from covering the full lot or site. Therefore in 1961 a new zoning resolution was created, that 'introduced a new type of space: privately owned public space, located on private property yet … physically accessible to the public at large' (Kayden 2000: 11). The authorities were now acknowledging not only that light and air were needed in the streets but also that public urban space was required. The incentive to developers to provide this space was the offer of a greater maximum floor area for a building if ground-level plazas and arcades were accessible to the public at all times.

These public accessible areas were given as of right regardless of design as long as the spatial thresholds were upheld. However, Kayden notes that this was abused by many developers who used the plazas as loading areas or garage entries, or just built barren empty spaces. As a result, over the following decades incremental amendments were made to the ordinance to ensure that the system was not abused, and that spaces were usable and attractive. Further corporate space typologies have been added to the official city list through legislation, making a total of twelve public/private space typologies: Plaza (1961), Arcade (1961), Elevated Plaza (1968), Through Block Arcade (1969), Covered Pedestrian Space (1970), Open

Air Concourse (1973), Urban Plaza (1975), Sidewalk Widening (1975), Residential Plaza (1977), Through Block Connection (1982), Through Block Galleria (1982), and Other Spaces.

Each typology has design standards that were pioneered through the work of William H. Whyte (1988) in the 1970s through his research on how public spaces function (see Chapter 1). This introduced standard design elements that developers could pick such as foliage, benches, cafés, drinking fountains, sculpture, etc, and prohibited the spaces being used for car parking, loading, and other uses not benefiting the public (Kayden 2000: 21–45).

The diversity of typologies has meant a diversity of such spaces, for example the Sony Plaza that combines a plaza and covered pedestrian space to produce an entertainment mall (Figure 2.16). Zukin (1995: 3), however, cites this as an example of the negative side of this increasing 'privatisation' of public spaces; an example where a covered pedestrian space intended for the public has controlled access and contains mass advertising for Sony Products. By 2000, over 500 public/private spaces had been delivered using incentive methods. But, as will be discussed in Chapter 3, when the owners of such spaces are large commercial corporations, design and management approaches are sometimes used to reinforce a desired corporate image that leads to the exclusion of individuals or groups who do not fit.

Downtown space 2: entertainment space

Returning to 1900, the trends to specialisation and zoning of uses across American downtowns increasingly resulted in another type of public space, this time for entertainment. The evolution of New York's famous Times Square can be used to understand the commercialisation of this type of public space throughout the twentieth century, and forms the subject for more detailed analysis and discussion in Chapter 9.

In fact, the term 'square' is somewhat of a misnomer as Times Square is merely the junction where Broadway intersects with 7th Avenue, forming a 'bow-tie' shaped area of two triangles: to the south from 43rd to 45th Street, then a second wedge from 45th to 47th Street to the north. As early as the 1860s the area began to develop into an entertainment district as open public space was reduced and vaudeville, theatres and brothels became the main attractions (Taylor 1991). For the first time in nineteenth-century America the lines were blurred between 'respectable and dissolute spaces' (Berman 1997: 76). In the approach to the twentieth century, American cities commercialised at an ever-increasing pace, with New York leading the way. One outcome was the spread of mass advertising, particularly following the electrification of the city in the 1880s.

The themes of theatre, illegal activity, and advertising feature throughout the history of the Times Square district, and are parallelled in London's Piccadilly Circus. Times Square too seemed to symbolise the vitality of the new century.

In 1904 the district gained its present name when the New York Times newspaper relocated its offices to the Times Tower on the southern end of the bow-tie (Figure 2.17). Times Tower was the second tallest building in the city in 1904, raising the prominence of Times Square through its moving-light news zipper and its 'ball lowering' New Year's Eve event. In that same year the subway station at Times Square opened, a major intersection of several lines. This brought people from all over New York to Times Square. Hotels and restaurants such as the Astor and the Knickerbocker opened at the turn of the century, adding a sense of class to the vibrant surroundings. Advertisers soon recognised the potential commercial gain of placing billboard signs around the area and the lights became multicoloured (Figure 2.18). In 1916 a new zoning ordinance permitted full-scale giant billboards in the area (Sagalyn 2001: 32–43). As such, Times Square quickly became a symbol of American free-market values.

In trying to summarise the social and architectural strands of the space Ada Huxtable (1991: 360) commented:

> Size and anonymity make it both a private and a public place, where offbeat or offcolor desires can be openly or secretly satisfied, but where New Yorkers can also gather at moments of crisis

2.17 Times Square looking south towards the Times Tower in the 1920s

2.18 North Times Square at night, showing signage and advertising in the late 1930s

2.19 Times Square at night showing election results on Times Tower in 1952

2.20 Times Square in the 1970s, perceived as dangerous and seedy territory (from Sagalyn 2001: 17)

and triumph; to celebrate the end of war, to wait for and share important news.

As such there is a duality to the square, offering a civic space (Figure 2.19) and also a consumer space.

Huxtable (1991: 358) observes that the square changed from 'news as advertising to advertising as entertainment', but also gradually declined as movie theatres replaced the theatres and real-estate values dropped after the Second World War. Increasingly the area gained a reputation as a sordid district of drug dealers and prostitutes, with the theatres being used for peep shows. The area was perceived to be the domain of the ethnic minority male, stereotyped as pusher and pimp, while 42nd Street became dubbed the 'dangerous deuce' (Figure 2.20) (Sagalyn 2001: 44–52).

Throughout the 1970s and 1980s various 'regeneration' schemes were created for the Times Square and 42nd Street district, only to be thwarted by real-estate slumps. In 1992 the Times Square Business Improvement District (BID) was created. Major corporate players were courted and the Times Square district now contains converted theatres with a Disney musical and megastore, a Madame Tussaud's museum, a Warner Brothers studio tour, an MTV store, a New York Yankees store, a Planet Hollywood restaurant and a Hilton Hotel. This is in addition to four high-rise office buildings (www.timessquarebid.org).

Reichl (1999) is scathing about the commercial and political motivations behind the regeneration of Times Square. Approaching the redevelopment from a cultural and racial standpoint, Reichl (1999: 171) observes that 'cultural symbolism' is being used as a vital component for including and excluding certain cultural groups in Times Square, and explains,

> cultural symbols inscribed in the urban form serve to establish and demarcate control over urban spaces. … Race and class are fundamental characteristics expressed in these cultural codes.

Reichl suggests cultural symbols are used through management and design to explicitly and implicitly control access to, and behaviour in, public space. This relies on users' perceptions and interpretations of urban public space and 'the other', this usually being a certain social group, within the space.

Once user perceptions of Times Square changed from those of an ethnic ghetto to those of a safe white-collar entertainment district, then the social as well as the physical regeneration was complete. Yet, for some, this 'success' has been achieved at a high social cost. The case demonstrates both close parallels to experiences across the Atlantic, but also that it is the ongoing management of spaces, often long after their original development, that determines how they are experienced by different users, and how, as a result, their character and clientele can change (often dramatically) over time.

2.21 Modernist functional space

Modernism and beyond: positive and negative urban space

This brief historical review of public space would not be complete without some reference to modernist urban space, and the post-modern reactions to this movement, both of which have had profound impacts on how space is managed.

Modernism and public space

Modernism saw the city as a machine, with form following function, and treated urban public space as an undifferentiated whole, with a concern for light and ventilation uppermost, and seen as decisive benefits for health. Social and psychological needs were generally eschewed by the modernists, and therefore the function of public space was never fully considered. As such the large areas of open public space found in many modernist projects typically had no prescribed social activity or function (Figure 2.21). Madanipour (2003: 202) notes how these open spaces were also unconnected:

> What resulted was vast expenses of space which could have little or no connection with other spaces of the city and could be left under-used, only to be watched from the top of the high rise buildings or from car windows. In this sense such space can be considered 'negative', in that its role is entirely subservient to that of the buildings in which the 'life' of the city is deemed to take place.

By contrast, 'positive' urban space can be seen as a container of public life, which, as the discussion in this chapter has shown, has been the dominant view of public space throughout history. Indeed, writing in the late nineteenth century, long before the modernists began their work, Camillo Sitte can be viewed as one of the first critics of the modern approach to city building. Sitte (1889: 53) eulogised historic spaces for their random and artistic city aesthetic (Figure 2.22), and instead attacked the uniformity and 'the artless and prosaic character of modern city planning'. His work was to be an inspiration for future critics of the modular regularity of the modernist city.

With reference to open space, Sitte criticises the power of the engineer and hygienist in determining design; the tendency of open space to be the unconsidered remainder of a site after a building has been placed upon it, the unenclosed open nature of modern streets and plazas, and the regularity of spaces. The importance of Sitte's work is that many of his criticisms are still relevant to contemporary public space, despite what some have characterised as a highly selective reading of the evidence (Bentley 1998). Sitte observed a convergence in urban public space designs that no longer had any link to the diverse artistic or cultural identity of man. Public space to Sitte was too often an afterthought.

SOCIAL CRITIQUES

Contemporary critics, by contrast, have tended to focus on social critiques for the failure of modernist public space. Sennett (1990: 4–5), like Sitte, eulogises past civilisations, particularly the ancients, in his case with reference to participation in public life. He argues that modern public life is too personalised, and it is modern society's obsession with personalities that has created a society where the majority of people have no real public role.

> The Ancient Greek could use his or her eyes to see the complexities of life. The temples, markets, playing fields, meeting places, walls, public statuary, and paintings of the ancient city represented the culture's values in religion, politics, and family life. [By contrast] it would be difficult to know where to go in modern London or New York to experience, say, remorse.
>
> (Sennett ,1990: xi)

Sennett (1977: 12) blames modernism for creating 'dead public space' where spaces are isolated and isolating and makes the criticism of many that modern public space is too often a space to move through rather than a place to be. He recognises that the city itself is an amalgamation of strangers and alludes to the problems the postmodern city dweller has in taking pleasure from the urban experience, particularly when space is divorced from context and sociability. He observes that the stranger is a necessity of the city, but 'The stranger himself is a threatening figure, and few can take pleasure in that world of strangers' which is the cosmopolitan city (Sennett 1977: 3).

Many critics ascribe the failure of modernist space to the poor definition between public and private particularly with reference to crime. One of the most vehement was the influential writer Jane Jacobs who blamed modernist urban design for disrupting stable social relationships. Thus her classic critique discusses public space with reference to safety on sidewalks and lists three qualities a public street should have for handling 'strangers':

2.22 Artistic space

First there should be a clear demarcation between what is public space and what is private space. ... Second, there must be eyes upon the street. The buildings on a street equipped to handle strangers ... must be orientated to the street. And third, the sidewalk must have users on continuously, both to add to the number of effective eyes on the street and to induce the people in buildings along the street to watch the sidewalks in sufficient numbers.

(Jacobs 1961: 45)

Jacobs' text repeatedly cites the 'stranger' within public space, with reference to those who are not local residents she is familiar with. This term creates an element of suspicion and danger within public space, and moves the social argument through to a psychological one: the perception of danger or crime.

In another classic text, the anthropologist Edward T. Hall examined the psychological impact of urban space. With reference to modernism, Hall also criticised the mass urban renewal schemes of his native US which separated people from their cultural context, particularly blacks and latinos (Hall 1966: 155–8). Hall instead argued for public space that embraced the numerous cultural strands. 'One of man's most critical needs', he argued, 'was 'for principles for designing spaces that will maintain a healthy density, a healthy interaction rate, a proper amount of involvement, and a continuing sense of ethnic identification' (Hall 1969: 157).

In summary, critiques of modernist urban public space are numerous and diverse, and argue that the movement led to a homogenisation of spatial types, ignoring the social and psychological needs of an increasingly diverse city. The imposition of a uniform aesthetic vision produced space that divorced its users from history and culture, and too often rendered urban public space as functionless while disrupting social relationships and creating suspicion of strangers within it. The movement demonstrated both the fundamental impact that design can have on the use and viability of public space (in this case often negatively), but also, as a consequence, that an aesthetic vision of public space, to the exclusion of other factors, can be a very dangerous thing.

The return to positive urban space

In the postmodern world, with the spread of an increasingly universal set of urban design principles (see Table 1.1), a general return to traditional urban space has been witnessed. Advocates argue that such urban space has the potential to support a range of complimentary social, economic and physical characteristics, such as the universal positive characteristics suggested in Table 1.2.

To achieve this, however, the modernist experiment has shown that it is first necessary to get the physical container correct, in order that the activities within can thrive. This is not to make a physically deterministic argument that the shape of the space will determine by itself the quality of the 'place' that emerges and the degree and type of human interaction, but it is to argue that some forms of space make it virtually impossible for meaningful human interaction to occur, and therefore for a strong (or any real) sense of place to emerge. Conversely, the right physical container will greatly increase the potential for a liveable local environment to be created and sustained (Bentley 1999: 125; 184).

Led by Le Corbusier who eschewed the use of traditional streets as 'oppressive' and constricting (quoted in Broadbent 1990: 129), the modernists rejected urban systems based on perimeter blocks (buildings defining spaces – Figure 2.23), and instead favoured freestanding buildings sitting in space. This allowed the buildings, rather than the public spaces, to take centre stage – 'object' rather than the 'ground' – and over time, through repetition of object-oriented building forms, shattered the urban block system. Lefebvre (1991: 303), described this as a 'fracturing of space' and concluded that the resulting disordering of elements was such that the urban fabric itself was also torn apart. Trancik (1986: 21) recognised that modernism itself had worthier ideals, but 'Somehow – without any conscious intention on anyone's part – the ideas of free flowing space and pure architecture have evolved into our present urban situation of individual buildings and isolated parking lots and highways' (Figure 2.24).

Other worthy, if often misguided, intentions were reflected in the proliferation of public health and planning standards throughout the second half of the twentieth century specifying road widths, density thresholds, land-use zoning, space between buildings and almost every aspect of public space. Ben-Joseph (2005) describes these as the 'hidden language of place making' arguing that today they still dictate much of the form and function of urban space around the world. In doing so, he argues, often the original purpose and value of such standards are forgotten, as the bureaucracies put in place to implement them increasingly do so in a manner that has little regard to their actual rationale, and even less to the knock-on effects of their existence. The results have been universally criticised for the bland, repetitive and sanitised public spaces that an over-emphasis on non-place specific standards can deliver (Figure 2.25).

2.23 Perimeter block urban systems

2.24 Modernist free-flowing space

BACK TO COMPLEX URBAN SPACE

By contrast, recent urban design has moved away from object architecture, arbitrary zoning and standards and above all from free-flowing space towards buildings as background defining 'positive' object spaces; typically streets punctuated by occasional squares. This return to streets follows the debunking of a further modernist tenet, the separation of vehicles and pedestrians. Carmona *et al.* (2003: 79) argue that 'sustainable urban design … requires patterns of development able to accommodate and integrate the demands of the various movement systems, while supporting social interaction and exchange'. Therefore, whilst tensions often exist between the use of public space as movement space for cars and other vehicles, and its role as connecting and social space for pedestrians, multi-purpose public space should only separate the two where absolutely necessary.

Numerous authors accept and support this idea of public space as both connective tissue and social milieu (Appleyard 1981; Engwicht 1999; Hass-Klau *et al.* 1999; Jacobs 1993; Moudon 1987; etc.). The Project for Public Space (2001), for example, suggest that good public space should provide good access and linkage alongside a sense of comfort and image, viable uses and activity, and strong sociability. Lang (2005: 370), following the most wide-ranging analysis of 50 international case studies spread across 50 years concludes 'The major clash in urban design paradigms has been over the way streets are considered. Are they seams or edges? … As seams they join blocks together, as edges they divide districts'. It follows that as seams streets focus primarily on bringing activities together, whilst as edges their role is dominated by movement. For Lang, we need to ask 'How comfortable and safe should

we strive to make the world?' when the 'uncomfortable bustle of streets is more popular'.

This requires the combining of street roles and avoiding separation without good reason. In turn, rather than the idealised simple, separated, and 'logical' forms of modernism, it creates and/or perpetuates the infinitely complex stage for management that is the modern, 'traditional' city.

Conclusions

This chapter has provided a rapid tour through the historical evolution of Western public space from antiquity to postmodernity. It demonstrates how many of the issues facing the management of public space today are not new, and often relate to constraints imposed by ownership public or private, and the range of public/private variants in between.

Pre-modern European urban public space had multiple functions, themes, and meanings, which have been repeated through history and are still relevant today. Pre-modern urban public space was discussed in relation to commerce, democracy, community participation, social hierarchy, access, civic obedience, informal social interaction, individual well-being, the power of the state/church, the display of status and wealth, and art and aesthetics. All these are salient topics in the current debates surrounding public space, as are the four primary functions that could be identified in the production, use and management of public space in pre-modern Europe:

2.25 Standards-dominated space

- to facilitate commerce
- to project power, sacred or political
- to display the status and wealth of the ruling class
- to foster civility and community.

The production and use of public space has been both formally planned from above and generated organically from below. In practice, however, both are subject to the unending urban cycle of change and conflict, dissolution and regeneration. Management therefore needs to adapt to competing and ever changing public space functions and demands.

The three London studies illustrate that public and private urban space have derived from a wide variety of ownership, access, and functional patterns. Now even the highest profile central spaces are increasingly subject to commercialisation pressures and increasingly this is generating new forms of management to eliminate perceived elements of 'disorder'. Discussion of the English marketplace demonstrated, however, that marketplaces have long been regulated, for commercial motives, whether by public or private owners, whilst still maintaining their civil and community functions. Civic and residential spaces have also been carefully managed, and after a period of universal decline in the quality of public space, as attention switched to the needs of the motor car, a new realisation has dawned that new modes of management may be required, not least to enhance the image of London in the global tourist market.

New York, like London, has struggled against increasing decay and disorder in public space, and in many cases has chosen privatisation and/or regulation as a means of addressing this, both of which are more prevalent in New York than London. Sometimes it is the City of New York

that is regulating private interests through legislation, for example, the use of zoning ordinances to give some order to the provision of public space. Elsewhere government intervention is acting to facilitate the private sector in the provision and management of public space.

The differences between the political systems of the UK and the US help explain why the New York studies are historically and currently more dominated by commercial considerations. New York City receives very little federal funding compared to London's financial dependence on central government, and therefore has to constantly seek global commerce in an effort to stay financially stable (Fainstein 2001: 82–4). The city almost went bankrupt in the 1970s, and now has to raise much of its income from business tax. This explains the financial drives behind the New York studies, particularly Times Square.

The chapter has demonstrated how the production, use and management of public space is shaped by the changing dominant forms of power, wealth and ideology. The discussion of London and New York show, however, that a diversity of historic public spaces types, shaped by different regimes, have increasingly converged in the age of globalisation. Contemporary postmodern public spaces are increasingly characterised by links to global commerce and to leisure and entertainment, and by the intensive management required to maximise financial returns and user satisfaction. Following the short-lived cul-de-sac that was modernist urbanism, increasingly the 'traditional' 'positive' forms of space that characterised earlier times have also been embraced. These eschew the simplistic overly logical physical forms of modernism, but, as future chapters will show, have not yet moved beyond highly compartmentalised modes of management.

3.1 Neglected public space

Chapter 3

Contemporary debates and public space

This chapter draws on different scholarly traditions including cultural geography, environmental psychology, urban design, and urban sociology to highlight the key tensions at the heart of the contemporary public space debate. In Chapter 1 it was argued that critiques of public space can often be placed into two camps, those who argue that public space is over-managed, and those who argue that it is under-managed. This, of course, greatly over-simplifies a complex discourse on public space that this chapter aims to further unpack. In fact there are a series of discrete but related critiques of the contemporary public space situation that the first part of this chapter identifies and organises. In so doing it also reveals a range of public space types that are used in the second part of the chapter to suggest a new typology of public space.

Critiques of contemporary urban public space

A range of recurring critiques characterise discussions of public space, ranging from the prosaic to the abstract. Most are based on a view about what public space should offer, often predicated on an idealised notion of public realm as an open and inclusive stage for social interaction, political action and cultural exchange. Although each of these qualities has distinct historical antecedents, as discussed in Chapter 2, it is also probably true to say that public space has rarely, if ever, achieved such a utopian state. Not least this is because the 'public' in public space is not a coherent unified group, but instead a fragmented society of different socio-economic (and, today, often cultural) groups, further divided by age and gender. Each part

of this fragmented society will inevitably relate to pubic space in different complex ways.

In that sense, today's critiques may be nothing new, although that should not diminish the critiques themselves as each have broad support in the literature, and the concerns they relate to are all too real. They begin with the notion that the public space, and therefore the public realm, is experiencing a physical decline.

Neglected space

Writing in the 1980s and commenting on the state of the urban environment, Francis Tibbalds' now classic polemic *Making People Friendly Towns* bemoaned the decline of public space across the world. Using the UK as an example of where a once rich public realm was declining, Tibbalds (2001: 1) argued that public space is too often:

> littered, piled with rotting rubbish, covered in graffiti, polluted, congested and choked by traffic, full of mediocre and ugly poorly maintained buildings, unsafe, populated at night by homeless people living in cardboard boxes, doorways and subways and during the day by many of the same people begging in the streets.

Tibbalds quoted Douglas Adams' *Hitchhiker's Guide to the Galaxy* when he said that the public realm is a 'SEP' (someone else's problem). Not only, he suggested, do the general public expect someone else to clean up after them, but so do the numerous organisations with a formal role in the creation and management of public space (Figure 3.1).

Like many urban designers, Tibbalds advocated the use of good design as a means to reverse the problems of a threatening and uncared for public

3.2 Lost space

realm, although unlike many others, he also recognised the vital role of public space management: 'Looking after towns and cities also includes after-care – caring about litter, fly-posting, where cars are parked, street cleansing, maintaining paved surfaces, street furniture, building facades, and caring for trees and planting' (Tibbalds 2001: 7). For him, after-care mattered every bit as much as getting the design right in the first place.

Empirical evidence that backs up claims that there has been a decline in the way we care for the urban environment (at least in the UK) is provided at the start of Chapter 5. The implications of this neglect are now widely accepted. Through their influential 'Broken Windows Theory', for example, Wilson and Kelling (1982) graphically demonstrated what a failure to deal with minor signs of decay within an urban area could bring – a rapid spiral of decline. They showed how a failure to repair broken windows quickly, or to deal promptly with other signs of decay such as graffiti or kerb crawlers can lead to the impression that no one cares, and quickly propel an area into decline.

Lost spaces

Other writers have written about certain types of contemporary urban space that make the management of public space a particular challenge. Loukaitou-Sideris (1996: 91), for example, writes about 'Cracks in the City'. For her, cracks are defined as the 'in-between spaces, residual, under-utilised and often deteriorating'. She argues that poor management is also to blame for the state of many corporate plazas, car parks, parks and public housing estates, 'where abandonment and deterioration have filled vacant space with trash and human waste'.

Trancik (1986: 3–4) has used the term 'lost space' to make similar arguments. For him, lost space is a description of public spaces that are 'in need of redesign, antispaces, making no positive contribution to the surrounds or users'. Examples of lost spaces are 'the base of high-rise towers or unused sunken plazas, parking lots, the edges of freeways that nobody cares about maintaining, abandoned waterfronts, train yards, vacated military sites, and industrial complexes, deteriorated parks and marginal public-housing projects' (Figure 3.2). He argues the blame for creating lost

spaces lies squarely with the car, urban renewal, the privatisation of public space, functional separation of uses, and with the modern movement.

However, not all writers are critical of these neglected spaces. Hajer and Reijndorp (2001: 128) suggest that:

> The new public domain does not only appear at the usual places in the city, but often develops in and around the in-between spaces. … These places often have the character of 'liminal spaces': they are border crossings, places where the different worlds of the inhabitants of the urban field touch each other.

They quote a broad group of supporters for the idea of 'liminality' (Zukin 1991; Shields 1991; Sennett 1990), each arguing in different ways that such spaces can also act to bring together disparate activities, occupiers and characters in a manner that creates valuable exchanges and connections. Worpole and Knox (2007: 14) have termed such spaces 'slack' spaces arguing that they should be regulated with a light touch. For them, urban areas need places where certain behaviours are allowed that in other circumstances might be regarded as anti-social.

However, responsibility for the state of these types of public space seems to rest with the fact that it is rarely clear who should be managing them after they are built, or after they have declined. As a consequence, they are universally neglected, with Hajer and Reijndorp (2001: 129) arguing that much greater attention needs to be given to such transitional spaces.

24-hour space

Other forms of space are not neglected in the sense that 'lost' or 'slack' spaces are, but have nevertheless also taken on some of the characteristics of liminality. Roberts and Turner (2005) argue that the increasing emphasis on the evening economy and support for 24-hour city policies has brought with it forms of behaviour that even the perpetrators would feel is unacceptable in their own neighbourhoods. In such places the conflicts often revolve around the needs of local residents versus those of the revellers and local businesses serving the evening economy. Leisure and entertainment destinations such as London's Soho are of this type.

In the UK, the 24-hour city and concepts of the evening economy became a major trust in the regeneration efforts of towns and cities throughout the 1990s, and the government-led deregulation of the drinks industry that followed stoked this heady mix, turning many urban centres into what have been termed 'youthful playscapes' (Chatterton and Hollands 2002). For some, these spaces may not have been neglected, but they have nevertheless been abandoned to market forces and to a clientele of the young with disposal income to burn (Worpole 1999), in the process

3.3 Car-reliant space: the American strip

3.4 Invaded public space

deterring other users from these previously shared spaces. For Roberts and Turner (2005: 190), the solution is the need for more active management and more sophisticated planning controls. Without suitable controls, they argue, the original ideals of a 'continental ambience', so admired by the original proponents of the 24-hour city, will not be achieved.

Invaded space

Perhaps the most universal derision is reserved for the impact of the private car which Gehl and Gemzøe (2000) have described as leading to invaded public space. They argue that in old cities and urban areas where car traffic has gained the upper hand, public space has inevitably changed dramatically with traffic and parking gradually usurping pedestrian space in streets and squares. 'Not much physical space is left, and when other restrictions and irritants such as dirt, noise and visual pollution are added, it doesn't take long to impoverish city life' (Gehl and Gemzøe 2000: 14).

The critique is nothing new, and manifests itself in four primary problems. Lefebvre (1991: 359), first, describes how urban space is often 'sliced up, degraded, and eventually destroyed by … the proliferation of fast roads' so that 'Movement between the fragments becomes a purely movement experience rather than a movement and social experience' (Carmona et al. 2003: 75). Buchanan (1988: 32), second, argues that the remaining public space itself is too often dominated by traffic and has lost its social function as a result. Thus even when the number of car users is greatly outweighed by the numbers of pedestrians using a street, the space given over to road space far exceeds that dedicated to footpaths.

A third problem relates to the ease with which car owners can move from one unrelated place or event to another – 'The in-between spaces simply fly past' (Hajer and Reijndorp 2001: 57). In such a context physically distant places can be compressed into a single space, whilst others (in between) can be ostracised and allowed to deteriorate because of their perceived reputation or absence of attractors. Hajer and Reijndorp (2001: 53–61) characterise this as an 'archipelago of enclaves' and argue that unless these parts of the city also develop an attraction value, the new network city will ensure that they continue to be ignored.

A fourth impact can be seen in the range of exclusively car-reliant environments that have spawned across the Western World, particularly in North America, where, in the same locations, external public space does not exist at all, at least not in any traditional form, but is instead replaced by a series of disconnected roads and car parks (Figure 3.3). This phenomenon is extensively covered in the literature (see, for example, Garreau 1991; Ford 2000; Duany et al. 2000; Graham and Marvin 2001), and although such developments are sometimes placed within landscape settings, these landscapes are typically designed to be experienced from the car, and rarely attract pedestrian traffic.

> Such cities are not intended for walking. Sidewalks have disappeared in the city centres as well as residential areas, and all the uses of the city have gradually been adapted to serve the motorist.
>
> (Gehl and Gemzøe 2001: 16)

Gehl and Gemzøe (2001: 14) argue that invaded space is generally impoverished space, and that most of the social and recreational activities that did or would exist, disappear, leaving only the remnants of the most necessary, utilitarian functions. In such places, people walk only when they have to, not because they want to. Collectively the invasion of private cars have led to a dramatic reduction in the space available to pedestrians, a reduction in the quality of the space that remains, significant restrictions to the freedom of movement for pedestrians both within and between spaces, and the filling of spaces with the clutter and paraphernalia that conventional wisdom has determined the safe coexistence of cars and people requires (Figure 3.4):

> This panoply is generally owned and managed by different bodies. At worst, there is no co-ordination and the only functional considerations are engineering-led and car-oriented. The pedestrian is ignored or marginalised. Some of these items are introduced on the grounds of 'pedestrian improvements', yet the 'sheep-pen' staggered pedestrian crossings and guard rails impede pedestrian movement while allowing a free run for the car.
>
> (Llewelyn Davies 2000: 102)

For Shonfield (n.d.) the solution can be found in a radical and somewhat utopian extension of the public realm to all spaces and buildings that can not specifically be identified as either home or work; for example to the places used for travel, caring activities, 'mind–soul servicing', 'body servicing', or in democratic pursuits. Built on a right to roam and a right of access, this would go hand-in-hand with a reclaiming of streets from the car. She argues it could deliver 'A city were each and every activity outside the home and work, promises the experience of democracy, the experience of freedom and the experience of security' (Shonfield n.d.: 13).

Exclusionary space

Rather than extending public space into realms where it has never existed, most commentators focus on preserving the quality and rights to public space that already exists. A number of the most influential figures in urban design, including Jane Jacobs (1961); Jan Gehl (1996), and William Whyte (1980; 1988), have argued that the use public space receives is directly related to the quality of that space. Therefore, if space is poorly managed and declines either physically, or in the opportunities and activities (social, cultural, political, economic) it offers, then a vicious cycle of decline may all too easy set in:

> If people use space less, then there is less incentive to provide new spaces and maintain existing ones. With a decline in their maintenance and quality, public spaces are less likely to be used, thereby exacerbating the vicious spiral of decline.
>
> (Carmona *et al.* 2003: 111)

Although the physical quality of public space will be important to all who choose to use it, for some it will be more important than for others. For some, particularly the disabled, those with young children in pushchairs, or the elderly, simple physical barriers can present major obstacles to their use of public space, often completely excluding them from certain areas as a result (Figure 3.5). Hall and Imrie (1999: 409) argue, for example, that the disabled tend to experience the built environment as a series of obstacle courses. For them, most built environment professionals have little awareness of the needs of those with disabilities, and the public space that results is itself disabling when it need not be (Imrie and Hall 2001: 10). Moreover, because disability is associated with wheelchair use when in fact only a very small percentage of the population with disabilities are wheelchair users (four per cent in the UK), the manifold ways in which the environment can be disabling is rarely appreciated (Imrie and Hall 2001: 43).

For Carmona *et al.* (2003: 43), addressing environmental disability involves:

- understanding social disability and the ways in which the environment is disabling;
- designing for inclusion rather than for exclusion or segregation;
- ensuring proactive and integrated consideration, rather than reactive 'tacked-on' provision.

In other words, because what is good for those with disabilities is generally good for all (making the environment more accessible and easier to use for everyone), the needs of less physically able users of the built environment should be considered as an integral part of processes that shape and manage the built environment. Likewise, the psychological barriers to accessibility may need to be tackled. These include fear of crime (see below), or simply a concern that the streets are unsafe for certain users (particularly children) because of their domination by fast moving traffic.

SPACE AND AGE

For Loukaitou-Sideris (1996: 100):

> the fragmentation of the public realm has been accompanied by fear, suspicion, tension and conflict between different social groups. This fear results in the spatial segregation of activities in terms of class, ethnicity, race, age, type of occupation and the designation of certain locales that are only appropriate for certain persons and uses.

Lofland (1998) describes such spaces as 'parochial' because they are appropriated by particular groups, so whoever wanders in feels either like a stranger or a guest, depending on how they fit in. Loukaitou-Sideris (1996: 100) describes users of contemporary public space as having suspicion of the stranger but, as opposed to the single undifferentiated spatial type of the modernist public space, there is now segregation into distinct spatial types and users.

The combined result of physical barriers, and concerns for the safety and well-being, in particular of the old and the young, means that life-cycle stage is amongst the most significant determinants of environmental accessibility and equity (Lang 1994:269). The reluctance of parents, for example, to let their children play in the street or walk to school has been widely reported, and linked to associated health and obesity problems amongst children unable to get enough exercise, as well as to a decline

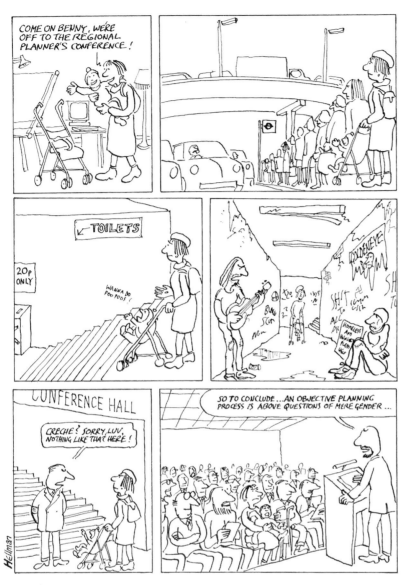

3.5 Exclusionary public space

of the overseeing role of children by adult strangers, and to a growing tendency to see the presence of children in public space as a threat to public order (Shonfield n.d.: 11). The development of car-dominated urban form may be partly to blame and has been extensively criticised, not least by 'New Urbanists' who argue that suburban environments too often dictate that only one lifestyle is possible; to own a car and to use it for everything (Duany *et al.* 2000: 25). But the way that existing environments are managed is likely to be just as culpable, and means that space for the pedestrian has increasingly been cut back and starved of investment in favour of space for cars.

Moreover, some heavy users of public space have been very actively denied access to it, or parts of it, prominent amongst which are the poor, homeless, and teenagers. Exclusion because of fear or an inability to consume are discussed below, and teenagers are excluded for both these reasons. But teenagers are also excluded because of their pastimes, the

most written about being skateboarding which is regarded by some as anti-social because of the conflict it creates with other groups and because of the damage it does to street furniture (Johns 2001).

Rather than actively designing for and managing such activities, the strategy is more often to banish such uses to dedicated spaces, and to design, or if not manage, them out of shared spaces. However, as Malone (2002: 165) has argued, 'It has become obvious from research that skate ramps and other youth-specific spaces on the margins of city centres are less than appealing places for young people (especially for young women)'. In such places teenagers experience problems of safety and security and feelings of exclusion, whilst what they desire in a public space is 'social integration, safety and freedom of movement' (Figure 3.6).

3.6 Dedicated teen space

RECLAIMING SPACE

Campaigners such as David Engwicht have written about the need to reclaim street space from cars to once again make it available as social space, available to the full range of users of all ages. He argues that 'the more space a city devotes to movement, the more exchange space becomes diluted and scattered. The more diluted and scattered the exchange opportunities, the more the city begins to lose the very thing that makes a city: a concentration of exchange opportunities' (Engwicht 1999: 19).

Urban designers have typically interpreted these ideas in terms of 'walkability', in other words, if a city is walkable, opportunities for social interaction also develop; opportunities that will be particularly pronounced by reclaiming the existing urban fabric, as well as by reflecting these principles in the design of new areas. Prescriptions abound, but one of the simplest is Llewelyn Davies' (2000: 71) 'Five Cs':

- Connections – good pedestrian routes that connect places where people want to go;
- Convenience – direct routes and crossings that are easy to use;
- Convivial – attractive routes that are well lit and safe and that offer a varied streetscene;
- Comfortable – an adequate width of footway without obstructions;
- Conspicuousness – easy to find and follow routes with surface treatments and signs that guide pedestrians.

They conclude that the best streets are designed for low vehicle speeds that allow all users to mix safely – cars, pedestrians and cyclists. They also take the most direct routes, and therefore do not separate modes of travel in order to get users from A to B (Figure 3.7).

Consumption space

In *Variations on a Theme Park* edited by Sorkin (1992: xiii-xv) it is argued that a new corporate city has emerged heralding an end to traditional public space. This new space is a global space, where economic phenomena cross over to society and culture. In the introduction to the book, Sorkin describes a world dominated by multinational companies, producing a standard departicularised urbanism where public space is for consumption. He argues public space is being heavily managed with an obsession on security, and that public space is at the forefront in creating a city of simulation where spaces are defined by pseudo-historic links to the past.

3.7 Shared space

3.8 Exclusive space

Hajer and Reijndorp (2001: 49–50) note an unprecedented increase in the deliberate consumption of places and events as a consequence of the dramatic expansion and domination of the middle classes in developed countries.

> A phenomenon that has mushroomed in recent years concerns the desire of the ordinary citizen to have 'interesting' experiences. Leisure experts talk about an 'experience market'. Where all kinds of events are offered that can excite people for a short time, from factory sales to art biennials. ... Cities and organisations compete with other places by producing experiences.

Boyer explores the question of simulation further, and how postmodern cities contain layers of history and symbolism that can be manipulated and exploited as an instrument of late capitalism:

> In Europe as well as in America, the postmodern return to history and the evocation of past city tableaux ... can be viewed as an attempt by political and social authorities to regain a centered world. ... [V]isual memories ... codified as fashionable styles and images ... could be manipulated to release the tensions that social changes and political protests, uneven urban and economic development, had wrought.
>
> (Boyer 1994: 408)

Boyer observes that districts in cities may be carefully designed, but do not cater for all in society. Other districts in the same city are neglected leftover pieces of public space containing the realism of social decay.

FINANCIAL EXCLUSION

Although design and management strategies can be used to explicitly exclude certain groups (and encourage others), other forms of exclusion can be practised through financial means. This might be explicit, for example charging an entry fee, tied to a series of codified rules and regulations often specified on the ticket. Many internal public spaces – museums, underground railways, etc. – adopt such a strategy. A more subtle practice

involves establishing visual cues that communicate that only those with the ability to pay are welcome, and that those who fall outside this category will be treated with suspicion, or even physically barred. For those who enter, it is necessary to advertise their right of entry through a separate set of visual cues, for example the clothes they wear (Carmona *et al.* 2003: 127). Many shopping arcades fall into this category, outwardly welcoming all, at least all with the ability to consume (Figure 3.8).

By the same token, Loukaitou-Sideris and Banerjee (1998: 291) argue that although public space in traditional cities serves as a venue for political debate, this is explicitly discouraged in the consumption space that characterises the new downtowns of America. 'Owners and developers want their space to be "apolitical". They separate users from unnecessary social or political distractions, and put users into the mood consistent with their purposes' – to consume.

Mattson (1999: 135–136) discusses this trend in the context of the ubiquitous American suburban shopping mall. He argues that many shopping malls are examples of what sociologists call a 'total institution', in which the outside world is intentionally locked out so as not to divert shoppers attention from their primary responsibility, to shop (Figure 3.9). However, as malls have increasingly become the only central gathering place in many communities, 'the activities of regular citizens who leaflet, protest, or otherwise use malls as public space have resulted in a number of contentious court cases'. In the US, many states have come down on the side of protecting private property rights over the constitutional rights to free speech, with only a minority validating the view of malls as public spaces.

> Whatever the specifics of the debates, they always centre on the core issue of public space and democracy in America's suburbs. Citizens have made clear that they need places where they can interact with fellow citizens and try to persuade others of their viewpoints. Malls, they have argued, must serve as these places, simply because they focus public interaction within a defined arena. In making the argument, these citizens have recognised a key weakness in the contemporary suburban landscape – a lack of public space and the insidious impact of that lack on democracy.
>
> (Mattson 1999: 136–137)

Privatised space

In the US and the UK, debates over the management of public space have increasingly highlighted concerns over privatisation and related security issues in recent years. Low and Smith (2006), for example, highlight the increased security and regulation in the US, especially post 9/11. However,

3.9 Shopping malls as consumer space

they also note that public spaces in the US were undergoing significant increases in security during the neo-liberal era of the 1980s and 1990s as well.

> The clampdown on public space ... is not simply due to a heightened fear of terrorism after 2001, and it has many local as well as national-scale inspirations. Many public uses of space are increasingly outlawed and policed in ways unimaginable a few years previously, but these rights were already under concerted attack well before 2001 (Low and Smith 2006: 2).

Low (in Low and Smith 2006: 82) makes the links with the privatisation of public space by corporate or commercial interests, arguing that:

> during the past 20 years, privatisation of urban public space has accelerated through the closing, redesign, and policing of public parks and plazas, the development of business improvement districts that monitor and control local streets and parks, and the transfer of public air rights for the building of corporate plazas ostensibly open to the public.

The argument is now widely accepted that urban public spaces in the US are more highly managed and policed due to the increasing private ownership of public space and the consequent spread of private management strategies. Ellin (1999: 167–8) argues that this privatisation is both a cause of the decline of public space, but is equally a consequence of it, as the desire to control private space has grown. For him, the move of facilities and amenities from public city centres to privatised

suburban locations, and their reincarnation as inwardly focused fortresses surrounded by moats of car parking, epitomises the problem. It represents an appropriation of public space by private corporations.

Madanipour (2003: 215–16) notes a further cause of privatisation inherent in the urban development processes that give rise to many new urban spaces. As development companies have grown in size and complexity, small locally based companies with links to local decision makers have increasingly given way to companies whose centre of operations typically resides outside the locale. Hand-in-hand, the financing of projects and ownership of commercial properties are increasingly the responsibilities of national and multi-national companies. The result is a growing disconnect between those responsible for development and the locality. Therefore, '[i]f particular developments had some symbolic value for their developers in the past, it is now more the exchange value in the market that determines their interest'; space becomes a mere commodity. In such a climate, a safe return (the investor's primary interest) will most easily be guaranteed through responding to the needs of occupiers, whilst those of the wider community will be a low priority. In the absence of strong planning controls to rectify the situation, and a general unwillingness of public authorities to take on the responsibility and cost of managing new spaces themselves, privatisation is the inevitable result.

Boyer (1993: 113–14) recognises a 'city of illusion', arguing that it is inappropriate to call something public space when in fact it is not. In central areas, she suggests, the emphasis is firmly on the provision of luxury spaces whilst ignoring the interstitial places between. Loukaitou-Sideris and Banerjee (1998: 280) agree, arguing that postmodern design eliminates unwanted and feared political, social and cultural intrusions:

> Space is cut off, separated, enclosed, so that it can be easily controlled and 'protected'. This treatment succeeds in screening the unpleasant realities of everyday life: the poor, the homeless, the mentally ill, and the landscapes of fear, neglect, and deterioration. In the place of the real city, a hyper-real environment is created, composed by the safe and appealing elements of the real thing, reproduced in miniature or exaggerated versions.

For them, the subjugation of public space to market forces is a recent phenomenon. Thus, in the US, downtown urban design, because it is determined by private interests, has become reactive and opportunistic rather than proactive. By contrast, the public sector typically reacts to the initiatives of the private sector for downtown building.

Increasingly the new downtown has come to be at odds with the traces of the old downtown; the Main Street of yesteryear. The public life of the Main Street downtown is vestigial at best and has

3.10 Privatised corporate space

3.11 Café-creep

been totally transformed by the culture of the poor, the homeless, and the new immigrants.

(Loukaitou-Sideris and Banerjee 1998: 288)

Their analysis not only revealed a lack of macro-scale strategic direction to steer investment into older parts of the city where the public realm was in decline, but also a series of micro-scale design strategies that deliberately foster exclusion: high blank walls, impenetrable street frontage, sunken plazas, hidden entrances (to new spaces), de-emphasised doorways and openings onto the street, no retail, etc., etc. The 'privatised' spaces inside can be seen as a series of spectacles or themed environments that can be packaged and advertised (Figure 3.10).

STATE PRIVATISATION

In the UK, Minton (2006) describes the shrinking local government model whereby the local council acts as enabler as opposed to provider, with private–public spaces not managed by the police but by private security. Often the process happens through public-led urban regeneration initiatives, with resulting developments being owned and managed by a single private landlord. As Minton notes, this is effectively a transfer of power for the management of public space from the state to private individuals:

> In terms of public space the key issue is that while local government has previously controlled, managed, and maintained streets and public squares, the creation of these new 'private–public' places means that ... they will be owned and managed by individual private landlords who have the power to restrict access and control activities.
>
> (Minton 2006: 10)

Minton uses the examples of Canary Wharf and Broadgate in London as examples of this phenomenon, whilst the redevelopment of Liverpool city centre has involved Liverpool City Council leasing out 34 streets to a developer to build and mange for 250 years. Graham (2001) notes an altogether more subtle and pervasive privatisation of the streets, in this case

through the move in the UK (and elsewhere) from publicly owned urban infrastructure, to privately owned. Although the phenomenon has not yet extended (new motorways and bridges aside) to the roads themselves, most of the infrastructure beneath the street has now been privatised, with associated rights transferred to these companies to obstruct, dig up and reinstate public space more or less at will.

A related issue, in common with the US, is the recent rise of business improvement districts (BIDs). BIDs amount to a group of businesses paying an extra financial levy in order to create an attractive external consumer environment (see Chapter 10). The relevant legislation to allow the creation of BIDs was approved in 2004, and by April 2006 there were 27 BIDs in England. These Minton (2006: 17) describes as 'private–public' spaces where private management tightly monitors and controls the public space. For him, BIDs are 'characterised by a uniformed private security presence and the banning of anti-social behaviours, from skateboarding to begging'. The evidence suggests that the UK is experiencing a similar set of changes to public space and public space management to that experienced by the US over the last 20 years: a shrinking local government; changes in land ownership; increasing private ownership of public space; increasing private control and management of public space; and an increased focus on cleanliness and security.

However, citing the impact of the 2001 Patriot Act in the US in evidence, Low and Smith (2006: 12) conclude that 'the dilemma of public space is surely trivialised by collapsing our contemporary diagnosis into a lament about private versus public'. For them, the cutting edge of efforts to deny public access to places, media and other institutions is occupied by the state, and the contest to render spaces truly public is not always simply a contest against private interests. At a less dramatic level, critiques of the instigation and spread of BIDs are based on similar concerns, of the state effectively passing aspects of their responsibility for publicly owned space to private interests. Kohn (2004) identifies another dimension of these same trends in what she characterises as a creeping commodification of public space. In this category she places the renting out of space by local government for commercial events, the sales of advertising space in and around public space, and 'café-creep', or the spread of commercial interests across the pavements of public spaces (Figure 3.11).

3.12 Gated communities in China

3.13 Active participants in the drama of civilisation

Segregated space

Trends in the privatisation of public space are not confined solely to corporate space, but extend also to the home environment. Contemporary and worldwide trends of physically gating communities, for example, have been well documented (see, for example, Blakely and Snyder 1997; Low in Smith and Low 2006; Webster 2001), and reflect the long-established desire of affluent groups in many societies (see Chapter 2) to separate themselves from the rest of society, often reflecting a fear of crime, or simply a desire to be, and to be seen to be, exclusive. In essence, the gates turn the space inside into a private space, accessed on the basis of relative wealth, whilst the residents turn their backs (the walls and gates) on the space around. Increasingly this is a global phenomena (Figure 3.12).

These trends may be an extension on what Sennett (1977: 5–15) has described as a decline in public life brought on by an increasing emphasis on the private relations of individuals, their families and intimate friends, driven by the rise of secularism and capitalism. By contrast, he argues, public life has increasingly been seen as a matter of dry, formal relations, whilst the introspective obsession on private life has become a trap, absorbing the attention of individuals rather than liberating them. The consequence is that the venues of public life, the streets and squares, have increasingly been replaced by the suburban living room, whist the spaces that remain become movement rather than sociable spaces.

THE IMPACT OF CRIME

Crime, or often, more correctly, the fear of crime, remains a major cause of this retreat from the public realm for those with choice (Miethe 1995), whether behind gates, or simply away from urban locations into suburban ones. Boddy (1992), for example, contends that people feel exposed and vulnerable when outdoors, and conversely safe and protected when inside, a fear that results in the increasing spatial segregation of activities by class, age, ethnicity and occupation – communities for the elderly, ethnic areas, gated communities, skid row, etc.

As well as explicit segregation strategies, policing (public or private) and surveillance strategies can also be used to a similar effect. Indeed, the fear of victimisation is real and a major factor in how the contemporary urban environment is both designed, and managed (Oc and Tiesdell 1997). Crime and incivil behaviour can quickly undermine the quality and experience of public space, encouraging users to manage the perceived risk by avoiding using places and in turn contributing to their further decline. Although men are statistically at greater risk of crime then women, and young men at greatest risk of all, the fear of victimisation is felt more acutely by women, no doubt helping to explain Whyte's (1980) observation that a low proportion of women in public space generally indicates that something is wrong.

A huge literature exists around approaches to crime reduction, with arguments around the extent to which environments can be made more safe through various combinations of defensive design, surveillance, street animation, active control, and social and educational approaches to crime reduction. Although prescriptions vary, most commentators would agree with Jane Jacob's basic prescription that public peace is kept primarily by the network of voluntary controls that most individuals in society subscribe to and which is (typically) codified in law. In this sense, as Jacobs (1961: 45) argued, users of the public space and occupiers of the surrounding buildings are 'active participants in the drama of civilisation versus barbarism' (Figure 3.13). By its very nature this requires users to be actively engaged in the process of civility, and a perverse consequence of the privatisation of residential environments may simply be the withdrawal (behind their gates) of many law-abiding participants from this role (Bentley 1999: 163).

Domestic space

Another aspect of this balance between private and public realms concerns the idea that the very notion of a public life is under threat from the spread of new technologies and new private venues for social exchange. Ellin (1996: 149), amongst others, notes, how many social and

3.14 The third place setting for public life: hairbraiding salon

civic functions that were previously – by necessity – conducted in the public realm, have increasingly transferred to the private. Entertainment, access to information, shopping, financial services, and even voting, can increasingly be undertaken from the home using modern technologies, in particular the internet. This, on top of increasingly dramatic rises in personal mobility, has in many places led to decline in the 'local', 'small-scale' and 'public' and to a growth in the 'regional', large-scale and 'private' as venues for public life. Thus Sennett (1977) has long argued that individual lives are increasingly private and that, as a result, public culture has declined.

This tendency may simply necessitate a broadening of the definition of public space, to incorporate some of the new forms of semi-public space that have been emerging. Banerjee (2001: 19–20), for example, has suggested that urban designers should concern themselves with broader notions of public life rather than just physical public space, reflecting the new reality that much public life exists in private spaces 'not just in corporate theme parks, but also in small businesses such as coffee shops, bookstores and other such third places'. For him, these spaces support and enable social interaction, regardless of their ownership.

This notion of 'third places' was originally advanced by Oldenburg (1999) who argued that because contemporary domestic life often takes place in isolated nuclear families, and work life, with the spread of new technologies, increasingly in a solitary manner, people need other social realms to live a fulfilled life. For him, this 'informal' public life, although seemingly more scattered than it was in the past, is in fact highly focused in a number of third place settings – cafés, book stores, coffee shops, bars, hair salons and other small private hangouts (Figure 3.14). These places host the encounters from the accidental to the organised and regular, and have become fundamental institutions of mediation between the individual and society, possessing a number of common features. They are:

- neutral ground, where individuals can come and go as they please;
- highly inclusive, accessible and without formal criteria of membership;
- low profile and taken for granted;

- open during and outside of office hours;
- characterised by a playful mood;
- psychologically supportive and comfortable places of conversation, and therefore also of political debate.

One might argue that these features also characterise (or should characterise) public space, but also that these third spaces are, again, nothing new; the British pub, French café, or American bar providing examples from the past that remain significant third places in the present. Today these have been supplemented with other forms of third place; the shopping centre, health clubs, video rental stores, and a surfeit of new leisure spaces.

VIRTUAL SPACE

What is new is the growth of virtual spaces – chat rooms, virtual worlds, radio phone-ins, and the like – that some have argued will supplant our need to meet and interact in traditional public space, and will eventually lead to new forms of urbanism (see discussion in Aurigi 2005: 17–31). Leaving on one side the most extreme predictions of the 'techno-determinists' of an end to urban life, some of the most thoughtful writers in the field have concluded that the nature of cities as we understand them today will be challenged and must eventually be reconceived, especially as '[c]omputer networks become as fundamental to urban life as street systems' (Mitchell 1995: 107). Others have argued that the new technologies, rather than undermining traditional cities, actually act to reinforce their role as IT applications are largely metropolitan phenomena, whilst those who work in these fields increasingly wish to live and work in places that bring them into contact with others in the field, and which meet their quality of life aspirations (Graham and Marvin 1999: 97).

Conversely, therefore, the quality of public space may become more rather than less important. In reality, the true impact of the new technologies on city form and public space has yet to be seen, but the fact that face-to-face communication remains the preferred mode of interaction for business as well as for private activities suggests that public space may not be as threatened by the new technologies as was once thought (Castells 1996; Sassen 1994). The expanded role of third places seems to confirm this.

Invented space

Some of the most frequent critiques of the new forms of public space are associated with the perceived loss of authenticity and growth of

3.15 Baltimore Inner Harbor

3.16 Manchester's Gay Village

'placelessness'. Various writers have discussed the components of place, typically focusing on the sum of three elements: physical form, human activities and meaning or image (Relph 1976; Canter 1977; Punter 1991; Montgomery 1998). Others have focused on the qualities of successful places, such as Carr *et al*.'s (1992) view that space should be 'responsive' to five needs:

- comfort, encompassing safety from harm as well as physical comfort;
- relaxation, allowing a sense of psychological ease;
- passive engagement, with the surroundings and other people (e.g. people watching);
- active engagement, that some people seek out, but which is often spontaneous if the situation allows;
- discovery, reflecting the desire for variety and new experiences.

However, these very qualities help fuel the desire for, and spread of, entertainment spaces where, without effort, participants can indulge in leisure activities. At the same time, the spread of globalisation processes, mass culture and the loss of attachment to place (Carmona *et al.* 2003: 101–2), has led to a repetition of certain formulaic responses across the world, a classic example being Baltimore's Inner Harbor, which, since its regeneration in the 1970s and 1980s, has spawned copycat leisure spaces across the globe (Yang 2006: 102–27, see Figure 3.15).

Although many settlements have at some time been 'invented' by their founders, increasingly techniques borrowed from theme parks are being used to re-invent existing places, with the danger that elements of continuity and character that might have been part of the distinctive qualities of a place can be lost. Wilson (1995: 157) takes Paris as an example, arguing that the Parc de la Villette, despite its international reputation, is 'designed for tourists rather than for the hoarse-voiced, red-handed working men and women who in any case no longer work or live there'. Thus in cities around the world, 'not only is the tourist becoming perhaps the most important kind of inhabitant, but we all become tourists in our own cities'.

Sometimes the process involves the creation of difference as a means to distinguish between places, for example the use of place marketing strategies to distinguish one city, neighbourhood or place from another (Figure 3.16). Sometimes the process involves the deliberate creation of sameness, copying a successful formula that has worked elsewhere – for example the emergence of formulaic China towns in many cities across the world, or the cloning of high streets with the same national and international brands (New Economics Foundation 2004). Criticism of such places is now widespread. Sorkin (1992: p xiii), to name but one, reserves particular bile for such places, arguing that America is increasingly devoid of genuine places, which are instead gradually being replaced by caricatures and 'urbane disguises'.

However, although such places can be criticised for being superficial and lacking in authenticity, all such places necessitate a considered and careful design process. Thus as Sircus (2001: 30), talking about Disneyland, argues, 'It is successful because it adheres to certain principles of sequential experience and storytelling, creating an appropriate and meaningful sense of place in which both activities and memories are individual and shared'. Zukin (1995: 49–54) agrees that Disneyland and its like represent one of the most significant new forms of public space from the late twentieth century, although she identifies different factors for its success:

- visual culture, through an aesthetic designed to transcend ethnic, class and regional identities;
- spatial control, through a highly choreographed sequence of spaces, allowing people to watch and be watched, and to participate without embarrassment;
- private management, aimed at controlling fear – no guns, no homeless, no illegal drink or drugs, promising to 'make social diversity less threatening and public space more secure'.

3.17 The creep of the private security industry

MANUFACTURED PLACE

This manufacturing of place occurs in a wide range of contexts, as do Zukin's factors for success, with the creation of entirely fictitious theme parks at one end of a spectrum, to ubiquitous shopping centres featuring specific place references (e.g. Milan's Galleria), to the reinvention of historic urban quarters at the other. At all scales there is one over-riding objective, 'to attract attention, visitors and – in the end – money' (Crang 1998: 116–117). In this sense, such places are undoubtedly popular, and invariably full of human activity. Returning then to the components of place, one might conclude that 'placelessness' is not a product of the lack of activity or carefully considered physical form in the places that lack authenticity, but instead an absence of place-derived meaning. For Sircus (2001: 31) even this is not a concern. He argues:

> Place is not good or bad simply because it is real versus surrogate, authentic versus pastiche. People enjoy both, whether it is a place created over centuries, or created instantly. A successful place, like a novel or a movie, engages us actively in an emotional experience orchestrated and organised to communicate purpose and story.

Ultimately, therefore, the challenge may not be to create authentic or invented places, but simply to create 'good' places, recognising that to do that, many factors over and above the original design will be of concern.

Scary space

Kilian (1998: 129–131) argues that restrictions can be broken down into power relationships of access and exclusion, and that it is these relationships that are the important factors in space. For Kilian, urban spaces contain three categories of people: inhabitants, visitors, and strangers; and each group has different rights to access and exclusion:

- Inhabitants, the controllers; these are often seen as the state/ government, but are frequently the private sector such as a large corporation. Inhabitants have rights to access and exclusion.
- Visitors, the controlled; these are the users of public space, with rights to access for certain 'purposes' and no rights to exclusion.
- Strangers, the 'undesirables'; they have no rights to access and are excluded by definition.

He freely admits that these are fluid categories that are controlled by the subjective definitions that inhabitants give to visitors and strangers, and concludes that the debate over the loss of public space relates to the processes of social relationships that control the function of urban public space.

For Minton (2006: 24), fear of crime (rather than actual levels of crime) have often been the driver of moves to privatise parts of the public realm, segregating communities in the process. She argues, however, that whist the ubiquitous reporting of crime in the media has undoubtedly driven much of the increased fear (at a time when actual crime is consistently reducing), processes of polarisation and the associated atomisation of communities also drive a heightened fear of 'the other' (strangers), and a further withdrawal of those with choice from public space. Research in the US, for example, has revealed that the perception of crime is linked to the presence of visibly different groups with mutual suspicions of each other sharing the same space, such as the presence of homeless people in public space (Mitchell 1995).

Minton (2006: 2) describes the potential for social exclusion in terms of 'hot spots' of affluence and 'cold spots' of exclusion. 'Hot spots' – such as urban regeneration areas or BIDs – are characterised by having clean and safe policies that displace social problems. On the other hand, 'cold spots' are characterised by the socially excluded who are unwelcome in the hot spots. By this analysis, public space management is actively creating socially polarised urban public spaces. Minton (2006: 21) also identifies the slow creep of the private security industry in the UK, effectively supplanting the role of the publicly funded police force in those areas that can afford it (Figure 3.17). On this issue, she quotes Sir Ian Blair, the Commissioner of the Metropolitan Police who has described Miami where despite 19 per cent of streets being policed by private security, the city remains the murder capital of the US. For her, 'private security does not equate with safety', but it does represent a further degree of privatisation of public space, and a further withdrawal of the state from this, its traditional territory.

Murphy (2001: 24) highlights how exclusion practices are not always the work of the private sector through processes of privatisation, but are increasingly supported in public policy aiming to counter undesirable social activities. The 'exclusion zones' that result vary, but control factors such as smoking, skateboarding, alcohol consumption, begging, use of mobile phones and driving. This raises concerns about personal freedom

3.18 Hard controls

versus personal and collective responsibilities. Returning to Jane Jacob's (1961: 39) assertion that society acts together to establish and police norms of behaviour, and in doing so controls what she described as 'street barbarism', the question arises, are such zones any more than the codification of these rules in areas where the voluntary controls have broken down? Are they therefore a delimitation of person freedoms, or simply a statement of the freedom of others to use public space in a manner that reflects societal norms?

In this regard, Ellickson (1996) has argued persuasively that if users of public space are not able to enjoy a basic minimum level of decorum in public spaces, then they will be all the more likely to flee to the privatised world of suburban shopping malls, gated enclaves or the internet. He makes the seemingly controversial argument that to avoid this, those who transgress societal norms should be confined to zones set aside for their use – in other words the skid row model of social control. In fact, as Kohn (2004: 169) contends, this is no more than codifying what already happens in many cities where the homeless and other 'undesirables' are tolerated in some areas – red light districts and the like – but herded out of others, including shopping and commercial districts. Davies (1992: 232–3), points to the danger of such a strategy, arguing that the no-go environments that result merely exacerbate rather than solve the problems, with the resulting problems inevitably spilling over into surrounding urban areas.

Carr et al. (1992: 152) argue that freedom with responsibility necessitates 'the ability to carry out the activities that one desires, to use a place as one wishes but with the recognition that a public space is a shared space'. The question of management, and what is appropriate and what is not, is therefore a matter of local judgment and negotiation. Lynch and Carr (1991: 415) establish that this involves:

- distinguishing between 'harmful' and 'harmless' activities, controlling the former without constraining the latter;
- increasing general tolerance towards free use, while stabilising a broad consensus around what is permissible;
- separating – in time and space – the activities of those groups with a low tolerance for each other;
- providing 'marginal places' where extremely free behaviour can go on with little damage.

HARD AND SOFT CONTROLS

Loukaitou-Sideris and Banerjee (1998: 183–5) identify two basic options, hard or soft controls. Hard controls are active and use a variety of private security, CCTV systems, and regulations; the latter either prohibiting certain activities or allowing them subject to control (permits, scheduling

or leasing). Soft controls are passive, using a range of symbolic restrictions that passively discourage undesirable activities or make others impossible through removing opportunities. Much of the concern in the literature over a perceived loss of freedom and a resulting change in character of public space relates to a view that the former set of controls are increasingly being favoured over the latter by those with responsibility for managing public space – both public and private (Figure 3.18).

Fyfe and Bannister (in Fyfe 1998: 256), for example, point out that:

> Responses to the fortress impulse in urban design, and the broader 'surveillance society' of which it is a part, range from optimism at the discovery of potential technological fixes to chronic urban problems, to despair at the creation of an Orwellian dystopia. Laying between these extremes, however, is a middle ground characterised by a profound ambivalence about the impact of increased surveillance.

They quote Ellin (1996: 153) who argues that while gates, private policing and CCTV will contribute to give some people a sense of greater security, for others, they will simply raise the levels of paranoia and distrust that they feel.

Extensive research in the UK reveals that the actual impact of CCTV on reducing crime is in fact very low, whilst the popularity of such systems grows at a seemingly exponential rate (Welsh and Farrington 2002). In such a context, Fyfe and Bannister (in Fyfe 1998: 265) conclude that:

Under the constant gaze of CCTV surveillance cameras, Boddy's (in Sorkin 1992: 123) claim that streets 'symbolise public life with all its human contact, conflict and tolerance' will be difficult to sustain.

Atkinson (2003: 1840), by contrast, in surveying British urban space policy, notes that although it is possible to see a 'revanchist'[1] strand at the extremes of public space policy in the UK as a coercive attempt to clear certain groups in order to protect the majority – zero-tolerance policing, ASBOs (anti-social behaviour orders), child curfews and exclusions zones, etc. – at the same time other 'more compassionate ideas and initiatives can also be detected, including neighbourhood wardens, policing without the police', etc. Moreover, coercive policies may simply be viewed as attempts to empower communities by tackling the most severe problems in order to re-claim streets for the silent law-abiding majority. For him, the direction of travel is still not clear.

Homogenised space

The discussion above strongly suggests that urban public space shapes and is shaped by society; its fears, power relationships and priorities. Edward T. Hall (1966) recognised the significance of culture in increasingly diverse cities while others, notably Loukaitou-Sideris (1996) and Fainstein (2001), note how contemporary urban public spaces have become increasingly contested and fragmented as those within them compete for spatial identities. The argument goes that as communication between groups is often misunderstood and differences cannot be resolved, users are willing to accept a homogenised vision of urban public space that neither fosters civility nor community.

In addition, global economic changes have meant that urban public space is now recognised as a valuable commercial commodity, and global business in partnership with city governments has re-ordered the historic functions of public space through the production of new forms of public space that bring together those in society who can afford to consume. As cities increasingly compete for investment at a national and international level, they need to create environments that are seen as safe, attractive and which offer the range of amenities and facilities that their (increasingly white-collar) workers, and the tourists that they hope to attract, expect (Madanipour 2003: 224).

As has been argued, this new public space is linked to the move to late capitalism and mass consumption. This is significantly different from previous historic periods as described in the previous chapter, or the economic systems in place at the start of modernism, and can be generically described as globalisation. These forms of contemporary public space use symbolism in design as described by Boyer (1994) as a wider part of postmodernism's referencing to history and culture. Symbolism, when combined with entertainment, can be viewed as populist, as described by Light and Smith (1998), or lacking the public sphere nature of public space as described by Sennett (1990).

Being an important global commodity, the owners and/or managers of urban public space ensure that visitors to public space perceive and interpret it as being safe. Therefore the multicultural and pluralistic nature of public space has meant that fear of the stranger is now dispelled by management and surveillance. The increasingly contested and fragmented nature of public space has increased this necessity, and, as Madanipour (2003: 217) notes, 'A combination of the need for safe investment returns and safe public environments has lead to the demand for total management of space, hence undermining its public dimension'. Moreover, in order that visitors interpret public spaces as safe, strangers are increasingly being removed through the use of semiotic codes in space as described by Goldsteen and Elliott (1994).

The combination of these traits produces Sorkin's (1992) departicularised urbanism or a form of homogenised public space. The trends, it is argued, are exacerbated by a further impact of globalisation, the speeding up of ideas and influences around the globe. Today designers, developers, and clients in both the public and private sectors are no longer tied to particular localities, but operate across regions, states, and increasingly on an international stage. The result is that design formulae are repeated from place to place with little thought to context. At the same time, in order to influence the design agenda locally, the public sector has increasingly adopted a range of standards, guidelines and control practices that in many cases merely parrot 'generic' 'globalised' design principles that may or may not be appropriate locally, or which are applied rigidly by de-skilled local government officers, again without thought to context. These pressures to standardise the design process have been extensively documented in the case of British residential (Carmona 2001b) and other (Bentley 1999) environments and produce both a homogenised public realm and associated architecture.

There has also increasingly been a reaction to the perceived 'compensation culture', as a result of which public authorities have been attempting to design out any risks in public space as a means to manage their liabilities in case of accidents and other dangers (Beck 1992). Although recent evidence in the UK suggests that the existence of an actual compensation culture is much overstated, the impact on the design and management activities of local government (and private developers) is not, and has often led to the creation of safe, but bland and uninspiring public space.

3.19 Homogenised public space in China

It can restrict innovation, leading to more standardised designs and less interesting places ... It is [therefore] easier for those engaged in making decisions about schemes, especially clients, to justify a decision that avoids risk than a decision that uses risk creatively.

(CABE 2007: 1)

Arguably, therefore, homogenisation is the product of both contemporary design and development processes, and of the impact of all the concerns discussed above (Figure 3.19).

Towards a typology of public space

Decline or revival?

On the face of it, the critiques are damming of contemporary public space, but despite this, some authors argue that the reported decline in public space is much exaggerated (Brill 1989; Krieger 1995; Loukaitou-Sideris and Banerjee 1998). Instead, they argue, public space was never as inclusive, democratic and valued as many commentators would have us believe. Jackson (in Fyfe 1998: 176), for example, concludes that:

In lamenting the privatisation of public space in the modern city, some observers have tended to romanticise its history, celebrating the openness and accessibility of streets. ... Various social groups – the elderly and the young, women and members of sexual and ethnic minorities – have, in different times and places, been excluded from public places or subject to political and moral censure.

Hajer and Reijndorp (2001: 15) argue that too much of the discussion about public space has been conducted in terms of decline and loss, something that in their opinion is both unsatisfactory and misplaced. For them, the pessimism of many commentators is founded on an artificial dichotomy that is established in many writings between the centre and periphery, the latter, seen as replacing the former with impoverished forms of space. Instead, they suggest, 'if we regard city and periphery as a single urban field then we discover countless places that form the new domains that we are seeking'. However, 'The urban field is no longer the domain of a civic openness, as the traditional city was, but the territory of a middle-class culture, characterised by increasing mobility, mass consumption and mass recreation' (Hajer and Reijndorp 2001: 28).

The way in which 'the market' – the economy, globalisation, 'new-liberal hyper-capitalism' – threatens or even destroys the 'authenticity' of the historic meaning of local 'places' has often been a topic of discussion. These viewpoints have little consideration for the creation of scores of valuable new places. The possibility of these being created by 'the market' seems to be peremptorily dismissed. Privatization and commercialization are considered irreconcilable with the concept of public domain, but that discrepancy is less absolute than it might seem.

(Hajer and Reijndorp 2001: 41

For them, the fact that something is private rather than public, suburban rather than urban, or civic rather than commercial does not determine either its quality as a place, or its potential role as part of the public realm. The consequence is that we should no longer associate public space solely with the streets and squares of the historic city core, but should instead embrace the new urban network of dissociated places. They conclude that now, as in the past, the quintessential character of public space is determined by those who occupy it, and society has long been fragmented into groups with a knock-on fragmentation of spatial types (Hajer and Reijndorp 2001: 85).

These observations are strongly supported by a body of research in the UK supported by the Joseph Rowntree Foundation. In summarising this research, Worpole and Knox (2007: 4) argue that 'Contrary to conventional assumptions, public space in neighbourhoods, towns and cities is not in decline but is instead expanding'. So, whilst concerns are frequently expressed that open and uncontrolled public spaces have been increasingly privatised and made subject to controls and surveillance, the evidence for this is not widespread, and anyway results from a tendency for commentators to confine their notions of public space to traditional outdoor space in public ownership. Instead, it is important to reframe debates to reflect how people actually use spaces, and the fact that to members of the public, ownership and appearance do not define the value of space, rather the opportunities it provides for shared use and activity. If this broader notion of public space is accepted, they argue that despite the tendency towards privatisation, opportunities for association and exchange have increased. 'Gatherings at the school gate, activities in community facilities, shopping malls, cafes and car boot sales are all arenas where people meet and create places of exchange' (Worpole and Knox 2007: 4).

Reflecting on the new forms of space, Light and Smith (1998: 4) suggest that the average American does not want to spend time with strangers, and cite a range of authors to support this view, including Robert Venturi, who described the plaza as 'un-American'; J.B.Jackson, who observed that American public space is designed for 'the public as an aggregate of individuals'; and Roberta Smith who describes Americans as consuming public spaces like french fries, 'thoughtlessly and without ceremony'. They observe that the American public prefers spaces that are entertaining and not collective, educative, or political; and cite the revulsion of the middle class from the dangerous urban public space of the Modernists, and the increasing competition of other forms of entertainment such as cinema, television, and the worldwide web. Instead they note that large corporations increasingly compete for consumers through 'sensation, sentiment and nostalgia' in urban public space, and quote Venturi's description of Disneyland as 'nearer to what people really want than anything architects have ever given them' (Light and Smith 1998: 5).

Banerjee (2001: 14–5) continues the argument claiming that an important function of public space is enjoyment: 'The sense of loss associated with the perceived decline of public space assumes that effective public life is linked to a viable public realm … where the affairs of the public are discussed and debated in public places … But there is another concept of public that is derived from our desire for relaxation, social contact, entertainment, leisure, and simply having a good time'. For him, 'Reinvented streets and places' seek 'to create a public life of *flanerie* (the activity of strolling and looking) and consumption'; and 'whether it actually takes place in a public or private space does not seem to matter'.

Lees (1994: 448–9) concedes that contemporary public spaces still contain important aspects of urban life, and although many of these primarily commercial public spaces lack wider civic functions, we should remember that commercial space has always been built into public space and vice versa. 'The core of city life – exchanges of goods, information, and ideas – still has a strong grounding in space … the design, accessibility, and the quality of such urban space can and ought to be criticised, but its existence must be recognised'. For others, such commercialised public spaces are at least 'profoundly ambivalent'. Goss (1996: 221), for example, examines the waterfront festival marketplaces which have been developed in several American cities since the 1970s, and acknowledges that simulation and nostalgia, as described by Boyer (1993), are used for mass consumption. Yet Goss asserts that there is no longer a general public in such a divided society:

3.20 Reconquered cities: Copenhagen

Critics must, of course, consider whether private ownership and the pursuit of profit compromises the claim of festival marketplaces to provide a new model of public space … however, they are wont to sound churlish … to blame festival marketplaces for failing to provide equal access to all members of a mythical 'general public' – which does not and cannot exist in an ethnically and class-divided society – and for failing to provide the context for authentic public interaction and transactions – which does not exist in a mass-mediated society – is to repeat precisely the impossible bourgeois desire for a genuine public sphere that the festival market articulates.

(Goss 1996: 231)

Others, have anyway noted an improvement and reinvestment or return to the traditional forms of space, with a consequential improvement in the quality of public space and a resurgence in public life. Gehl and Gemzøe (2000: 20), for example, examine 39 public space exemplar projects from across the world, and conclude that:

In a society in which increasingly more of daily life takes place in the private sphere – private homes, at private computers, in private cars, at private workplaces and in strictly controlled and privatised shopping centres – there are clear signs that the city and city spaces have been given a new and influential role as public space and forum.

They argue that examples of such reconquered cities can be found across the world, particularly across northern Europe (Germany, Netherlands and Scandinavia see Figure 3.20), and– standing out as notable exemplars in the Americas – Portland in the US (Figure 3.21) and Curitiba in Brazil. Carr *et al.* (1992: 343) suggest that new forms of public space are only to be expected as cultures and societies develop and new uses need to be housed. For them, this is a sign of life, rather then death.

3.21 Reconquered cities: Portland

Classifying public space

If nothing else, this discussion confirms that the nature of contemporary public space is directly affected by the complex socio-economic context within which it is generated. Public space is a political arena, and in the most extreme cases has been actively fought over by groups with seemingly irreconcilable ideological visions concerning the nature and purpose of public space – a place of free access and interaction, unconstrained by the control of commercial and/or state forces, or, a space for particular defined purposes, subject to behavioural norms and control over those who are allowed to enter (Mitchell 1995: 115). But it is too simple to put the nature of public space down to these factors alone. In fact, public space as experienced today will be a result of:

- historical trends and norms that go back to the ancient world;
- the diverse modes of governance, regulation, legal dominion, and investment under which it is created;
- cultural traditions, that vary even across the Western world;
- political priorities and the particular lifestyles they support;
- the balance between political and market forces the increasing complexity of public space, and the limitations on professional skills and responsibilities to tackle this (see Chapter 1).

So, although much of the literature points to a homogenisation in the experience of public space, to its physical decline, and to trends in privatisation, commercialisation and exclusion, it is also true to say that much of the literature comes from a narrow academic perspective, and critiques certain types of public space, whilst not necessarily recognising the sheer diversity of space types that constitute contemporary cities, or the very different development models that often predominate around the world.

Reflecting the diversity, many attempts have been made to classify public space according to a range of characteristics, often inspired by the different academic traditions from where they derived:

1 From a sociological perspective – Wallin (1998: 109) defines much contemporary urban public space as 'dystemic space', a space of impersonal and abstract relationships, and as a deliberate antithesis to Hall's (1966) 'proxemic' spaces that are controlled by culture. Instead, the dystemic is 'a community of strangers' who inhabit public space. This is the world of the shopping mall, television, or worldwide web: the culture of capitalism where society is 'incessantly kept in a passive, voyeuristic, consumeristic state of mind and emotion'.

2 Focusing on the experience of space – Gulick (1998: 135–41) defines three types of public space that he claims many critics are confusing with each other:
   - 'public property': the traditional definition where the government or state formally owns space;
   - 'semiotic': made up of 'spatial identities' that encourage competition for, and segregation in, urban space (Fainstein 2001: 1);
   - 'public sphere': the community space, where citizens can interact socially or politically.

3 In terms of power relationships – Kilian (1998: 115–16) argues that all spaces are expressions of power relationships containing both the public and the private. He identifies two urban public space types, public space as the sites of contact, and public space as the sites of representation (respectively Gulick's public sphere and semiotic public spaces), and argues that critics of both types of space are concerned with both public and private space. In fact, he suggests, all spaces are both public and private and contain restrictions, whether of access or activity, explicit or implicit.

4 As a journey from vision and reality – Lefebvre (1991: 39) distinguishes between 'representational space' (appropriated, lived space, or space in use) and 'representations of space' (planned, controlled and ordered). In this sense, space is seen as a chronology, developing and changing over time. Thus space typically begins as a representation of a particular type of space, with a particular range of uses, but is appropriated over time by other uses and activities.

5 By means of control – Van Melik et al. (2007: 25–8) argue that the design and management of public space has in recent years responded to two trends: 'On the one hand, a rising anxiety about crime induced people to avoid the public domain of the city and retreat into the private sphere. Yet, the appeal of urban entertainment also grew, inducing people to indulge in fantasy and new experiences outside the home'. For them, these represent two sides of a tendency towards greater control, but produce two distinct types of public space:
   - secured public space – characterised by measures to create a sense of safety, through CCTV, enforcement activities, and exclusion of unwanted groups.

- themed public space – aims to create ambience and stimulate activity in order to attract more people to public spaces and thereby encourage their self-policing.

6 In terms of their adaptability in use – Franck and Stevens (2007: 23) argue 'The looseness and tightness of space are related conditions, emerging from a nexus of the physical and the social features of a space'. Thus loose space is adaptable, unrestricted and used for a variety of functions, ad-hoc as well as planned. Tight space, by contrast is fixed, physically constrained or controlled in terms of the types of activities that can occur there. For them, although these qualities are adjustable and relative, existing along a continuum from tight to loose, the new types of space that have emerged are often more restrictive in nature than they have been in the past and actively discourage the kinds of unplanned activities that lead to looseness.

7 Through their exclusionary strategies – Flusty (1997: 48–49) distinguishes between five types of space, each designed to exclude to different degrees:

- 'stealthy space', which is camouflaged or obscured by level changes or intervening objects, and which therefore cannot be changed;
- 'slippery space', which is difficult to reach because of contorted, protracted means of access or missing paths;
- 'crusty space' to which access is denied due to obstructions such as walls, gates and checkpoints;
- 'prickly space' which is difficult and uncomfortable to occupy, for example seats designed to be uncomfortable and discourage lingering, or ledges that are sloped and can not be sat upon;
- 'jittery space' that is actively monitored and which cannot be used without being observed.

8 Reflecting degrees of inclusion – Malone (2002: 158) adapts Sibley's (1995) notion of open and closed spaces to define spaces according to their acceptance of difference and diversity. Thus open spaces have weakly defined boundaries and are characterised by social mixing and diversity (e.g. carnivals, festivals, public parks), whilst closed spaces have strongly defined boundaries and actively exclude objects, people and activities that do not conform (e.g. churches, some shopping malls, schools). The latter are also strongly preoccupied with boundary maintenance and definition.

9 By their clientele – Burgers (1999) classifies space as a series of landscapes that form the domains of various social sectors or interest groups:

- erected public space – landscapes of fast-rising economic and government potential;

- displayed space – landscapes of temptation and seduction;
- exalted space – landscapes of excitement and ecstasy;
- exposed space – landscapes of reflection and idolisation;
- coloured space – landscapes of immigrants and minorities;
- marginalised space – landscapes of deviance and deprivation.

10 In terms of how users engage with space – Dines and Cattell (2006: 26–31) use social engagement with space and perception of it as a means to identify five categories, although these are not necessarily mutually exclusive:

- everyday places – the range of non-descript neighbourhood spaces that make up much of the public realm and the everyday venues for interaction;
- places of meaning – that differ from person to person and that relate to particular associations and meanings attached to particular spaces, both positive and negative;
- social environments – that through their design and uses actively encourage social encounters between users, both fleeting and more meaningful;
- places of retreat – that offer a chance for people to be alone with their thoughts or to socialise in small groups of friends;
- negative spaces – where some experience aspects of antisocial behaviour, including racism and disruptive activities that are often perceived as threatening.

11 Through their physical / morphological character – from Sitte's (1889) deep and broad squares, to Zucker's (1959) closed, dominated, nuclear, grouped and amorphous squares, to the Krier brothers attempts at more sophisticated typological classifications for urban space (see Papadakis and Watson 1990).

12 And, by function – for example Gehl and Gemzøe (2000: 87) classify 39 'new' city spaces into five types: main city square, recreational square, promenade, traffic square, monumental square, whilst Carr et al. (1992: 79) identify eleven types of space:

1 public parks
2 square and plazas
3 memorials
4 markets
5 streets
6 playgrounds
7 community open spaces
8 greenways and parkways
9 atrium/indoor marketplaces
10 found spaces/everyday spaces
11 waterfronts.

Table 3.1 Urban space types

Space type	Distinguishing characteristics	Examples
'Positive' spaces		
1. Natural/semi-natural urban space	Natural and semi-natural features within urban areas, typically under state ownership	Rivers, natural features, seafronts, canals
2. Civic space	The traditional forms of urban space, open and available to all and catering for a wide variety of functions	Streets, squares, promenades
3. Public open space	Managed open space, typically green and available and open to all, even if temporally controlled	Parks, gardens, commons, urban forests, cemeteries
'Negative' spaces		
4. Movement space	Space dominated by movement needs, largely for motorised transportation	Main roads, motorways, railways, underpasses
5. Service space	Space dominated by modern servicing requirements needs	Car parks, service yards
6. Left-over space	Space left over after development, often designed without function	'SLOAP' (space left over after planning), modernist open space
7. Undefined space	Undeveloped space, either abandoned or awaiting redevelopment	Redevelopment space, abandoned space, transient space
Ambiguous spaces		
8. Interchange space	Transport stops and interchanges, whether internal or external	Metros, bus interchanges, railway stations, bus/tram stops
9. Public 'private' space	Seemingly public external space, in fact privately owned and to greater or lesser degrees controlled	Privately owned 'civic' space, business parks, church grounds
10. Conspicuous spaces	Public spaces designed to make strangers feel conspicuous and, potentially, unwelcome	Cul-de-sacs, dummy gated enclaves
11. Internalised 'public' space	Formally public and external uses, internalised and, often, privatised	Shopping/leisure malls, introspective mega-structures
12. Retail space	Privately owned but publicly accessible exchange spaces	Shops, covered markets, petrol stations
13. Third place spaces	Semi-public meeting and social places, public and private	Cafés, restaurants, libraries, town halls, religious buildings
14. Private 'public' space	Publicly owned, but functionally and user determined spaces	Institutional grounds, housing estates, university campuses
15. Visible private space	Physically private, but visually public space	Front gardens, allotments, gated squares
16. Interface spaces	Physically demarked but publicly accessible interfaces between public and private space	Street cafés, private pavement space
17. User selecting spaces	Spaces for selected groups, determined (and sometimes controlled) by age or activity	Skateparks, playgrounds, sports fields/grounds/courses
Private spaces		
18. Private open space	Physically private open space	Urban agricultural remnants, private woodlands,
19. External private space	Physically private spaces, grounds and gardens	Gated streets/enclaves, private gardens, private sports clubs, parking courts
20. Internal private space	Private or business space	Offices, houses, etc.

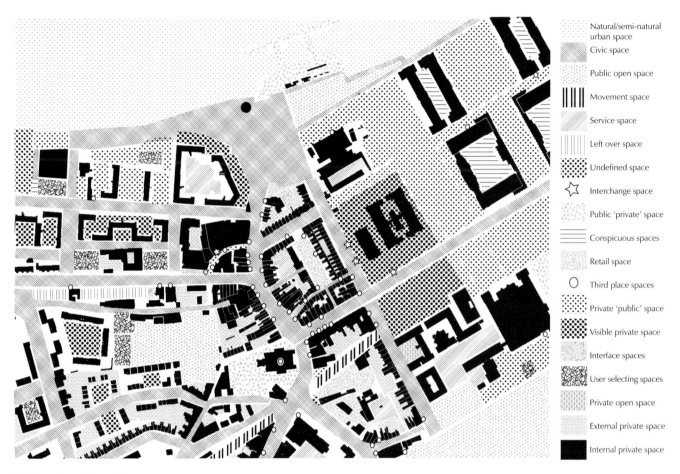

Legend:

- Natural/semi-natural urban space
- Civic space
- Public open space
- Movement space
- Service space
- Left over space
- Undefined space
- ☆ Interchange space
- Public 'private' space
- Conspicuous spaces
- Retail space
- ○ Third place spaces
- Private 'public' space
- Visible private space
- Interface spaces
- User selecting spaces
- Private open space
- External private space
- Internal private space

3.22 Space types: Greenwich

The reality is that public space can be classified in all these ways and more. Thus Kohn (2004: 11–12) concludes that the term public space is a cluster concept in that it has multiple and sometimes contradictory definitions. She identifies three concepts to distinguish between spaces: ownership, accessibility and intersubjectivity (whether it fosters communication and interaction), but concludes that a categorisation is becoming increasingly difficult as public and private realms are increasingly intertwined.

Nevertheless, as much of the contemporary public space 'problem' revolves around a failure to understand public space and its multiple dimensions, arguably it may be more by accident than design that public space has deteriorated. With this in mind it is useful to conclude with one further typology, one that specifically addresses the concern on which this book focuses; the management of public space (see Table 3.1).

Reflecting the discussion in this chapter, and developing Kohn's three-part classification, this new typology uses aspects of function, ownership, and perception to distinguish between space types. Twenty types are identified in four overarching categories, reflecting a continuum from clearly public to clearly private space.

Table 3.1 demonstrates both the wide range of space types that a typical urban area would possess, but also how many of these are in one sense or another ambiguous in that their ownership and the extent to which they are 'public', or not, is unclear. Some of these have always been so, for example private shops that are nevertheless publicly accessible. Others, for example the forms of internalised 'public' space, are relatively recent phenomena, or are simply becoming more dominant in the urban areas.

Figure 3.22 and 3.23 illustrate how for two different Thames-side town centre contexts in south-east London, the balance of space types varies, but also that each is made up of a patchwork of different public space types and, consequently, different management requirements and responsibilities. In Greenwich, a World Heritage site, the historic urban grain remains largely intact, and although conflict exists between vehicles and people, space remains largely public. There, however, the naval history of the town has left behind a large number of institutional buildings in grounds that bring with them their own restrictions on public access, and a fragmented pattern of ownership. Erith, by contrast, offers a fragmented landscape, where private interests have been allowed to buy up and now manage much of the town centre in their own narrow interests. The result is a car-dominated and controlled landscape, where the former 'public' parts of the town have been left to decline, and are now eschewed by much of the local population. No public life of any significance remains in the traditional public spaces of the town.

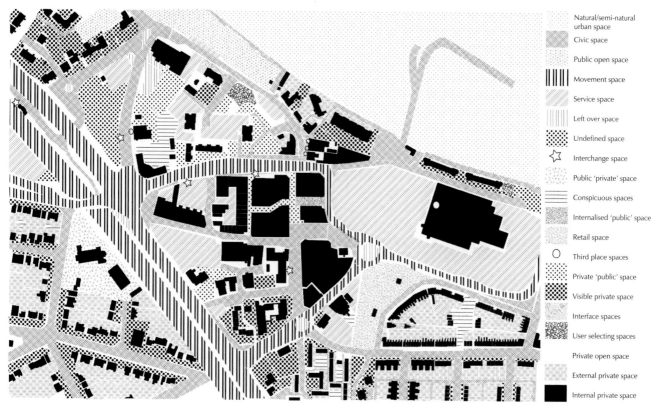

Natural/semi-natural urban space

Civic space

Public open space

Movement space

Service space

Left over space

Undefined space

Interchange space

Public 'private' space

Conspicuous spaces

Internalised 'public' space

Retail space

Third place spaces

Private 'public' space

Visible private space

Interface spaces

User selecting spaces

Private open space

External private space

Internal private space

3.23 Space types: Erith

Conclusions

Richard Sennett (1977: 21–2) has argued that the public space of the modern city has always represented a hybrid of political and commercial forces. At the root of many critiques, however, is a perceived increasing severance between the two.

Whether these critiques are any more or less pertinent today than during any period in the past are open questions. As discussion in Chapter 2 demonstrated, there has always been a strong historical link between commerce and urban public space, and strong exclusionary tendencies amongst those with management and ownership responsibilities. Nevertheless, the concerns of those who criticise trends in contemporary public space design and management are powerful and all too relevant to the way public space is used and perceived today. On the other hand, one might argue that it is hardly surprising that corporate interests are determined to take responsibility for their own public spaces, or for spaces that directly impact on their businesses, when the public sector has so often done such a poor job in managing the spaces for which they are responsible; spaces that still make up the large majority of the public realm, and that have long been perceived as in decline.

As the typology of public space demonstrates, the management context, reflecting the patchwork of public space types, is perhaps more complex now than ever before. Although some of the literature takes a relatively sanguine view about the nature and quality of contemporary public space, the majority offers a more pessimistic view, arguing that how

urban areas are managed today is increasingly undermining the 'public' in the concept of public space. In the future, if the critiques themselves are to be consigned to history, then public space management strategies will need to be sensitive to the full range of space types, and not just the selected few – the historic, affluent or private.

Note

1 Smith's (1996) notion of revenge against minorities and the affirmative action directed at them, including, for example, asylum seekers, beggars, and young people.

Chapter 4

Models of public space management

The first part of this chapter discusses the concept of public space management and its evolution in a context of wider changes to urban governance. Public space management is taken as a sphere of urban governance in which conflicting societal demands on, and aspirations for, public space are interpreted through a set of processes and practices. Four interlinked dimensions for public space management are proposed: the coordination of interventions; the regulation of uses and conflicts between uses; the definition and deployment of maintenance routines; and investment in public spaces and their services. Within this conceptual framework, the chapter looks at recent changes in public space management and in a second part suggests the emergence of alternative models of management. These are based on the roles ascribed to the state, to private agents and to community organisations, and on different approaches to dealing with the four management dimensions. Although the discussion shows that these models are more than just abstract formulations, and have been used to deal with a variety of public space problems, an important purpose for the chapter is to provide an analytical framework through which to examine emergent practices in the management of public space and their potential consequences.

The nature and evolution of public space management

The recent urban policy focus on issues of sustainability, social exclusion, economic competitiveness, place image, culture, gender and ethnicity, reveals an increasing awareness of the multidimensional nature of the challenges facing cities, their managers and inhabitants. This has also permeated our understanding of the roles of the built environment in general, and public spaces in particular, partly explaining the renewed global policy interest in the quality of public spaces. From civic, leisure or simply functional spaces with an important but to some extent discrete part to play in cities and urban life, public spaces have become urban policy tools of a much wider and pervasive significance.

Within this context, the broadening concern with public space and its quality, from the iconic parks and gardens to the ordinary streets and squares, reflects a more complex view of the relationship between the local physical environment and the social and economic well-being of its inhabitants (see Gospodini 2004). This goes well beyond the more mechanistic formulation of that relationship which characterised modernist planning and design. As a result, urban policy instruments have emphasised the potential roles of public spaces, variously as weapons in the arsenal of global and local inter-city competition, as catalysts for urban renewal, as potential arenas for community revitalisation and participatory local democracy, as well as fulfilling their more traditional functions as a source of amenities and connecting tissue between the private spaces of the city (Hall 2000, Fainstein and Gladstone 1997, Smyth 1994, Low and Smith 2006).

This wider understanding of public space and its urban policy role has also led to a closer attention to the processes through which its qualities and its ability to fulfil all those functions are created and maintained, and through which rights and obligations are established. Therefore, the concerns with design issues that have informed the planning literature, or those with ownership and rights that have dominated much of the geography debate on public spaces, are gradually incorporating a more explicit critical attention to the management regimes shaping public spaces and their uses. The key issue is whether the regime for public space governance and management consolidated in the middle years of the twentieth century in most Western countries is still the most appropriate way to realise all the roles ascribed to those spaces. This clearly means a critical appraisal of the traditional forms of management of public spaces, and an understanding of the meaning and implications of emerging management forms.

As Chapter 3 indicates, there is a considerable literature on what has been happening to public space over the last quarter of a century, much of which centred on the implications of a key element in the recent changes: the retreat of the state and the privatisation of public space provision and governance. As with other public sector activities, this process is linked to broader changes in the nature of contemporary governance of cities, in the relationship between civic society and the state, and in the economic and social context in which governance takes place.

What those changes come to is an on-going re-arrangement of urban governance mechanisms which is leading to important and sometimes painful changes in the organisational structures and practices through which traditional state functions are delivered, including the provision and management of public space. The evidence in this book suggests that new organisational forms have emerged, and that responsibilities, power and resources have been redistributed within and beyond government structures.

Whether or not this redefinition of rights and responsibilities in the management of public spaces is socially desirable is an open and contested issue. However, any critical analysis of what is going on with public spaces requires a historically well-grounded perspective of what public space management has come to mean as a public service, and how patterns of provision and management gradually built over a long period are coping with current demands. This chapter discusses these new forms of public space management that have emerged recently, using England as its focus, and dwells on their significance for the debate on the future of such spaces and their governance. This theme will be re-examined in later chapters through a more detailed examination of current practice in England and elsewhere around the world.

What is public space management?

All public spaces, no matter how inclusive, democratic and open require some form of management so that they can fulfil their roles effectively. Chapter 1 introduced the broad range of functions that public spaces of all sorts have to accommodate. Linked to these various roles are a wide array of stakeholders who are concerned that public spaces meet their own requirements as, for example, providers of infrastructure, motorists, pedestrians, retail operators, park users, etc. The potential for conflicts of

interests in the daily usage of public space is therefore quite significant, and, in a sense, inextricably linked to the very 'publicness' of such spaces. Public space management is therefore:

> The set of processes and practices that attempt to ensure that public space can fulfil all its legitimate roles, whilst managing the interactions between, and impacts of, those multiple functions in a way that is acceptable to its users.

This is a very broad definition, and there are clear issues here concerning who legitimises the different roles of public space, what is acceptable and what is not, and who decides; as well as with who are the users – the owners, defined groups, or wider society. This reflects some of the discussions in Chapter 3, and will be returned to in connection with the case studies in Part Two of the book.

Public space management is anyway the governance sphere where stakeholder demands on, and aspirations for public space are articulated into sets of processes and practices. Given the multifunctionality of public space, the variety of stakeholders whose actions contribute to shape its overall quality and the plurality of elements that constitute it – the 'kit of parts' discussed in Chapter 1 – it is clear that the management of public space is a complex set of activities, that often goes well beyond the remit of those organisations, public or private, formally in charge of delivering it.

For the purposes of this book, the management of public space is conceptualised into four key interlinked delivery processes:

1   The regulation of uses and conflicts between uses: the use of public spaces and the conflicts between uses have always been regulated, either formally through byelaws, and other prescriptive instruments, or informally through socially sanctioned practices and attitudes (see Ben-Joseph and Szold 2005 and Madanipour 2003). Regulation sets out how public spaces should be used, sets a framework for solving conflicts between uses, determines rules of access and established acceptable and unacceptable behaviour. How regulation is conceived, adhered to, and how it adapts to changing societal needs is a vital dimension of public space management.

2   The maintenance routines: these ensure the 'fitness for purpose' of the physical components of public space. Public spaces and the infrastructure, equipment and facilities vested in them need to be maintained in order to perform the functions that justify their existence. This concerns anything from ensuing that public spaces are usable, uncluttered, clean and safe, maintaining the surfaces of roads, street furniture, lighting, vegetation and facilities of all sorts;

to removing anything that might deface or offend the symbolism invested in civic spaces; to occasional capital intensive replacement of parts of the public realm.

3   The new investments into and ongoing resourcing of public space: regulating uses and conflicts and physically maintaining public spaces requires resources, financial and material. The degree to which regulatory instruments and maintenance routines can be effective is linked to the amount of resources devoted to those activities. Moreover, resources can come from several sources, each of them with a different combination of limitations and possibilities. This involves both ongoing revenue funding, for day-to-day management tasks, but also significant capital funding from time to time as and when significant re-design and re-development is required.

4   The coordination of interventions in public space: because regulation, maintenance and resourcing are likely to involve directly or indirectly a wide array of people and organisations, there is a necessity for coordinating mechanisms to ensure that the agents in charge of those activities pull in the same direction. This need for coordination applies equally to units within an organisation, such as departments of a local authority, as it does to different organisations. As some of the case studies in Part Two will show, the need for coordination has been made all the more pressing by the fragmentation of the 'command and control' state and the emergence of 'enabling' forms of urban governance (Leach and Percy-Smith 2001: 29).

These four dimensions apply whether public space management activities are undertaken primarily by public-sector agencies, by voluntary bodies or community organisations, or by private-sector companies (Figure 4.1). However, as the historical overview of public spaces in Chapter 2 has shown, even if the key dimensions of management are broadly constant, management responsibilities change and there is no final of definitive state for them. Therefore, it is not possible to refer to an ideal pattern of responsibilities over public space as these are invariably the result of messy governance arrangements resulting from the historical evolution of social practices and urban governments. What might intuitively appear as the normal or 'natural' form of public space management, defined by direct state ownership and management, captures only one moment in the history of that set of practices, freezing in time what is essentially a dynamic process.

Discussion moves on now to explore these issues, taking the example of the history of public space management in the UK in order to identify the current changes shaping the new practices and approaches.

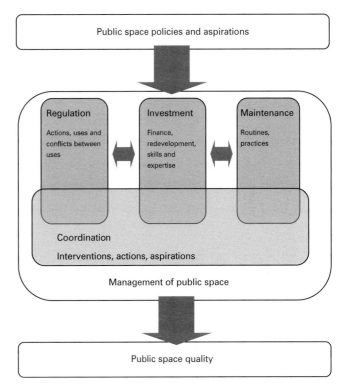

4.1 Public space management and its key dimensions

Public space management, a public good?

The idea of public space and public space management are normally associated with the public sector, and more specifically with local government. There are strong reasons for this coming from history and also from the economic dynamics of modern societies.

In a capitalist economy, goods and services tend to incorporate the character of commodity; something with value and a price, traded in the marketplace. It follows that provision is determined to a large extent by demand and supply relationships between buyers and sellers competing in the market. Some argue that the history of capitalism is the history of an ever increasing part of social life being subsumed under the category of commodity (Watts 1999, Thrift 2000).

However, not all goods and services are equally suited to the commodity character and to market relationships, even if they are vital to the functioning of the economy and society. The provision of such goods cannot therefore depend entirely on markets, and relies instead, at least partly, on alternative forms of provision, often involving the state. Public space is of this type, as it exhibits the characteristics of what economists call 'public goods' (see Cornes and Sandler 1996). Just like clean air, defence or policing, public spaces are goods that, once produced, can be enjoyed by more than one consumer simultaneously without affecting the utility derived by any of them. It is difficult and/or onerous to exclude from consumption anyone who wishes to benefit from those goods and, therefore, it is equally difficult to charge at the point of consumption.

This possibility of free consumption makes market provision of such goods unlikely as there is no incentive for it, even if demand is high. As with other public goods, public spaces have been historically provided and managed by philanthropy or collective organisations – as opposed to private, profit-seeking ones – and more recently the state through general taxation. This public-goods character of public space underpins much of the history of state involvement in its provision and management in modern societies.

In most Western countries, the progressive codification of the roles of the state during the twentieth century, and its takeover of the roles of previous collective and philanthropic organisations, led to the provision and management of public spaces becoming a public service, along with health, education, social housing and welfare. Vital functions performed by public spaces (linkage between places, traffic corridors, leisure, meeting and ceremonial spaces, health enhancing, etc.) became accepted as key to the well-being of modern societies and thus part of the array of goods and services whose adequate provision should be secured by the state. In most countries, the essentially local character of most public spaces and the functions they perform have resulted in their management becoming the responsibility of local government.

THE UK: THE RECENT HISTORY

In the UK, the development of local government as provider of public services resulted from the consolidation of multi-purpose, elected local authorities, a process that started in the early nineteenth century, gained impetus in the early twentieth century and reached its apex in the post-war years. During the Victorian period, the growing demand for infrastructure, health, education, poverty alleviation, and so forth, produced by rapid industrialisation was generally met through the piecemeal increase of state intervention, replacing or, more often than not, functioning side-by-side with a plethora of voluntary bodies, private companies, charitable organisations or private philanthropy that had traditionally provided for those needs (Leach and Percy-Smith 2001: 48). This was also the case with public parks provision and maintenance, and services related to road and waterway infrastructure, from street lighting to maintenance, waterworks, drainage, etc. (Southworth and Ben-Joseph 1997). Simultaneously, local

government became gradually more democratic, moving away from the business-dominated municipal corporations and coming to embody a wider array of local interests (Leach and Percy-Smith 2001: 51).

Multi-purpose, elected local authorities, as the principal provider of a series of public services in UK were a product of reforms in local governance structures from the late nineteenth century to the 1930s. This established the two-tier system that still characterises local government structures in parts of the UK, as well as the central government/local government dualism that has dominated local politics, given the absence until recently of a regional sphere with any practical meaning for the delivery of public services. Thus, for most of the twentieth century, the local single-purpose private, voluntary or charitable bodies that were so prevalent in the Victorian period, almost disappeared as public service delivery organisations. For more than half a century, local governance appeared synonymous with local government (Leach and Percy-Smith 2001: 50). In this context, public space management has been provided through local government's hierarchy of operational structures, and has been responsive to users' needs through the same means that render all local government's actions accountable to citizens, the ballot box.

PROFESSIONALISM OR SILO MENTALITY?

From the middle of the twentieth century, the growth in importance of local government as part of the welfare state machinery contributed to the transformation of local authorities into large multi-purpose organisations, with a high degree of internal specialisation and professionalism (Goldsmith 1992; Leach and Percy-Smith 2001). This meant the formation of large, self-contained service delivery units organised around specific areas of welfare policy (e.g. housing, education) or particular services (e.g. traffic and highways management, street cleansing, parks maintenance). This is at the heart of what is now decried as the 'silo mentality' that came to dominate the strategic thinking and the delivery of public services, characterised by an exclusive focus on one particular service and an inability to understand the connections and linkages across services and policy areas (Richards et al. 1999).

In the case of public space management, although the activities that make it up were for the most part located within local authority service delivery structures, it was not in itself the focus of public services, simply the context in which the service happened. Moreover, given their utilitarian origins, ordinary streets and squares were not viewed primarily as pubic space until very recently, and their management was focused on the functions and activities that used those spaces, not, in a holistic sense, on the spaces themselves. Public space as a concept tended to be limited to parks and iconic civic spaces, and this was expressed, for instance, in the existence of parks departments within local authority structures, which mirrored the independent, comprehensive management structures of historic parks. Even as late as 2004, a survey conducted for the research reported in Chapter 5 showed that the vast majority of English local authorities did not have an operational definition of public spaces that went beyond parks and a few iconic squares.

Therefore, as analysis in Chapter 5 confirms was still true at the time of the research, by and large, care for the majority of public spaces in England over the last half century has been dealt with as an implicit part of the general environmental management responsibility of local authorities. The professionalisation and compartmentalisation of public service delivery structures and the lack of a specific focus on public space – with the exception of park management – meant that public space management was carried out by a fragmented collection of agencies, very often located in functionally different departments and with a focus on narrowly defined services that happen to take place within public space.

This approach to managing public space, prioritising the delivery of discrete tasks without an overall strategy encompassing all forms of public space, lasted relatively unquestioned until very recently. Its utilitarian rationale suited policy priorities in the expanding urban economy of the 1950s and 1960s, with its large share of centralised state control, and it was not until problems of urban decline and economic and state restructuring in industrialised economies were acknowledged towards the end of that period that the need for a more strategic view of public space and its management started (very slowly) to emerge.

That approach is now being challenged by alternative models that imply a shift in public space management away from local government structures, and towards an increased involvement of other stakeholders (other public sector agencies, the private sector, community organisations, interest groups, etc.). This sits together with an increased awareness of public space management as a public service in its own right. If the key dimensions of coordination of interventions, regulation of uses, definition of maintenance regimes and investment and resourcing were subsumed into the management priorities of services that were peripherally concerned with public space, they are now slowly coming to the fore.

The drivers behind current changes in public space management

So what are the reasons for the current changes and key elements shaping them? Today the concern with the vitality and viability of town and city centres – and the public spaces within them – is now well consolidated in British and European urban regeneration (see for instance Urban Task Force 1999; DETR 2000). Similarly, the roles of parks and green spaces in the quality of urban life, and in the urban economy are now widely recognised. Therefore, part of the reasons underpinning changes in public space management are linked to an evolution in the thinking about urban regeneration and its aim of bringing sustainable vitality and viability to urban areas, and to the role of public space quality in this process.

In a related but separate process, the evolving understanding of the role of public space in social and economic life has directed attention to the ability of local service delivery agencies to meet more ambitious challenges. In the UK, for example, from the early 1980s, with the curbing of powers and spending of local authorities by an incoming Conservative government, there was a steady decline in funding for public space maintenance, a trend reversed only very recently (DTLR 2002a; Audit Commission 2002a). Emblematic of this process was the decline of park management systems and the disappearance of the park keeper. Park keepers were responsible for the care and management of individual parks and represented the continuity of localised and dedicated care mechanisms dating from Victorian times, but fell victim to rationalisation and cuts in public services as park maintenance was 'rationalised' and incorporated into spatially undifferentiated maintenance routines. As awareness of the importance of public space quality grew, so did the concern with the ability of a poorly funded and neglected system to meet the new demands on public space (DTLR 2002a).

GOVERNMENT TO GOVERNANCE

These factors – recognition of the key role of public space in urban policy and the need to raise levels of funding to public space services back to what they had been – although important, are not in themselves enough to fully explain the more recent re-thinking of public space management and its emergence as a policy concern in its own right, both in the UK, and elsewhere. For that, it is important to understand the general context in which the changes are situated.

First, as a public service, public space management has not been immune to considerable changes affecting state and public services in general over the last 15 to 20 years. Drawing on the policy theory and public policy literatures, recent trends can be situated within the political, cultural and institutional context of contemporary urban governance (Hajer and Wagenaar 2003; Kooiman 1993; 2003; Andersen and van Kempen 2001). Changes in the relationship between central and local government, society and government, the economy and government triggered by deeper transformations in the economy and society (globalisation, the move to a service-based economy, affluence, fragmentation of social life and changing lifestyles, etc.) have challenged hierarchical, 'command and control' forms of government. In turn this has lead to the rethinking of public sector cultures, structures and procedures (Goss 2001; Leach and Percy-Smith 2001; Pierre and Peters 2000).

Moreover, an increasing public policy focus on problems that seem intractable, persistent and not amenable to simple solutions such as environmental quality, social exclusion, sense of safety (i.e. the 'wicked issues' of the literature – Clarke and Stewart 1997), has strengthened the case for collaborative forms of making, managing and delivering policy (Sullivan and Skelcher 2002: 33). Recent trends in the management of public spaces are therefore part of the process whereby 'government' – an analytical and practical focus on the formal structures of government and on the state as the central governing actor – is being replaced by 'governance' – a focus on the process of governing and on the multiple state-society interactions that constitute it (Kooiman 2003: 4).

What is happening to public spaces and their management is therefore a re-shaping of the specific sets of institutional arrangements in a context of more general change in the way urban governance takes place in an increasingly diverse, fragmented and complex society, and in which no single social actor has the solutions for the policy problems at hand, or the power to implement them (Hajer and Wagenaar 2003: 7). Changes in public space management are therefore a reflection of wider changes in the relationship between government, especially local government, and society, which have concrete manifestation in the management of most public services (Goss 2001: 24).

RE-DISTRIBUTING POWER

For public services in the UK, this has meant a substantial redefinition of how they should be funded and delivered, what type of standards should guide delivery and how they should respond to changing user needs. An increasing emphasis on cost effectiveness, competition among providers and on consumers' choice has underpinned a retreat of government from direct service provision, the transfer of public management responsibilities to private and community stakeholders, an increasingly complex trade-off between service quality and public control, and so forth. In this regard, the provision and management of public spaces seems to fit a general pattern followed by other public services in which forms of collaboration between different sectors and jurisdictions have become widespread (Sullivan and Skelcher 2002; Bailey 1995). This is at the core of the notion of the 'enabling' local authority, whose main role as far as public services are concerned, is to 'stimulate, facilitate, support, regulate, influence and thereby enable other agencies and organisations to act on their behalf' (Leach and Percy-Smith 2001: 162)

Public sector and local government reforms in the 1980s and early 1990s were translated into privatisation, agencification and the flowing of power to subsidiary bodies within and outside the formal boundaries of the state (Rhodes 1997, Stoker 2004). This has led to a multiplication of agencies with a stake in the delivery of public services, and it has been no different with public space services. Managing public space might now involve a plethora of privatised public sector bodies, utility providers, area-based urban regeneration organisations, local authority departments, semi-public delivery agencies and so forth, all responsible for parts of the space, or for different services, or different operations within the same service.

The spread of contractual relationships in service delivery has added to the fragmentation. For example, compulsory competitive tendering (CCT) for all local public services in England was progressively introduced in the 1980s to increase competitiveness and efficiency, but it added another layer of fragmentation as it separated the client and contractor functions (Leach and Percy-Smith 2001: 162–3 – see a more detailed discussion in Chapter 5). With services such as refuse collection, grass mowing, street cleaning, grounds maintenance and so forth having to be contracted out to the lowest bidder in a competitive tendering process, integration between service design and delivery became dependent on increasingly complex contractual or quasi-contractual arrangements between the client commissioning the services and the contractor delivering it. This occurred whether or not the contractor was the in-house delivery arm of the client organisation or a private company. Although this practice ceased towards the end of the 1990s, it contributed towards shaping attitudes to service delivery and management as well as to the service delivery structures which are still in place.

CHANGING CONTEXTS

At the same time, demographic and cultural changes have put new diversified and often conflicting demands on public spaces with corresponding new pressures on management systems. In English town and city centres, for instance, the emergence of a young, alcohol-based sub-culture providing the mainstay of the evening economy, and vital for the economic viability of those areas, has created a context for the use and management of public spaces dominated by conflicts between uses and rights of use across space and time; in this case between night-time and day-time users, or between different age groups (see Chapter 3). Similar conflicts between the needs of a growing, sometimes gentrified and increasingly important leisure economy and the resident population have been reported elsewhere around the world (McNeill 2003; Smith 2002).

Conflicts in the use of public space are, of course, nothing new and it was precisely the coexistence of highly polluting industries and housing in the nineteenth-century industrial cities in Europe and the US that led to the birth of urban planning. What is new is that such conflict should happen in spite of the highly regulated urban environment, suggesting an inability of urban management tools to deal with the new context. Moreover, these conflicts and their consequences are being addressed in an increasingly risk-averse culture (Giddens 1999), which implies a more direct and expanded liability of the state for the services and facilities for which it is statutorily responsible. The increased level of liability facing public space managers, for example, has had an influence on the design of spaces and the equipment they contain, including the suppression of equipment and facilities deemed to increase the potential for law suits (Kayden 2000). It also impacts on the deployment of management routines and thereby on the nature of the relationship between providers and users of public space (CABE 2007).

In addition, as public space is perceived as a vital component in strategies of urban regeneration, city marketing, place identity, neighbourhood renewal, social inclusion, and so forth, it has been required to accommodate an increasingly complex range of expectations. The potential conflicts associated with this plurality of functions require management structures that can cut across specialised remits and understand the cumulative impacts of apparently unconnected activities, in the process mirroring the scope of urban policy objectives. For example, streets are increasingly expected to provide a focus for community life, provide a distinctive identity for an area, be a safe space for all, vibrant and vital at all times, and at the same time provide an efficient corridor for public and private transport (see Audit Commission 2002a, ODPM 2002).

The cumulative results of these contextual demands on public space and its management have exacerbated the shortcomings of public space services as traditionally delivered. They have made more acute the need to re-think the very notion of public space and its place in urban policy and challenged the exclusive focus on iconic civic spaces and parks and lack of focus on the variety of more ordinary public spaces.

A GLOBAL PHENOMENON

If the process as described above refers centrally to the UK, it is far from being unique. Indeed, there is plenty of evidence of similar processes elsewhere (see Chapters 7 and 8). The need to find ways of funding and operating public services in the context of a globally competitive economic environment, the challenges to traditional forms and practices of the welfare state by economic groups and by citizens, the multiple demands arising from differentiation of lifestyles, growing social fragmentation, city competition and so forth are global phenomena, which are impacting on the management of public services, and public space services across the world. The rhythm and intensity of those pressures and the responses to

it might differ, but even here the similarities are more noticeable than the differences.

The evidence from the literature and research reported in this book suggests that pressures for changes in the way public spaces are managed are bringing about a different understanding of what is public space, which management activities should be prioritised, how they should be resourced and implemented, and how they should be accountable to users. Consequently, new ways of dealing with the management of public space have emerged, which try to address issues of fragmentation, responsiveness, and quality over a broader range of public space types. However, this is an ongoing process, and as Part Two of the book will show, traditional and new ways of dealing with public space issues coexist, and are being combined to tackle the challenges found in localities.

It is also the case that a re-thinking of public space management has not affected equally the different services that make it up, or even the totality of public spaces. English historic parks, for example, have benefited from a long-standing tradition of coherent management structures, even if recently partly dismantled and starved of funds, as a basis on which to address the kind of problems discussed earlier. More than two decades of town centre management schemes have also provided a good starting point for their streets and squares. The challenge remains far greater with the range of ordinary spaces that make up so much of the urban realm.

The next section explores the emerging alternative approaches to public space management and their implications.

The management models

The literature, recent trends in the UK, and the empirical research reported in Part Two of this book all point to three emerging models of public space management (i.e. three different ways of addressing the issues of coordination, regulation, maintenance and investment). One represents a modified version of the current framework of public provision of public-space services, with public agencies playing the roles of coordinators, regulators, maintainer and funder. The second involves partial or complete delegation of those roles to private-sector organisations through contractual arrangements and reciprocal agreements. The third is similar to the second, but roles are devolved to voluntary and community-sector organisations as part of a move to reduce the distance between user and provider of services.

These are not mutually exclusive, and places and services have used a combination of them, depending on policy priorities, the relative strength of the various social agents with a concern for public space, and

	State-centred Public service ethos, accountability, separation provision-use, separation public-private	**Market-centred** Delegation, value for money and profitability, contract relationship, overlap provision-use, separation client-contractor, overlap public and private,	**Community-centred** Delegation, civic spirit, co-production of services, overlap provision-use, overlap public-community, overlap client-contractor
Coordination	• Hierarchies • Organisational restructuring • Consultation and user feedback	• Contract specification • Partnership design	• 'Compact', agreement and partnership design • Contract specification • Stakeholder engagement
Regulation	• Legislation and enforcement • Performance management	• Contract enforcement • Partnership performance management	• Contract enforcement • Partnership design • Institutional support • Capacity building
Maintenance	• Separation delivery-use • Technical expertise • Standards setting • Consultation and user feedback	• Overlap delivery-use • Separation client-contractor • Contract drafting • Outcome specification	• Contract drafting • Standards setting • Institutional support • Local x general standards
Investment	• Budget allocation • Rationalisation and efficiency gains	• Alternative sources • Value for money and competition • Stakeholder identification and involvement • Vested interests	• Alternative sources • Stakeholder identification and involvement • Commitment • Local knowledge • Capacity building

4.2 The three models of public space management

on the nature of the management challenges at hand. However, each of these approaches has its own dynamic and its own implications, and it is important to look at each in more detail (Figure 4.2).

The state-centred model

The first model centres on the state-centred provision of public space management, which was the dominant form of public space services in most countries for most of the twentieth century. It relies on public-sector institutions to plan and deliver the array of services that make up public space management, with minimum use of external input from either private contractors or the voluntary sector. Its key characteristics are:

- hierarchical structures of planning and delivery;
- clear vertical lines of accountability both upwards to policy makers – the politicians who set up public space policy whether explicit or implicit – and downwards to service users;
- clear separation between service and use;
- a public-service ethos based on the impartiality of officers and a commitment to the public interest.

In some cases this model can be regarded as inertial, a mere continuation of public space management practices and cultures developed over decades. This carries on despite the challenges posed by contemporary demands on public space and its quality and despite the sort of problems widely associated with this model, including: service specialisation caused by strong departmental cultures and professionalisation; clear separation of policy conception and service delivery leading to a fragmentation of

the different components of public space management; rigidity in dealing with varying contexts, including the ability to deliver fine-tuned variation of basic services; a disjuncture between, people's perception of issues and those of specialised service deliverers; issues of costs and cut-backs; and a lack of responsiveness to changing needs and demands (Audit Commission 2002a, ODPM 2004). It was precisely the growing realisation of those negative consequences of the traditional model of public space management that raised the need to re-think management systems.

However, this model can encompass attempts to tackle those negative aspects of traditional practice in ways that still retain the positive elements of state-controlled public service delivery with its public-service ethos and democratically accountable system. Indeed, the main strength of this model is that it is based on visible and widely acceptable lines of accountability, as service planning and delivery are directly subject to established mechanisms of elected local democracy. Moreover, it maintains clear lines of demarcation between the public and private spheres and therefore sets a clear, easily understood framework of responsibilities, of property rights, ownership, and of public rights and duties. Also, as discussed in Chapters 7 and 8, in many other countries the pressures to reform public services management and delivery have not been as intense as in the UK, local services funding has not been so eroded and the costs of this traditional model have not as yet offset its benefits to the point of demanding radical change.

COORDINATION

This is still, by far, the dominant management model throughout the world, requiring that efforts to tackle the issues of bureaucratic rigidity, fragmentation, excessive specialisation, lack of responsiveness, insensitivity to context and so forth are made within a public-sector service framework. For example, the key issue of coordinating the actions of agents whose actions impact on public space will to a large extent imply the coordination of public-sector services, either horizontally within and among local authority departments, or vertically among agencies at different levels of government, from the neighbourhood scale upwards.

Hierarchical structures to secure horizontal and vertical coordination will be very important in this model. This can mean the creation of clear lines of management and responsibility for public space services at local authority level, or formal agreements linking the performance of, for example, national and regional agencies to the service delivery strategies of local authority departments. As discussion in Chapter 6 will show, in England the effort to better coordinate public space interventions has often meant restructuring local authorities to create 'cross-cutting', more strategic structures that can focus on several dimensions of public space and are not limited by the narrower remits of specific services. 'Task forces' and working groups that can oversee and harmonise the actions of different agencies have been another common way of securing multi-agency coordination in public space management.

As this state-centred model maintains the separation between service providers (the public-sector agencies) and service users (public space users), an important issue for coordination is how the different aspirations, demands and actions of users are factored into public space management. The normal participation channels of parliamentary democracy are obviously important as public space users can express their views, on the quality of their public space when they elect local government. However, this might not be sensitive or flexible enough to respond to changing demands or contextual variety. This need for more responsive ways of coordinating the aspirations and actions of users requires the development of consultation and reporting mechanisms with effective feedback to users and linkages to service delivery agencies. It is likely to be a challenge in a complex multi-level, multi-agency institutional context.

REGULATION

The hierarchical nature of many of the coordination initiatives in this state-centred model means that a regulatory framework for public space management has two sides to it. One is straightforward legislation on uses and their impact on public space, on how users should relate to public space, and so forth, accompanied by enforcement action to secure compliance with legislation. This is clearly associated with the law-making and policing roles of the state and addresses the relationship between public space users and the state, framed by accepted rules, norms and customs.

The second refers to the regulation of relationships between public space service providers and is about securing compliance with public space policy aims and objectives and service commitments among public-sector agencies at different levels. Coordination initiatives in this model seek to organise roles and responsibilities among agencies so that public space policy can be achieved, but this needs mechanisms to ensure that those agencies commit the effort and resources required to an area that in many cases is poorly understood and, as a result, is seen as marginal.

This is to some extent secured by the hierarchical nature of the state apparatus, but the fragmentation, restructuring and withdrawal of the state over recent years (see Chapter 5) have weakened traditional command-and-control hierarchical structures. New forms have emerged to regulate performance of public sector organisations which rely less on hierarchical lines of command and more on performance management (Hill 2000, Leach and Percy-Smith 2001). In England, for example, as part of a drive for efficiency in local government, there has been a sustained effort to implement a performance measurement culture based on target setting for public services and auditing of results, with sanctions imposed on agencies that miss their targets and rewards given to those who perform well (Leach and Percy-Smith 2001; Audit Commission 2002a). As a consequence, regulation of public-sector agencies actions as regards public space is now done through the setting of targets at national, regional and local levels, measured through officially approved indicators (e.g. on street cleanliness, park quality, user satisfaction, and so forth).

MAINTENANCE

In this model maintenance routines are primarily technical and budgetary exercises, confirmed by political sanctioning in policy instruments and public consultation to secure support when necessary. This is public space management in the narrowest sense, which in this model is typically conducted by specialised departments of local government and other public agencies. However, as discussed in Chapter 1, there is an increasing awareness of the importance of public space maintenance, for example concerning the appropriateness and contextual sensitivity of maintenance routines. This has put the spotlight on how these routines are defined and what rationale underpins them, and indeed whether or not their deployment is an exclusive public sector affair.

In this context, key to the maintenance dimension of public space management are the mechanisms that secure the involvement of policy makers and users in designing maintenance routines, while at the same time maintaining the separation between service delivery and use. Chapters 6 and 8 illustrate some of the ways in which public space users have been involved in the definition of, for example, cleansing or tree pruning routines, on the basis of the general aspirations they have about their public space. This is incorporated into the technical deployment of those routines; into more general policy instruments regulating public space quality; and, importantly, into budgetary considerations.

INVESTMENT

Finally, in this state-centred model, the fourth dimension of public space management, investment, is primarily about capturing an appropriate slice of public-sector budgets for public space services. This can in turn pay for the skills and equipment necessary for the delivery of the desired levels of public space quality. As resources come exclusively from within public sector service budgets, increases in the quantity or quality of public space services are linked to one of two processes.

On the one hand, increases are possible if budgetary allocations to public space services grow, because either the total budget grows, or those services manage to capture a larger slice of the total public-sector budget. The latter is a product of policy shifts in a context in which public space quality is valued more highly in relation to other public service goods, and can be instigated by pressure from public space lobbying groups, shifts in public appreciation of public space, shifts in central government policy, and so forth. On the other hand, those increases can be the result of rationalisation, for example through better use of existing human, technical and financial resources, introduction of new technologies in maintenance routines, reduction of duplication in activities through organisational restructuring, and so forth.

In recent times, given the growing importance of public space quality for securing a range of urban policy objectives, both of these processes have been at play. For example, increased budgets for park maintenance were documented in England as park quality became a key indicator of local service delivery within the new 'liveability' agenda (MORI 2002). At the same time, the re-organisation of local authority structures, with the merging of previously separate public space services, has created the potential for increased coordination and, by pooling budgets and reducing duplication, for more resources for service delivery (ODPM 2004, ODPM 2006). Therefore, within this model the issue of resourcing centres around the role of public space management vis-à-vis other public services and how budgets are shared among them, and on how service delivery can be rationalised so that existing resources can be used more productively.

Devolved models

The other two emerging models share the common characteristic that they imply the transfer of responsibilities for provision and management of public space away from the state and towards other social agents. More than a rearrangement of responsibilities, they suggest a redefinition of what public space is or should be, and how its public character should be kept. This is part of what are referred to in the literature as process of privatisation of public space (see Chapter 3). In practice it comprises widely differing practices that go from the provision and management of public space by corporate organisations as part of the process of securing control upon externalities that might affect the performance of their business, to the take over of public spaces by community organisations or interest groups, whose own interests become equated with the 'public interest'.

This retreat of the state from its responsibilities over public space should not be confused with, or restricted to, the transfer of ownership of public spaces, although it is certainly linked to it. The real issue for public space management is how 'devolved' public spaces are managed and maintained, which also has a bearing on how 'publicness' is defined. Thus spaces owned and maintained by the embodied representation of the public interest (i.e. the elected state machinery) are intuitively 'public' and belong to all citizens, whereas spaces owned by private agents and managed by them will have their public status secured through contracts, legal instruments and regulated practices and might feel (and actually be) less 'public', even exclusionary. These devolved models imply a definition of property rights over public space management, separate from the issue of ownership of such space.

Therefore, what characterises these models is not necessarily the transfer of ownership of public spaces such as those produced through private property development in the UK or the US (see Kayden 2000). It is rather the transfer of management responsibilities (i.e. those of coordination, regulation, maintenance and investment) to others away from the public sector; to a variety of collaborative arrangements with other social agents with a shared interest in their outcomes. These arrangements will vary from contracts, to partnerships, to looser networks. As such, the models embrace the process of collaboration between the state, private agents and communities in the delivery of public services, or co-production of services (Sullivan and Skelcher 2002).

The market-centred model

The first and more common model of the devolved type is the transfer of management responsibilities over public spaces, whether publicly or privately owned, to private entities. This involves the transfer of rights and obligations for managing public spaces, and in some cases the power to define management objectives. This is done either through straightforward service delivery contracts, or as part of a development agreement in which private provision and/or management of public space results from negotiations around the conditions for, and outcomes from, private property development. The contracting out of street cleaning or park maintenance services, common in the UK, are examples of the former, whereas the public–private spaces in the US are examples of the latter. In both cases, these arrangements involve a business, profit-making logic on the part of the contractor (the agent), either directly profiting from a management/maintenance contract, or indirectly profiting from the performance of the development of which the public space is a part, and, in part, because of it.

Contracts in one form or another are an essential part of this process, and are more clearly expressed in terms of a principal–agent or client–contractor relationship (Sullivan and Skelcher 2002: 82–4). In these, one part – usually a public-sector agency – defines the services to be delivered and sets the standards of delivery, policy obligations and legal requirements. The other – normally a private agent – delivers those services in return for financial gain. For the private sector, even when not imposed by planning, zoning or other urban policy regulations, such collaborative relations can be justified by the characteristics of public space and public space management as commodities from which profit can be made and, given the externalities created by public space, by its potential to maximise the utility derived from ownership of surrounding property. For the public sector, they represent a way to fund public services by means other than the public purse. The rationale here is the same one underpinning the development of public–private partnerships (see Bailey 1995, Harding 1998):

- increasing public service budgets by tapping into private resources;
- bringing in skills and expertise not available to public-sector agencies;
- securing levels of service in excess of those normally provided by the public sector;
- creating more responsive, user-led management strategies for business-sensitive public spaces.

Although private management of public space is not a new phenomenon, its re-emergence as a practical policy option in post-welfare state societies runs contrary to many accepted notions of the direction of social progress. It is more established in the US, but it is rapidly gaining ground in other industrialised societies, especially in Europe, in spite of concerns about some of its implications. This is precisely the process denounced in the increasingly vast literature on the 'death of public space' (see Mitchell 1995, Sorkin 1994, Smith 1996, Kohn 2004 – see Chapter 3).

In the UK, this model came about as an extension of privatisation and the use of contracts in other public services, notably health (see Sullivan and Skelcher 2002). Service delivery through private contractors is now common in a range of services such as street cleaning, graffiti removal, verge maintenance, tree pruning, etc., as a way of buying-in expertise and lowering fixed operational costs. However, this is not only about the private delivery of public space services as planned by a local authority or another public-sector agency. Increasingly it involves the total design and delivery of services in particular areas, or even the private provision of a framework of design guidelines and service standards for public spaces that are privately owned and managed.

COORDINATION

This privatised delivery of public space management and its constituent public services, dominated by contractual relationships, has important implications for the key dimensions of coordination, regulation, maintenance and investment. Whereas in the previous model coordination was essentially a matter of devising better, and more integrated links between public-sector organisations at different levels, here this is compounded by the need to coordinate the outcomes of public–private arrangements and contracts. Therefore, besides the normal vertical and horizontal coordination mechanisms within the public sector, coordination requires considerable attention to contract specification and the negotiation of public–private agreements, as well as to their monitoring and enforcement. Hierarchical structures might secure adherence to commonly-agreed practices and objectives among public sector organisations, but clear and detailed specifications of outputs and outcomes and penalties for non-compliance are required in the case of contractual, multi-sector relationships (Sullivan and Skelcher 2002: 84).

Detailed contractual specifications might ensure that particular public space management tasks are carried out to pre-defined standards, frequencies and levels of outputs, as in the case of street cleaning or waste collection, thus securing the desired level of public space quality. Similarly, clearly drafted agreements on, for example, the use, access, opening hours and maintenance standards of a privately built and owned public space, can help to ensure that such spaces feel 'public' by their users. In most of these cases, coordination is about making sure that private contractors or developers conform to public space policy objectives. However, detailed contracts and agreements are not necessarily effective in dealing with situations in which great flexibility is required, or where public space management involves a wide range of private actors.

For this, partnership mechanisms between public agencies and private agents have been used as a way of coordinating actions across sectors. A good example of this are the town centre management schemes in England, and business improvement districts (BIDs – see Chapter 10) which are becoming an increasingly common arrangement for public space management in cities around the world. These partnerships provide a forum for achieving some degree of consensus on what is required for public spaces under their control, for distributing responsibilities and for agreeing on a framework of objectives upon which actual contracts and agreements can be drafted.

In this model, there is no separation in principle between the delivery of public space services and their use, as many of those managing public space on behalf of a local authority or other public body might also be users of the spaces with a vested interest in its quality. Their aspirations, demands and actions as public space users will be factored into public space management through their involvement in partnership boards, forums, compacts, panels and so forth. For others, the same system of consultation and feedback as used in the state-centred model will be required.

REGULATION

The regulation dimension of public space management in this model still typically depends on legislation and powers of enforcement vested in public bodies to manage conflicts between uses and usage patterns. Increasingly, however, private regulation of pseudo-public space that is owned and controlled by private interests has caused tensions that are reflected in the literature. These concern the potentially discriminative practice of private regulation and enforcement, but also the lack of a public-interest motivation in how authority over space is wielded.

It also depends on those mechanisms described earlier in this chapter to regulate and enforce cooperation and compliance with agreed public space objectives among service delivery agencies (i.e. the hierarchical lines of command within the public sector as well as the performance management systems that have come to dominate public sector management in recent years). However, given the extension of public–private contractual relationships defining this model, regulation also means the adequate regulation and enforcement of contracts and formal agreements; their outputs and outcomes.

This can happen directly through contractual dispositions and related penalties, but also indirectly through carrots and sticks embedded in planning and other regulatory systems. A case in point are conditions within a planning permission for a new development which require that the public space associated with that development be built and maintained at a particular standard of quality and with particular rights of access, even when that space will not be adopted by the relevant local authority and will remain a 'private public space'.

In the case of partnerships and other less structured forms of public-private collaboration, regulation can also mean the expansion of performance management regimes to all public space service delivery organisations, whether or not in the public sector. Chapter 6 explores examples of partnership performance measurement in the UK and the difficulties in creating effective mechanisms to regulate and monitor organisations with very different logics of accountability.

MAINTENANCE

The separation between client and contractor has fundamental implications for the task of defining and deploying maintenance routines. Whereas in the state-centred model both the definition and the deployment were undertaken within the same organisation, with the same ethos and rules of operation, in this case they are likely to be separated. The client, normally a local authority, will define the basic elements of routines such as frequency of services, coverage, and so on which will be specified in the contract, and it will be the contractor who will deploy them and will have to adapt them to conditions on the ground according to their own interpretation of the contract. This reinforces the importance of careful contract drafting so that the expected outcomes are actually produced, but also that some flexibility needs to exist to deal with changing demands on public space so that the expected outcomes are achieved even if under different conditions to the ones assumed when contracts were drafted.

In this context it may be better to base maintenance contracts on outcomes rather than crude measurements of process: for example a street cleaning contract that requires the service to be carried out if streets are dirty beyond a certain degree, rather than on a determined frequency whether or not it is needed, thus saving resources for more urgent actions elsewhere (ODPM 2004: 138). Outcome-based contracts have been used successfully in Groningen, in the Netherlands (see Chapter 8) in a range of public space maintenance tasks. This allows for more flexibility in contracted service provision that can adapt to any changes in demand and also to contextual differences, for instance in relation to parks or streets used more heavily and thus more susceptible to wear and tear. However, those experiences also show how difficult it is to specify outcomes (i.e. the conditions public spaces should be kept in) rather than quantities of services to be provided by contractors, in what is essentially a rigid legal instrument designed to define the limits of the legal obligations of both parties.

INVESTMENT

On the last of the four key dimensions, investment, there are significant differences compared to the state-centred model. Resourcing here is not exclusively about securing a slice of the public services budget, although very often this will still be important. One of the main elements of the rationale for privatisation of service provision is precisely the ability to draw resources, financial and technical, from outside the public sector. In some circumstances, resourcing decisions will imply determining whether or not private money and expertise are likely to be more effective at delivering a public space service, for example because there might be cost savings, better use of existing resources or access to particular skills. In others it may imply determining who has a stake in the fortunes of a particular public space and therefore a direct interest in its management in order to engage them financially in the process.

This may simply encourage contributions to a public-run pot of money to be spent on basic services. Alternatively it may allow private stakeholders to take over full responsibility for the management of such spaces. An example of the former are the sponsoring arrangements for parks and public gardens in which private organisations contribute to the costs incurred by local authorities in maintaining them. An example of the latter are BIDs through which a consortium of private organisations effectively takes over the public space management of an area of direct interest to them, and coordinates and supplements public sector expenditure in that area.

At the same end of this spectrum are the public spaces produced as part of development agreements through the planning process which remain in private ownership and are managed separately from the surrounding publicly owned spaces. In these examples, even through private management may supplement public funds and/or free up public resources to be spent on other areas, it also raises issues of the disparity in expenditure and levels or service between places. Moreover, as the willingness of private organisations to maintain public spaces is rarely dissociated from at least a degree of private control on how those spaces are used and by whom, it raises questions of freedom and exclusion. This last issue returns the discussion to questions of contract and agreement drafting and whether issues of control and exclusion can be adequately controlled by such instruments.

The community-centred model

The third model is perhaps the least developed of the three, although not necessarily the most recent. It constitutes another form of devolution of responsibility for the provision and/or management of public spaces and related services, but this time to community organisations, including associations of users of public spaces, interest groups organised around public space issues, and so forth. A fundamental difference from the previous model is that the organisations to which public space management is devolved are in principle not structured according to market principles of profitability and competitiveness. They do not exist to provide public space services for a fee or to maximise economic returns on investment in or surrounding public space, and instead have a direct interest in the quality of the public spaces and related services primarily for their use value.

In these cases, the 'public interest' dimension that characterises public services is not confined to one side of the devolved arrangement, although this coincidence of interests might be very localised. In real life these distinctions are more nuanced, and communities residing around a public space might have an interest in its quality also because it affects the capital value of their homes. However, this is unlikely to be the main or only purpose of the organisation, and even if it were, it would not operate according to market rules. These organisations do not belong either to hierarchical (the state) or market (private-sector) modes of social governance, and are more closely linked to 'network' governance (Rhodes 1997) in that they exert influence and pursue their objectives by developing formal and informal horizontal linkages with other similar organisations and with the public and private sector.

As with the previous model, this approach can be seen as a result of the retreat or 'hollowing out' of the state (Rhodes 1994), weakened by the reshaping of the economy and society since the mid-1970s It can also, and perhaps more positively, be explained by the trend towards the co-production of public services with their users (Sullivan and Skelcher 2002, DTLR 2001). The need for flexibility to match services to a variety of needs, for local knowledge to understand very localised demands, coupled with the effort to redefine the relationship between the state and citizens in mature democracies has led to an erosion of the separation between provision and use. Co-production (i.e. user engagement in the provision of public services) has been seen as the most effective way to tackle diversified and complex demands brought fourth by the increase in wealth and the variety of lifestyles and associated needs (Goss 2001). This applies to a whole gamut of public services, from health and education to social housing and urban renewal, as well as to public space management.

This model is also a rediscovery and extension of a long-established tradition of involvement of charities and the voluntary sector in welfare delivery, which pre-dates state provision and was never fully replaced by it. Charitable organisations have long been associated with the provision and management of public services. In the UK, for example, the recent re-emergence of this form of provision and management of public space has a number of key drivers:

- There has been a sustained effort to modernise the state, and local government in particular, to establish the 'enabling state'. This will include the search for more effective, responsive and cost-effective ways of delivering public services, but also the formulation of a new contract between citizens and the state by re-distributing responsibilities (DTLR 2001, DCLG 2006).
- At a more practical level, there have been attempts by government to reach sections of society normally at the margins of social programmes, such as some difficult-to-reach ethnic groups, teenagers in social housing estates and so forth through fostering their involvement in the provision of public services relevant to them (DTLR 2001).
- Specifically in the case of public spaces, there is plenty of evidence of problems of under-use and exclusion by particular groups within a community, which could be better addressed through the involvement of the relevant groups in design and delivery of solutions (DTLR 2002a, Audit Commission 2002a).

If contractual relationships defined the nature of devolved service provision to private-sector agents, given the variety of contexts in which public space management by communities has evolved, it is difficult to define a single set of characteristics for the relationship between the state and voluntary agents. In the UK, devolved service provision through community and voluntary sector organisations has also tended to take a contracts-dominated form, with the state acting as the principal, and the voluntary organisations as the agents. However, this has proved to be fraught with tensions because of the threat to the independence of those organisations created by their progressive transformation into public-sector contractors (Deakin 2001). As a result, there are moves now to replace conventional principal–agent, or client–contractor arrangements with more complex 'compacts' involving mutually agreed principles, practices and distribution of responsibilities.

Well-defined public space management contracts with voluntary organisations exist side-by-side with much less formal agreements with ad-hoc residents' groups centred on the management of particular spaces whose existence and survival depend both on government funding and on the capacity of the community in question for sustained collective action. An example of the former is the transfer of the management of social housing estates and its recreational and green spaces to housing associations, or the management of parks or open spaces by long-standing 'friends' associations. An example of the latter are the neighbourhood management schemes, funded by government neighbourhood renewal initiatives, in which communities in deprived areas are encouraged to manage their own public spaces (DCLG 2007). More recently, there have been a few examples of role changes in contractual relationships, in which organised communities have been able to produce public space management strategies for their areas and have them recognised by their local authority, effectively becoming the clients for public and private contractors (see Chapter 7).

COORDINATION

Like the state-centred model, the interventions of community-centred agents on public space also require better vertical and horizontal connections within public sector organisations. Given the contractual nature of many public space management agreements between the public sector and voluntary organisations, contract specifications are also important in establishing that what is being delivered by the contractor is what is required, and that it reflects broader public space policies. However, enforcement and sanctions that went hand-in-hand with specifications as means of coordinating contractual relationships between the public-sector and private agents are less effective here as not all forms of voluntary and community organisations will be affected by contract sanctions in the same way.

As the separation between the providers and users of public space services is even narrower than in the other models, partnership mechanisms are essential tools of coordination. Adequate partnership structures, with clear consultation, participation and decision making mechanisms can lead to the formulation of clear agreements about what outcomes should be expected, what is required from each partner, why they should comply with broader policy strategies, and how sustained engagement between partners will be maintained. The ability to negotiate with and engage partners is the key skill.

Coordinating the inputs from public space users into management is not an issue in itself in this model, as it is already implied in the involvement of users in management tasks. However, this involvement is mediated by the way in which voluntary and community organisations work, and it depends on how representative they are of their own constituencies, and how well they absorb and deal with the demands and aspirations of their members.

REGULATION

The regulation dimension in this model also relies on the law-making and policing roles of the state to deal with conflicts of uses in public space and patterns of usage, often in support of a less formal policing role played by the community itself. Contract enforcement mechanisms are also relevant to regulate devolved service provision, but less so than in the market-centred model.

Voluntary organisations are not necessarily in competition with one another for the same service, especially the more localised community groups, and the effectiveness of contractual sanctions is less clear. A more established voluntary-sector organisation delivering public space management services in a variety of locations, with assets to back their liabilities will react differently to contractual sanctions compared to a small, local friends group, which might simply dissolve under pressure. In the same vein, performance measurement systems setting clear targets for public space management are important to secure standards in a devolved approach, but are less useful as an enforcement tool for the same reasons. Moreover, they need to be linked to capacity building measures and thus to resourcing policies to secure that targets can really be met.

MAINTENANCE

As regards maintenance, the appropriate definition of routines, techniques and procedures is still the core of this management dimension. As with the market-centred model, there is a separation between the definition of standards and routines and their deployment; the first, the responsibility of the local authority, the second of the organisation undertaking the management task. This is especially so where contractual relationships are employed, and in these cases contract specifications are an important part of management; as they were in the previous model.

However, the gap between the definition and deployment of maintenance routines is not so clear when standards of public space and maintenance are agreed through partnership work and deployed by community partners. The key issues here are about setting standards of public space maintenance that are compatible with the capacity of the partnership or the community organisation to deliver. This may very well involve the provision of technical and institutional support to those organisations by the public sector so that the desired standards can be achieved.

Locally defined standards and maintenance routines are more likely to reflect local aspirations, be more responsive to local context, and benefit from a sense of ownership by local communities. However, they are likely to lead to differences in standards or maintenance across areas within the same local authority, as inevitably communities will have different aspirations as regards public space quality, and varying capacities to deliver them. In this model, therefore, the acceptability or otherwise of local difference, and the understanding by all parties of its implications are key issues in the maintenance of public space.

INVESTMENT

As with the previous model, resourcing is not primarily about securing a slice of the public sector budget for public space management but is instead about drawing resources from outside the public sector. In this case, this may not involve finding alternative sources of money or technical expertise, although that can be important, but instead involves drawing local knowledge into public space management by harnessing the active commitment that can be provided by public space users. Again this implies identifying who are the social actors with a stake in the fortunes of a public space, what resources they can add to its management, how these resources can be combined with those already available, and how those actors can be engaged in public space management.

However, even when contractual relationships are in place setting up rights and responsibilities, the nature of community involvement is such that those resources of knowledge, mobilisation and commitment can only be released if the right structures are established to make this possible. Therefore, in this model public space management resourcing is also concerned with building community capacity to act collectively, developing skills to form and manage partnerships, and about creating and fostering relations of trust; all of which create and sustain the basic conditions for those resources to be released. Indeed, experiences reported in this book suggest that releasing the kinds of resources communities can offer to the management of public space requires in turn a sustained effort to maintain commitment and a sense of purpose.

Conclusions

In this chapter three models of managing public space have been put forward which have emerged as a response to perceived problems of the more traditional approach. From the discussion it should be clear that although there are clearly identifiable rationales underpinning each model, in practice they do not constitute entirely separate approaches to public space management. The next chapters will show how public space management strategies use elements of these different models to tackle specific challenges and contexts, sometimes harmoniously, sometimes with contradictions. How they combine these models is determined by the nature of public space issues, political contexts, local social and economic factors, and so forth.

There is no moral or practical superiority of one model over the others. In both theory and practice approaches centred on state action, or on private sector effort, or in direct community participation, can all provide solutions to particular public space challenges in the particular contexts in which they are applied. These models have their own intrinsic advantages, from the clear accountability or the public interest ethos of the state-centred model; to the ability to draw resources from a much wider constituency and more sensitivity and responsiveness to changes in demand in the market-centred model; to the sensitivity to user needs and the commitment of the community-centred approach.

They also have their own potential disadvantages too, from the potential bureaucracy and insensitivity of the state-centred model, to the very real risk of exclusion and commodification of the market-led approach, to the fragmentation, lack of strategic perspective and inequality of a community-centred model. These issues and how they have played out in practice will be returned to in the chapters that follow in Part Two of the book.

Part TWO

Investigating public space management

Chapter 5

One country, multiple endemic public space management problems

Through a national survey of urban local authorities in England, this first chapter in Part Two explores the approaches of English local government to the management of external public space. It is the first of two linked chapters that explore current and developing practice in England from the perspective of what the public sector is actually doing to manage public space. As such, it focuses on what was referred to in Chapter 4 as the state-centred model of public space management, the model that is still dominant throughout the world today. The discussion begins by introducing the local government context within which public space is managed in England and briefly explores evidence for a decline in the quality of public space and the services charged with its management. Next the research methodology is discussed for this and the next chapter. The third and main part of the chapter reports on the outcomes from the national survey itself, whilst a fourth section links the findings to a related study to gauge the opinions of key user groups on the state of public space management in England.

The state of English public space, and its management

The empirical research upon which this book is based began in 2002 with a deliberately broad focus, examining the management of the full gamut of public space types encompassed in the typology in Chapter 3 (Table 3.1) and the definition in Chapter 1. In subsequent chapters the focus is narrowed to particular forms of public space from the typology; public open space (Chapters 7 and 8) and civic space (Chapters 9 and 10).

In England, responsibility for managing the wide range of spaces that fall under the adopted definition usually resides with local authorities. A national survey of local authority approaches and policy concerning the management of public space was therefore conducted in 2003, the aim being to establish a baseline of knowledge about what might be described as 'normal' practice across the county, whilst also seeking to uncover innovative practice that might point towards more effective public space management in the future. Chapter 6 goes on to examine in greater depth the views and experiences of twenty local authorities that exhibited such innovations.

Local government in England

Before discussing public space management specifically, it is first important to establish the broader context within which local government in England operated at the time of the survey. Since 1997 Tony Blair's 'New Labour' administration had been active in implementing what has been collectively described as a Modernising Local Government agenda. Thus 1999 saw the first of a series of Local Government Acts that formed the legislative basis for these changes. In fact this drive for 'modernisation' was not an isolated programme, but instead sat as part of a much larger tide of change worldwide characterised as the 'new public management' (NPM)[1].

Central to this agenda has been the idea that public services should be managed in a rational fashion, drawing lessons from private sector performance management which itself has roots in management accounting. The legislative programme arose from an analysis of local government that was highly critical of both political and managerial decision making and that was itself part of a much broader programme of reform in the public sector.

PRE-1997

The reforms actually began in the 1980s and 1990s which were also characterised by a flow of legislation, directives and regulations directed at local authorities in the UK. Discussing NPM in general and the period under the Conservative government in particular, Pollitt et al. (1999) divides public management reform into three phases. First, the period from 1979 to 1982 was characterised by a fierce but crude drive for economies. Second, the government moved to emphasise efficiency and there was a push towards privatisation of public services; this phase lasted until the late 1980s. Although the three 'E's of economy, efficiency and effectiveness were constantly referred to in this period, most of the procedures and national performance indicators concerned the first two – economy and efficiency. It was during this period that the Audit Commission was set up, in 1982.

The third and probably most radical phase was after the 1987 elections. The reforms in this period included: extended use of market-type mechanisms (MTMs); intensified organisational and spatial decentralisation of the management and production of services, (even some shifts from local authority control to independence), although not necessarily their financing or policy-making; and a rhetorical emphasis on service quality, exemplified by the launch of the Citizen's Charter programme. Rogers (1999) usefully summarises the themes that ran through the reforms from 1979 onwards:

- accountability – local government to central government, authority to citizens, services to users, managers to councillors, employers to senior management;
- the explosion of audit and inspection – the role of the Audit Commission in particular expanded from its responsibilities in relation to financial accountability to include inspection and determination of performance indicators;
- customer choice – the legislative provision of choice; even to 'opt out' of local authority provision; moving beyond limiting accountability mechanisms to elections, politics and complaints;
- competition and contractualisation – which was exemplified by compulsory competitive tendering (CCT), through which authorities were effectively forced to out-source certain specified services;
- centralisation and control of government – despite the increase in rhetoric about partnership;
- The Citizen's Charter – these proposals contained in a 1991 White Paper and intended to improve performance of public service organisations, included the principles of standards and targets publishing, user consultation in standard-setting and to ensure independent validation of performance to achieve value-for-money.

POST-1997

The publication of the 1998 White Paper 'Modern Local Government: In Touch with the People' proposed further local government reforms to strengthen the leadership role of local government within the community, whilst making it more accountable and providing better quality, cost effective services (Planning Officers' Society 2000). In his introduction to the 1998 White Paper the then Deputy Prime Minister outlined the scope for change:

> People need councils which serve them well. …There is no future in the old model of councils trying to plan and run most services. It does not provide the services which people want and cannot do so in today's world.
>
> (DETR 1998: foreword)

The comments reflected what central government saw as the old culture of local government, a culture not conducive to effective local governance and leadership in the modern context, a culture typified by:

- a paternalistic view from members and officers that it is for them to decide what services are to be provided on the basis of what suits the council as a service provider;
- the interests of the public coming second to the interests of the council and its members;
- more spending and more taxes seen as the simple solution rather than exploring how to get more out of the available resources;
- relationships between the council and its essential local partners being neither strong nor effective;
- local people indifferent about local democracy;
- overburdening of councillors and officers;
- a lack of strategic focus concentrating on details rather than essentials.

Change under the 'Modernising Local Government' agenda sought to recast the culture of local authorities, and to transform how authorities undertake their statutory functions – principally through delivering and monitoring 'best value'. In reality the modernising agenda represented a continuation of public-sector reforms already in motion before 1997, albeit with a change in emphasis, including the introduction of a comprehensive system of performance related incentives and disincentives and tougher requirements for community and local governance.

This was elaborated in the 2001 White Paper 'Strong Local Leadership, Quality Public Services' which stated that the government will provide support to underpin local community leadership building on new well-being powers (wide-ranging freedoms for local authorities to act in the well-being of their communities) and local strategic partnerships (LSPs), designed to bring the public and private stakeholders together to plan the future for their areas (DTLR 2001). In part this was to be achieved through the production of community strategies as the vision and coordinating framework for investment and public sector services. The over-riding emphasis became one of public–private partnership in the delivery of services, replacing the earlier regime that could be characterised as a gradual private takeover.

The government proposed to manage the whole reform process through a national framework of standards and accountability, by setting out the comprehensive 'best value' performance framework, accompanied by a substantial package of deregulation. The framework comprised:

- defined priorities and performance standards – the latter encapsulated in the national best value performance indicators;
- regular performance assessments – most notably the national best value reviews, with inspections undertaken of local authorities by the Audit Commission;
- coordinated incentives – rewards and tools which address the assessment of results, including publicised performance information; freedoms, powers and flexibility over resources; action to tackle failing councils; and national/local agreements over service standards.

The White Paper proposed to accompany the increase in responsibility and accountability with removal of restrictions on planning, spending and decision-making within high-performing local government departments, with a view to encouraging more innovation and improved quality. However, by tying the freedoms to performance, a system of 'carrots and sticks' was effectively created. It built on the foundations of the 1999 Local Government Act which outlined that 'from April 2000, the duty of Best Value will require local authorities to make continuous improvements in the way they exercise their function, having regard to a combination of economy, efficiency and effectiveness'.

For local authorities, a particular bonus of the new system was the removal of the former requirement to tender for, and if necessary contract out, their services under the auspices of the much derided compulsory competitive tendering (CCT). In its time, CCT undoubtedly drove down the costs of providing public services, but very often this was achieved at the expense of service and delivery standards. The public space remit provided a case in point, leading in the process to a decimation of local government capacity and capabilities in this area of responsibility.

The public space/management context

For this and other reasons, the recent story of public space management in England has not been a happy one. Some of the most graphic examples of a general failure to manage public space were captured by the joint CABE/BBC Radio 4 initiative 'Streets of Shame' which called for nominations for the UK's best and worst streets. Following thousands of nominations, the five best and five worst streets of 2002 were chosen (Boxes 5.1 and 5.2). The results and the comments from nominees were instructive and revealed that what was identified as good and bad by nominees usually represented two sides of the same coin (Figure 5.1).

They also confirmed that much of the perception that users form about space, and whether that perception is positive or negative, relates to how space is managed and maintained, rather than to its original design. Therefore, although all the qualities in Figure 5.1 (except the first) relate in some way or other to the original design and layout of the streets, all (except perhaps the last) correspond more strongly to the way streets are cared for following their original construction.

A comprehensive and objective assessment of the state of public space in England is not yet available, although a range of evidence gathered shortly after the start of the second Tony Blair administration, when the Prime Minister himself was backing action on this front (see Chapter 1), suggested that the challenge faced by public space managers was substantial.

First, on the quality of public space:

- Polling company MORI's ongoing work tracking the perceptions of around 100 local authorities revealed a falling satisfaction with the street scene as a whole and with street cleaning in particular. They argue, '[i]n longitudinal survey after survey, the trends are negative'; a trend that contrasts strongly with rising satisfaction in the 'big ticket' services that have benefited from targeted funds and strong inspection regimes (MORI 2002). The work has revealed that highways and pavements is the worst-rated local government service.
- Results from the first 'Local Environmental Quality Survey of England' undertaken by the environmental charity ENCAMS (2002) across 11,000 sites and 12 'land-use' classes revealed that 50 per cent of the local environmental elements surveyed were registered as unsatisfactory. These included litter, detritus, weed control, staining, highways, pavement obstructions, street furniture condition and landscaping (Table 5.1). Although there has been some improvement since, the improvement is often from a very low ebb.
- A self-assessment by 85 per cent of UK local authorities of their green spaces undertaken for the Urban Parks Forum (2001)

BOX 5.1 BRITAIN'S WORST STREET (OF 2002)

Streatham High Road, London – Concrete and metal barriers, 'wasting away in places, supposedly designed to protect pedestrians from the full force of the dual carriageway traffic, are used as an assault course by those determined to get from one side to the other

Cornmarket Street, Oxford – 'An example of small mindedness, inefficiency and ineptitude', 'filthy dirty', 'smelly' and 'an embarrassment'

Drakes Circus, Plymouth – 'The lack of diversity and the out dated office spaces mean it is unattractive to commercial and retail tenants and the threatening feel at night, with lack of activity and poor lighting, make this a no go area'

Maid Marion Way, Nottingham – 'Dubbed the ugliest street in Europe since its construction in the 1960s, municipal engineers are doing their best to maintain its position at the top of the premier league'

Leatherhead High Street, Surrey – 'An example of cheap and thoughtless pedestrianisation taking the heart out of a whole town'

BOX 5.2 BRITAIN'S BEST STREET (OF 2002)

Grey Street, Newcastle-upon-Tyne – 'The shop fronts may not be original but they are in keeping with the spirit of the original design and fit in very well with the scale of the buildings.' 'A street on a human scale with a grand vision'

High Pavement, Nottingham – 'Well maintained' and 'offers respite in what can become a busy street at weekends'

Buchanan Street, Glasgow – 'Well lit', 'clean', 'good public seating', 'attractive tree planting' area

New Street, Birmingham – 'The fact that people can now walk from Brindleyplace to the Rotunda without having to worry about fumes and traffic, with opportunities to sit in well designed seats and see an eclectic mixture of art and sculpture is a great achievement'

Water/Castle Street, Liverpool – 'The scale is human, there is light and life and a feeling of safety 24 hours a day'

LOCAL AUTHORITIES WITH NO INTEGRATED STRATEGY

The large majority of councils that responded to the survey (four-fifths) did not possess a truly integrated strategy for managing public space. In many cases the local authority responses described a strategy for managing public space, but this could not be considered 'integrated' for the purposes of the research for one of a range of reasons:

1. TOO LIMITED IN SCOPE

The local authority definitions of 'external public space' and 'public space management' were too limited in scope to fit the research team's definition of what constitutes an integrated strategy for managing public space. This was either because external public space was defined as a distinct type, or because management processes were clearly limited to particular narrow aspects of the agenda. Instead, public space strategies generally addressed either one of two distinct types of public space: green/open space or urban/city centre space. While some authorities addressed both, few authorities supplied details of an integrating strategy that linked the two public space typologies to each other, or to the wider public realm.

Examples of the former included green-space strategies, parks and open spaces strategies, trees and woodlands strategies, rights of way strategies and recreation strategies. Other place-specific or area-based strategies included management plans for individual parks or other open spaces under the control of the local authority. A typical example was Sandwell District Council's numerous green-space strategies covering, parks, playing fields, and trees. Examples of the latter included city or town centre management strategies, and management strategies for particular urban locations, such as commercial streets, nightlife districts, or residential areas. In some cases these were pilot schemes that may, if successful, be extended to mainstream practice. An example was Westminster City Council's Action Plan for Leicester Square, which, if successful was to be extended to other areas in the borough (see Chapter 10).

2. TOO GENERAL IN COVERAGE

In such cases, mention was made of integrating practice, but coverage was too general to provide a meaningful coordination strategy. Examples included local authority-wide policy documents such as corporate strategies or service strategies that were general in the extreme, but also statutory planning documents such as development plans, or community strategies that provided no more than strategic aspirations. An example was Cheshire County Council's County Structure Plan and Community Strategy, both documents that describe rights of way and open space, but do not mention details of a strategy for managing public space.

3. DIVIDED RESPONSIBILITIES

In many parts of the country, the responsibilities for public space remain split between tiers of local government, between the county council (the highway authority) and the district council, which retains responsibility for most other publicly owned spaces. For example Runnymede District Council had an integrated management strategy in respect to its parks and open spaces, but this does not extend to highways and street landscaping. In fact, in this case the district council recently lost responsibility for these areas which it used to manage on behalf of the county highways department.

4. IN PREPARATION

In a number of cases, authorities reported that an integrated management strategy was in preparation, often resulting from the best value review process, but failed to provide any evidence to substantiate their emerging approach. An example was the London Borough of Richmond who responded that they were about to embark on a Street Scene Best Value Review that would address the integration of services in the public realm. The range of these responses suggested that better integration of public space management services is increasingly on the agenda.

5. PARTIAL INITIATIVES IN PLACE

Some authorities had partial public space strategies in place that would indicate that there is a good degree of coordination in the local authority's management of public space, although not full integration of all public space responsibilities. Examples of such partial strategies included those that integrate public space management policy and delivery. For example, Cambridge City Council have a public space management strategy whose aim is to integrate street cleaning and grounds maintenance service delivery, although this does not extend to the whole of the city's public space network.

LOCAL AUTHORITIES WITH AN INTEGRATED STRATEGY

Only nine local authorities responded with anything close to an integrated strategy for the management of the public spaces in their area, although

Table 5.2 Public space integrated strategies

No.	Authority	Inspiration	Name of integrating strategy/ document	Details
1	Newcastle City Council	Best Value	Best Value Pilot for Integrated Environmental Services	Unified public space budgets and restructured the council so that the public space management processes are all covered in a single council green paper known as an Urban Housekeeping Plan, that unifies other public space initiatives
2	Dartford Borough Council	Best Value	Street Scene Best Value Review	Unites most public space types, but excludes parks. Mainly maintenance and regulation/enforcement based
3	Harlow Council	Best Value	Urban Landscape and Street Scene Best Value Review & Service Improvement Plan	Addresses most management processes. Excludes parks and highways
4	Lancaster City Council	Best Value	Maintaining the Environment Best Value Review Improvement Plan	Coordination document to improve integrated working, at early stages
5	London Borough of Waltham Forest	Best Value	Street Scene Best Value Review	Unites public space types, mainly maintenance based
6	Leeds City Council	Best Value	Parks and Countryside Best Value Review	Brings together different public space management processes. Green space biased
7	Westminster City Council	Internal	West End Public Spaces Report	Management plans including detailed analysis of several high profile public spaces and districts. Includes all four management processes
8	Bristol City Council	Internal	City Centre Strategy	Brings together different initiatives that while limited by area do cover several public space types and a range of management processes
9	Oxford City Council	Internal	Public Realm Strategy	The only document supplied with public realm in the title, with detailed strategies for designing and managing public space

even here strategies rarely covered the full range of spaces encompassed in the definition of public space adopted for the research. Table 5.2 summarises these integrated strategies which were of two types.

First, some local authorities have achieved an integrated strategy through changes in public space management structures or through specific public space initiatives. These tended to be internally inspired, mainly through local authority members and officers with a passion and dedication for the public realm. Most of these strategies were for the city centre public realm, and therefore covered a limited range of public space types.

Westminster City Council, for example, had commissioned the West End Public Spaces Report, which was one of the few documents to describe management plans covering investment, coordination, regulation and maintenance concerns. However, the report concentrated in detail on only a few high-profile public spaces, while not discussing lower profile or residential public spaces. Bristol City Council's City Centre Strategy was primarily urban design based and limited to central Bristol, but encourages the coordination of activities, and is regularly updated on issues of public space maintenance and investment. Finally, Oxford City Council have produced a Public Realm Strategy for the city centre that includes a history and analysis of Oxford's public realm. While the emphasis of this is also primarily on design, clear aims for maintenance and investment are established.

Second, following recent public space-specific best value inspection processes, six local authorities had something approaching a comprehensive strategy for managing public space. Of these, only one – Newcastle City

Council – have what can be described as a completely integrated strategy for their public space that has also been implemented. Amongst the rest, by way of example, Harlow Council could demonstrate that internal restructuring meant that public space issues were all covered within one directorate with responsibility for the range of public space management processes. The London Borough of Waltham Forest had adopted a holistic definition of public space, and referred to a wide range of public space management processes with a particular emphasis on urban design in their Street Scene Best Value Review.

For their part, and resulting from their choice as a national exemplar – as a Best Value Pilot for Integrated Environmental Services – Newcastle City Council have been able to unify public space budgets and restructure the council so that the public space management processes are all covered in a single green paper known as the Urban Housekeeping Plan. The plan takes a deliberately holistic definition of public space, and demonstrates how public space management services are delivered and what the council's future plans for public space are.

Local authority public space initiatives

Although most local authorities in England do not have an integrated strategy for managing public space, there are nevertheless a wide range of initiatives increasingly being adopted by local authorities to deliver the better management of their public spaces. Moreover, many of these

Table 5.3 A selection of coordinating local authority initiatives

Authority	Name of coordinating initiative	Focus of initiative
Cambridge City Council	Street Scene Project	Coordinating the street scene through a team comprising officers across several departments, workforce, unions, members and residents' representatives
East Riding Council Newcastle City Council St Albans City Council	One stop shop hotline	A dedicated telephone line/ one-stop shop for coordinating all public space services
Great Yarmouth Borough Council	Street Scene Working Party	Coordinating the delivery of services for streets and open spaces
Great Yarmouth Borough Council Harlow Council Liverpool City Council Newcastle City Council Northamptonshire County Council	New council structure	Restructuring so that one department covers most public space issues with an Executive member and a chief officer directly responsible
Leeds City Council	Green Space Implementation Group	Integrating the working of different open space departments at policy and delivery levels
Liverpool City Council	Grounds Maintenance Continual Improvement Group	Coordinating contractors, client departments and other stakeholders to improve public space management
London Borough of Greenwich North Tyneside Council	Clean Sweep	Coordinates officers and direct labour to improve the maintenance of public spaces across typologies
Newcastle City Council, Nottingham City Council Watford Borough Council	Environmental Ward Stewardship, Locality Managers, Area Committees	Coordinating and improving the responsiveness of public space management to local demand
North Lincolnshire Council	Neighbourhood Teams	Locally based teams responsible for the maintenance of the street scene and open space, as well as related policy

initiatives are focused on finding means to better coordinate public space management and thereby improve the quality of public space. Other initiatives were more limited in their aspirations, and focused instead on particular aspects of public space or its management.

INITIATIVES FOCUSING ON THE BETTER COORDINATION

The main category of public space initiatives, and the most diverse, were those that created new coordinating structures for the delivery of services on the ground; although they tended to focus on one type of public space or another. This continued sectoral thinking seems to be the biggest influence on local authority public space management structures, and also the biggest barrier to the integration of service provision.

Local authorities named numerous different public space initiatives that relate to the improved coordination of public space management. Table 5.3 provides a range of examples of these types of initiatives. Usually inspired by best value processes, coordinating initiatives typically start by focusing on a particular type of public space (e.g. streets or green spaces). Cambridge City Council and Great Yarmouth Borough Council, for example, started with the street scene, while Leeds City Council and Liverpool City Council began by examining green space management.

A few local authorities look to coordinate as many public space typologies and management processes as possible. Authorities such as Great Yarmouth and Newcastle City Council have restructured so that an

executive member and chief officer are directly responsible for all public space issues cutting across space types and management processes. Other authorities, such as East Riding Council, look to coordinate public space services to users – local residents and businesses – through a one-stop-shop service that can be contacted through a variety of methods (telephone, fax, email, video box, in person). Other local authorities look to make public space management processes more responsive to changing local needs through area-based maintenance teams.

INITIATIVES LIMITED TO PARTICULAR ASPECTS OF THE SERVICE

Some public space initiatives identified in the survey covered specific aspects of public space and management processes. By themselves these initiatives were quite narrow in focus, unless part of a broader management strategy. Table 5.4 offers a range of examples of these initiatives.

At the time of the research, many of these were very recent and still needed to be evaluated for their effectiveness. A minority had been around for much longer, including the use of design guidelines, or project-oriented approaches. The latter type tend to relate more to securing the initial quality of the public spaces rather than to their ongoing management, or to the processes by which management is delivered.

Management plans and strategies also have a longer pedigree in two key situations; open spaces/parks/countryside sites, and town centres; both of which are area-specific rather than council-wide. In these situations,

Table 5.4 A selection of individual local authority initiatives for public space management

Authority	Name of initiative	Focus of initiative
Birmingham City Council London Borough of Camden	Street Design Guidelines	These apply to highways, new roads in residential areas and rural lanes
Carlisle City Council Lancaster City Council London Borough of Bexley	Town Centre Management	Area based management of town centres, included strategies and partnerships also
Corporation of London Sandwell Borough Council	Individual management plans for open space, parks and green spaces	Clear plans that refer to the four management processes but are limited by typology
Corporation of London Newcastle City Council	Award schemes	Internal authority or national or international environmental competitions, (such as 'Britain in Bloom')
Dartford Borough Council London Borough of Richmond Newcastle City Council Spelthorne Borough Council	Sponsorship schemes	Sponsorship for the maintenance of public spaces, including schemes such as 'friends of…' or 'adopt an area'
Eastbourne Borough Council London Borough of Kensington & Chelsea London Borough of Waltham Forest West Sussex County Council	Individual public space capital investment projects	New public spaces or major enhancement of the environment; it applies both to streets, town centres and parks and open spaces.
Lancaster City Council London Borough of Haringey London Borough of Wandsworth Wycombe District Council	Warden schemes	Human management of environmental and/or antisocial behaviour. These apply to park patrols, neighbourhood wardens, street watchers.
London Borough of Camden	Boulevard Project	Introducing new standards for street scene management

efforts have often been made to create a relationship with the users of the public spaces, with schemes such as 'friends' who volunteer to improve or maintain these spaces – usually parks – or businesses who contribute to the public realm through sponsorship schemes – usually town centres.

TRANSFERRING RESPONSIBILITIES

However, there was still wide variation in how extensive the use of private contractors is, with some local authorities reporting several instances of such a practice, and others reporting very little. Contracts themselves have varied from short, well-defined instruments that aim to keep contractors responsive to local authorities' needs, to broader long-term arrangements, which shift management responsibilities away more decisively from the public sector. This transfer of public space management responsibilities has often been done through more or less formalised partnerships between the local authority and local businesses, focusing on the management of specific areas.

A similar process has shaped the involvement of community organisations and the voluntary sector in public space management, often spurred on by national neighbourhood regeneration policies. The research shows a general agreement with the principle that the public should have an active role in tackling public space problems, as the 'ownership' of public spaces by their users might be the most effective way to maintain their quality. This is so despite a tendency of some authorities to keep a clear distinction between their role as provider and that of the community as recipient of services.

In general there is a tendency to experiment with service integration and redistribution at the local level, often through pilot schemes, in order to make the delivery more responsive to users. For example, the interdisciplinary working of officers from different departments, or the establishment of working parties that include external partners. Increasingly structural changes that bring together the various sections within local authorities with a role in the management of public spaces are also being adopted.

Local authority public space documents

Of the 64 authorities that responded to the survey, 41 enclosed a total of 134 documents with their response. The different types of documents supplied are shown in Table 5.5, loosely grouped by subject area. This information was often of a high quality and useful in understanding particular approaches and practice. A huge variety of document types were supplied by local authorities and analysed during the research. The diversity of documents serves to illustrate the highly complex nature of external public space practice and local policy.

Because national best value processes were resulting in local authorities reviewing the way in which public space services are delivered, documents associated with the Audit Commission inspection process and with performance reviews featured strongly in the response. However, broader aims and key priorities of councils were expressed in their corporate/community plans or strategies or in statements from the leader or cabinet. These also provided the framework for the service plans and best value reviews.

The corporate plans set out the authority's vision and identify key issues. They revealed that the aspirational objectives associated with the management of external public spaces tend to reflect common themes for both processes and outcomes. Examples of process objectives included:

Table 5.5 Local authority public space policy and documentation

Type of document	Nos. received
Major corporate strategy, vision or policy	
Cabinet/Leader's Statement/Report	4
Corporate Plan/Strategy	9
Community Plan/Strategy	9
Crime & Disorder/Community Safety Strategy	6
Service Plan/Strategy/Review/Action Plan	9
Specific public space typology	
Open Space Strategy/Review/Management Plan	8
Plan/Management Plan for Individual Open Spaces	4
Trees and Woodlands	2
Environmental	1
Biodiversity Action Plan	1
Rights of Way	1
Town Centre/Public Realm Strategy	5
Street Scene Projects/Pilots	4
Transport Plan/Strategy	2
Neighbourhood	1
Best Value	
Best Value Inspection Service (Audit Commission)	3
Best Value Review	7
Best Value Service Improvement Plan/Action Plan	3
Best Value Performance Plan	8
Other government initiatives or their local variants	
Safer and Cleaner Environments: Local Government Association (LGA)	1
Neighbourhood Management Pathfinder/Delivery Plan/ Project	1
Local Strategic Partnership (LSP)/Partnership Agreements	1
Street Scene Local PSA Target	1
Single Regeneration Budget (SRB) Delivery Plan	1
BIDs or similar schemes	1
Miscellaneous	
Leisure & Culture/Tourism/Marketing	3
Design Guidance	3
Award Schemes (Tidy Britain/Clean City)	1
Public Information Leaflets/Booklets	1

- improve services through efficiency savings and improve essential services such as roads, pavements, street cleaning, lighting, crime reduction (Coventry City Council);
- provide quality, value for money services through setting high standards, supporting innovation and the integration and joining-up of services (Peterborough City Council);
- integrated and coordinated provision of services, with environmental issues and crime high on the agenda (London Borough of Greenwich).

Examples of outcome objectives included:

- a safer, cleaner, greener city: improve the quality of the local environment through high performing public services (Birmingham City Council);
- a safe, clean and attractive place to live (Carlisle City Council);
- clean, safe, attractive streets and open spaces (London Borough of Waltham Forest)

- a quality environment through cleaner streets, better air quality, improved open spaces (London Borough of Southwark).

Corporate objectives were subsequently elaborated in a variety of specific strategies including town centre/public realm strategies, open space strategies, safety/crime prevention strategies, and best value performance plans. The latter provide means through which authorities are looking to make improvements to the way in which their services are delivered. Best value performance plans typically start by examining how services are delivered, challenging the processes in place and proposing changes that would result in improved delivery. Examples of how this process operates illustrate the different means through which best value can improve external public space management.

In the case of Harlow Council, following a best value review of its street scene activities, the council 'challenged' the delivery of its services that had been based on compulsory competitive tendering (CCT), distinguishing in the process between statutory and discretionary activities. This resulted in changes to the organisational structure with a single service head to deliver all street scene services. The London Borough of Waltham Forest, as a result of their street scene best value review, took a comprehensive look at delivery in this area and related services that impact on the appearance of the borough. Grouping services allowed activities to be viewed from a cross-cutting perspective, with the review resulting in a structural reorganisation to improve joined-up working. The London Borough of Southwark's best value performance plan focused on priority environmental issues such as cleansing and enforcement, street lighting and highways maintenance and the need to exercise greater control over the statutory utilities, whilst Coventry City Council chose to draw up their best value performance plan with an objective to improve the quality of the services through joined-up working at the local level.

Examples revealed by the research suggested that this process was generating considerable fresh thinking about public space and about the way its management is organised. Many of these examples will be examined further as part of the exploration of innovative practice discussed in Chapter 6.

Stakeholder views?

A final survey stage involved gauging the views of a range of professional institutes, government agencies and amenity societies engaged in the management of public space in England. By this means it was hoped to better reflect a broader range of views from the complex array of

BOX 5.3 KEY STAKEHOLDER GROUPS INTERVIEWED IN THE UK

Association of Chief Police Officers

Association of Municipal Engineers

Audit Commission

Association of Town Centre Managers

British Retail Consortium

Commission for Architecture and the Built Environment

ENCAMS

English Heritage

Groundwork

Improvement and Development Agency

Institute of Civil Engineers

Institute of Highways and Transportation

Landscape Institute

Local Government Association

Living Streets

Royal Town Planning Institute

Secured by Design

SITA

stakeholder groups identified in Chapter 1 (see Tables 1.3 and Box 5.3). The 18 groups agreed that the absence of dedicated strategies on public space management was largely a symptom of two overarching problems:

1   Poor coordination – on a range of levels, between policy formulation and implementation; different local authority directorates; different services within the same directorate; different initiatives in local authorities; different fragmented funding streams; between local authorities and other public and private landowners; and between the public and private sectors.

2   A lack of resources – for public space and its management brought about by a general absence of investment in public space services; the complexity of funding steams; a focus on 'special' initiatives and on capital investment rather than on revenue spending; an inherited 'lowest cost' mentality from the days of compulsory competitive tendering; and projects featuring poor 'cost-cutting' design solutions.

Two further sets of problems were themselves seen as compounded by the lack of coordination and resources, but were nevertheless identified as discrete concerns in their own right:

1   Poor use of regulatory powers – because authorities did not adequately prioritise enforcement; utilise the patchwork of laws and byelaws available to them; have sufficient powers in key areas (e.g. to control busking, travellers, fly tipping, derelict sites, litter, anti-social behaviour, vehicle abandonment, street trading,

skateboarders, placarders, leafleters, etc.); make connections between regulatory regimes; and because they feared the costs and processes of litigation.

2   A low priority given to maintenance – resulting from problematic relations between local authority client and contractor functions; procurement problems externally; a failure to adequately define standards and routines; higher community expectations than could be secured within budgets; failures to design-in maintenance concerns from the start; and conflicts with public space uses and users.

The four problem areas map onto the four interlinked areas of public space management responsibilities suggested in Chapter 4. They suggest, that in all the key areas of responsibility the state and local government in England have been underperforming.

For the stakeholder groups, the four problem areas were in turn exacerbated by an increasingly complex set of pressures impacting on decision-making at the local level, some of which were discussed in Chapters 3 and 4 (Figure 5.2). They include: organisational pressures, because organisational structures were rapidly changing and therefore seen to be untried and tested, despite the benefits that might ensue; societal pressures, because society seemed to be increasingly anti-social (i.e. the alcohol culture) and less concerned with place and community (i.e. the litigation culture); legislative pressures, because new powers inevitably remain untried and untested until they are enacted, and sometimes have unintended consequences (i.e. EU fridge and electrical appliance legislation leading to dumping in public space); economic pressures, that have reflected an expanding national and international environmental agenda but with negative externalities locally (i.e. the impact of the landfill tax and low vehicle recycling values); local political pressures, encompassed in frequent descriptions of the lack of political will to take public space concerns seriously, and by a diversion of resources to other services; and spatial/physical pressures, brought about by the increasingly complex range of uses and infrastructure that public space is required to accommodate.

The groups argued for:

•   better coordination of activities – both in policy frameworks and delivery services;

•   a move away from the philosophy that 'cheapest is best';

•   more resources for space management, but also the better management of existing resources;

•   an emphasis on the importance of routine maintenance through enhanced revenue budgets, rather than solely on projects and capital spending

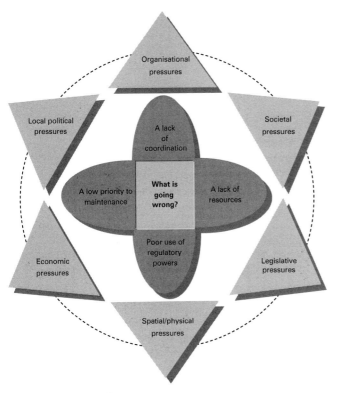

5.2 Generic problems and pressures

- maintenance as an act of enhancement of public space, i.e. a positive attempt to improve standards rather than to simply uphold them;
- good design to be factored in as a fundamental prerequisite for quality public space;
- management regimes to be extended to private space if perceived to be part of the public realm;
- better monitoring of public space quality, linked to more effective use of regulatory powers to better control public space;
- relations between the public and private sectors to be mutually supportive, whether the private sector is operating as sponsors, contractors or partners in managing public space;
- the community to be viewed as an untapped resource and to be more actively engaged in public space management.

The stakeholder groups concluded that public space remained a low political priority at the local level, and that a process of education might be required in order to raise it up the agenda. For them, the barriers between the traditional 'silo'-based professional disciplines needed to be overcome – both as part of the education process – and because key issues continue to fall between the gaps. Indeed, the groups argued that poor management skills dog public space services. Therefore, although stakeholders were remarkably consistent in identifying the important qualities of good public spaces – namely clean, safe, inclusive and robust space – they were also aware that the complex interactions remain poorly understood.

Conclusions

The evidence confirmed that in some places much is going on, even if, as yet, this practice was the exception rather than the norm. In this regard, it is hardly surprising that the public space literature and specifically the empirical evidence concerning public space quality in England reports a widespread deterioration, when local authority management practice seems so fragmented and partial, and lacking in vision about how to improve practice in the future. Indeed, the survey confirmed that the large majority of English local authorities did not have a dedicated and detailed strategy for the management of their public space, and instead, very broad 'motherhood'-style corporate objectives or individual strategies for parts of the external public space agenda were more common.

Although the provision of management services for external public space varies between councils, it continued to be divided on the traditional model between parks/leisure, planning/highways and street maintenance services. Sometimes these were under a single directorate, often they were under two, and sometimes three or more. Usually, however, there was little coordination between individual services that continue to operate along sectoral professional lines. As can be expected, the focus of these different services was not public space in itself. Their main concerns remained the tasks themselves, of road sweeping, tree pruning, controlling traffic and parking, and so forth, whereas public space was merely the context in which these tasks were carried out. The emphasis reflects criticisms raised in the literature explored in Chapter 4.

Nevertheless, two top-down influences had been inspiring changes. First, the national best value inspection process which has been challenging a number of local authorities to plan for cross-cutting public space services through the preparation of integrated best value plans. Thus best value processes seemed to be the driving force behind the use of integrated strategies, where they exist, often tied to changes in organisational structure. Best value reviews were also encouraging a number of initiatives to 'join-up' street scene services through special working parties and projects, whilst best value performance plans were often critical in challenging existing processes and proposing changes to improve delivery.

Second, more fundamental cross-authority structural reviews were leading some authorities to bring public space management services together. Typically these resulted from the rethinking of the structure and management of local authorities in light of the Local Government Act 2000, although perversely often at the cost of separating public space policy from delivery services. Nevertheless a wide range of initiatives now exist in local authorities across England, covering 20 types of initiative:

1   better cross-departmental corporate working arrangements
2   restructuring (e.g. merging departments into larger directorates)
3   IT initiatives to improve communication
4   working parties/stakeholder liaison (internal and external)
5   partnership arrangements (contacting, crime, etc.)
6   improving customer focus/care
7   setting new work standards/targets/guidelines/performance
8   devolution to neighbourhood level
9   coordination strategies – design, open space, transport, crime, etc.
10 capital investment projects/programmes/exemplar schemes
11 dedicated area management regimes
12 sponsorship schemes
13 warden schemes
14 award schemes
15 public space audits/indicators/health checks/monitoring
16 peer review schemes
17 training schemes (design, management, etc.)
18 byelaws (safety, litter, etc.)
19 community involvement
20 public space champions.

Many of these local initiatives suggest a degree of redefinition and redistribution of roles and responsibilities for public spaces within local government, and between them, the private sector, and community/voluntary sector organisations. Whether or not this redefinition of rights and responsibilities in the management of public spaces is socially desirable remains contested, in England, however, it seems to be inherently linked to the simultaneous redefinition of the very nature of local government (see Chapter 4). Nevertheless, these moves towards greater community or market involvement in the management of public space are (so far) typically tentative and do not amount to a wholesale move from a state-centred to either a market or community-centred model of management.

The research confirmed that this is an area of public sector responsibility in need of significant investment and reform, but also that top-down initiatives from national government are beginning to inspire a burgeoning range of bottom-up initiatives from below. In time, the initiatives could have a major impact on improving public space management responsibilities and structures and on delivering integrated strategies within local authorities. To do this, however, the problems identified by the stakeholder groups and associated with poor coordination and lack of resources, and the poor use of regulatory powers and low priority given to maintenance will need to be overcome.

For these groups, the limitations with the current state-centred delivery model were obvious, and many argued for a greater use of market and community-centred models as a supplement to state activities. They argued that the private sector and the community both have a long-term stake in, and responsibility for, the public realm, and therefore have an important contribution to make as part of the three-way partnership identified in Chapter 1 (see Figure 1.10). However, this should be a long-term mutually supportive relationship and not an exploitative one (in either direction), or one that furthers the 'us and them' mentality.

The approaches reported in the next chapter suggest how some local authorities are actively planning a way forward. Elsewhere, the reality is still often of too many hands all trying to do their best with limited resources, but with little coordination between efforts and with few attempts to overcome the pressures that limit the effectiveness of key public services. The result, it seems, continues to be a widespread deterioration in the quality of public space.

Notes

1   See Carmona and Sieh 2004 for a more comprehensive discussion of 'new public management' and performance management in English local government.
2   In this chapter and the next, local authority departments, directorates, or units will all be referred to as departments.

Chapter 6

One country, twenty innovative public space management authorities

In this, the second of two linked chapters exploring public space management policy and practice in English local authorities, detailed interviews with 20 local authorities provide a means to comment on the key challenges and opportunities facing public space managers. The chapter begins with a discussion of the national agenda to set the context. Discussion moves on in a second part of the chapter to examine the innovative practice. In turn this deals with local authority aspirations for public space, management structures and the coordination of public space management processes, stakeholder involvement in these processes, and the key challenges and solutions that the featured local authorities are engaging with. Conclusions recognise that although public space management remains a fragmented area of local government activity in England, a number of authorities are beginning to establish a bottom-up agenda that maps a way forward.

A burgeoning national agenda

Chapter 5 has already sought to describe and analyse the 'normal' approaches to public space management in England through a national survey of local authorities. The analysis concluded that it is hardly surprising that the literature and national surveys report a widespread deterioration in the quality of public space when the services responsible for its management remain fragmented, uncoordinated, and without a clear vision of how the situation can be remedied. By focusing on the 20 innovative local authorities identified through the national survey and associated key stakeholder interviews, it was hoped that clues would be revealed about how the management of public spaces could be improved in the future.

The research on which this chapter is based came at a time of growing national interest in issues of public space and its management, driven largely by an increasing national political awareness of the potentially decisive impact of such factors in voters' minds. Persuasive surveys from MORI (2002), for example, revealed that while people still think the 'traditional' measures of quality of life (i.e. jobs, education and health) make a good place to live, it is issues of 'liveability' (the day-to-day issues that affect people's quality of life at the local level) that they most want improved. Low levels of crime and road and pavement repairs score particularly highly in these surveys, as do activities for teenagers, reflecting the otherwise negative environmental impact of bored teenagers roaming the streets.

A poll for CABE (2002), for example, focusing specifically on what might improve the appearance of people's local environments identified general cleanliness, traffic management, roads/pavement/lighting maintenance, and the availability of local amenities as the four top concerns. 85 per cent of people asked believed that the quality of public space impacts on the quality of their lives and that the quality of the built environment directly impacts on the way they feel.

A policy concern

As a policy issue, much of the growing concern for public space management stems back to the impact of the Urban Task Force. Constituted to review the ills of urban areas in the light of increasing housing pressures, their influential report also put urban management issues on the national political consciousness. It argued, 'There is a shared sense of dissatisfaction and pessimism about the state of our towns and cities', and 'a widely held view that our towns and cities are run-down and unkempt' (Urban Task Force 1999: 115). They contrasted this with the fact that more than 90 per cent of the urban fabric will be with us in 30 years time, and it is therefore in these areas that the real 'urban quality' challenge lies, rather than with the much smaller proportion of newly designed areas created each year.

A flurry of initiatives from Government and other organisations followed, and led to an unprecedented array of research, reports and policy statements on public space (Urban Parks Forum 2001, Fabian Society 2001; DTLR 2002a, 2002b; Audit Commission 2002a; CABE and ODPM 2002; CABE 2002; Institution of Civil Engineers 2002; ODPM 2002; DEFRA 2002; London Assembly 2002; Civic Trust 2002; Improvement and Development Agency 2003; ODPM 2003a; ODPM 2003b; ODPM 2004; CABE Space 2004a; CABE Space 2004b; House of Commons 2004). Space does not permit an exposition of the detailed content of these reports. However, a range of common management solutions can be identified and classified into eight key types:

1   explicit public space management strategies, aiming to establish and deliver a clear vision for public space and its management
2   cross-departmental working structures and initiatives, aiming to better integrate public space management services – restructuring, coordination, devolution, champions
3   initiatives aimed at better liaising with and involving a wider range of stakeholders – public, private and community – in the management of public space
4   approaches aiming to redefine the standards required of public space management efforts – targets, guidelines, performance standards, specifications, training, award schemes
5   attempts to attract more resources to the public space management agenda, both public (i.e. regeneration) and private (i.e. sponsorship, planning gain, business contributions)
6   schemes aimed at establishing and setting long-term delivery standards, through exemplar projects that build in long-term maintenance regimes, or though taking new powers (i.e. new byelaws), or better using existing powers (i.e. enforcement powers)
7   initiatives that respond to the challenges of particular contexts, through dedicated area management regimes, personnel or designations
8   investment in monitoring public space changes and initiatives, in order to better focus resources and better enforce decisions – audits, indicators, health-checks, peer reviews.

Many of the themes were picked up and summarised in perhaps the most important document, the policy statement *Living Places: Cleaner, Safer, Greener* (ODPM 2002). This laid out a series of Government intentions and initiatives to tackle the problems associated with the decline of public space. The document argued that 'achieving high-quality spaces will require new thinking that better integrates the ways we design, create, manage and maintain our public realm'; and picked out four main challenges: 'public space is not a single definable service; local environmental problems can feed off each other; problems need to be tackled where they are worst; and circumstances can change quickly' (ODPM 2002: 12). It established a 'cleaner, safer, greener' agenda:

• cleaner – by improving how streets and public spaces are maintained and how services are management and delivered;
• safer – by improving how they are planned, designed and looked afte
• greener and healthier – by ensuring access to high-quality parks and green spaces.

A pragmatic delivery agenda

The policy agenda has since taken shape in a variety of national policy initiatives that have attempted to address the issues of public space and the quality of its management. These encompass: changes in legislation giving local authorities formal responsibility for environmental quality through their new powers to promote community well-being; the creation of an Urban Green Spaces Task Force to report and advise on green spaces; a public-funded organisation to champion good design and the management of public spaces (CABE Space); the adoption of auditing regimes for local authorities' street-related services, with rewards offered to those performing well; the institution of funding programmes to support community-based management of public spaces in deprived areas; the introduction of business improvement district (BID) legislation; and so forth.

Two things underpin and unify most of these initiatives. The first is a gradual shift in emphasis from a concern with initial design and implementation, to more attention to the life-cycle of public spaces in which long-term management and maintenance are seen as paramount (see for instance Audit Commission 2002a). Second, a widening of the definition of urban public spaces to encompass also the ordinary streets and squares that make up the living spaces of communities and neighbourhoods (CABE and ODPM 2002; Audit Commission 2002a).

The government argued that five components stand out as key factors in much of the work being undertaken concurrently on the management of public space: 'committed leadership, strong partnerships, active community involvement, the desire for quality and innovation; and better communicating of ideas' (ODPM 2002: 14). In so doing it confirmed that local government retains the decisive role in their delivery, effectively endorsing the state-centred model of public space management into the future (see Chapter 4). However, both government policy, and the range of research, reports and policy statements from government and non-government organisations (see above), universally reflected a pragmatic view on delivery, arguing the case for partnership and involvement from as wide a range of parties as possible, and effectively endorsing market- and community-centred models as viable alternatives (or supplements to the state-centred model), where appropriate.

From government, this pragmatic approach might be seen on the one hand in the rolling back of compulsory competitive tendering (CCT) requirements, that in the 1980s and 1990s had forced local authorities to contract out much of their public space management responsibilities to the private sector on the basis of lowest price, and almost regardless of quality (see Chapter 4). On the other, the enabling of BIDs through legislation can be seen as a leap forward in the rights of local business interests to manage their local environment in a manner that best suits their own private interests.

Living Places: Cleaner, Safer, Greener concluded that 'local government is vital to the creation and maintenance of good public spaces', thus 'many of the successful schemes to improve the quality of local environments across the country are driven by strong local political leadership, clearly defined local targets, successful local consultation and productive local partnerships' (ODPM 2002: 18). The research reported in the remainder of this chapter examines how this was being done.

Managing public space in England – what can be done?

The research methodology for the 20 case studies is briefly discussed in Chapter 5. Interview findings were recorded at length before summaries were prepared following a common structure to enable comparison. Broad subjects for discussion which also structure this section of the chapter included:

- aspirations for public space
- public space management structures and coordination
- stakeholder involvement in public space management
- challenges facing local authorities.

Aspirations for public space

Authorities' aspirations began with their conceptualisations of what constituted public space. From the national postal survey it was found that no local authority in England had a holistic definition of public space, and indeed many were anxious for central government to provide one. The 20 authorities did, however, cite different types of public space in their various policy documents, with some definitions combining two or more typologies to form a more holistic definition of public space. The best example was Newcastle which combined the management of the street scene, open space, and parks in its 'Urban Housekeeping Plan'.

Despite not having their own definitions, most of the authorities agreed with the definition of public space provided in the interview pro-forma, based on that offered in Chapter 1. However, several local authorities considered that public space did not always benefit from unrestricted access, citing temporal access restrictions through the day, week, or year. Examples include urban parks, many of which have railings and are closed at night; public/private spaces, such as those framed by large private institutions that own the external public space but provide public access during office hours/days of the week; and public/private interfaces, such as those between the internal private spaces of stations or shopping centres and external public space that can also be closed at night. North Tyneside also argued that any space that could be seen from a public environment – internal or external – was to some degree public space by virtue of its 'visual accessibility', adding a further dimension to the definition.

Most of the 20 local authorities argued that the critical element determining whether external space was 'public' was its relative ease of access, rather than its ownership or necessarily responsibilities for its management. For example, the Corporation of London described numerous external spaces in private ownership which it has either negotiated access to, or has agreed to manage on behalf of a private landowner. They described external routes through the City as containing patterns of ownership and management that are invisible to users, a characteristic that applies to many central urban environments, and to a lesser extent to rural environments through public rights of way. The key aspiration of some authorities has therefore been to create a seamless public space network, rather than necessarily a continuous management regime or continuous public ownership.

LOCAL AUTHORITY OBJECTIVES

A number of objectives for better public space quality were repeated across authorities, demonstrating that, at least amongst the 20 selected authorities, a clear idea about how they would improve public space management was emerging. This contrasted with the findings from the national postal survey from which it was clear that most local authorities have given little thought to a coherent vision for their public spaces, and relied instead on highly generalised and aspirational statements in their corporate plans and community strategies (see Chapter 5).

Best value processes usually provided the impetus for the 20 local authorities to review their public space objectives. For example, Harlow set out an objective for public space in its Street Scene Best Value Review for a 'town that is clean, green and safe for people to live and work'. The local authorities often had such generic corporate statements on public space, but these aspirations were backed up with other documents describing public space strategies and operations, often containing examples of exemplar spaces to further inspire practice. Birmingham for instance had a vision for 'high quality, accessible, pedestrian friendly, and attractive public spaces', operationalised through its Best Value Performance Plan and other strategic documents relating to public space. These in turn were related to examples of high quality public spaces that the council had delivered and now manages (Box 6.1).

BOX 6.1 BIRMINGHAM: STREETS AND SQUARES STRATEGY

Centenary Square, Birmingham

In the late 1980s Birmingham had to address the loss of its manufacturing base and reinvent itself. The city had inherited a highways-dominated environment, and the council through the Streets and Squares Strategy sought to restore the fractured environment and link the centre to the distinctive quarters surrounding the city's core. Political continuity and ongoing commitment to the strategy has enabled Birmingham to implement the wider vision after the initial impetus and early successes of Centenary Square and Victoria Square in the early 1990s. These early successes ensured that the initiative received budgetary priority driven by the long-term need to lever in new private investment into the city.

The initial commitment amounted to £5 million per year over five years, including money from the European Fund to prime pump the project and as a lever for private sector investment. Private-sector involvement in delivering the Streets and Squares Strategy began in the early 1990s at a time when the business community still lacked the confidence to locate in the city centre. Following the city's lead, the developer of Brindleyplace recognised the value of high-quality external

space as a showcase for the development and provided the public spaces before the rest of the development was delivered. The move proved to be a very successful marketing strategy.

In Brindleyplace the developer has built the external spaces to a very high specification and has set up long-term management structures to safeguard the initial investment as well as the environmental quality of the development. The high levels of maintenance have set a new benchmark for the rest of the city and show what extra resources can achieve, setting the scene for a future BID in Birmingham as a mechanism for raising revenue.

When devising their Streets and Squares Strategy, Birmingham City Council was in the exceptional position that much of the city centre had been allocated to the highways network of roundabouts, underpasses and elevated roads several lanes wide. This gave the council the opportunity to become a major player in the regeneration process, leading the transformation of the centre into a pedestrian friendly environment where safe streets link attractive squares which become civic spaces in their own right and the backdrop to events that contribute to the vitality of the city.

A recurring theme in public space objectives was the desire to better engage external stakeholders. An example was Westminster, which is well aware of the difficulties in reconciling public space activities between local residents, businesses and visitors, and has put plans in place to address all three stakeholder groups in proposals for Leicester Square and the surrounding area. Great Yarmouth, for its part, had an objective to actively pursue formal partnership working with the private sector to improve public space management, whilst Bristol had developed this further by stating its objective 'to respond to, and consult more effectively with, end user demands on and of public space'. In Sandwell, the objective was to actively engage local people in public space issues and to look for design-led solutions to public space problems.

Another prominent theme was the use of standards and indicators. In some cases, local authorities were very aware of how their public spaces stand comparison against international standards. In one case – Kensington and Chelsea – an explicit objective of the borough has been to deliver the best streets in Europe, an objective illustrated through its work regenerating Kensington High Street. Other local authorities use a more down-to-earth set of standards to establish their quality aspirations in the form of indicators developed by the environmental campaigning organisation ENCAMS. Others have developed their own standards for litter, crime, graffiti, and other public space issues, including Coventry through its public/private city centre company CV One.

Many local authority public space objectives covered context-specific issues. Examples include objectives to better address crime and safety, as well as user perceptions of crime and safety; to reduce street clutter; to improve maintenance regimes; to rejuvenate commercial areas (Box 6.2) and to increase the commercial opportunities provided by public spaces. Some local authorities described objectives to develop a fully integrated public space strategy, bringing together different public space types, strategy/policy and operations, and professional disciplines within councils. Leeds, East Riding of Yorkshire, and Bristol all share these objectives, with Leeds also adding an objective to maintain the uniqueness of its public spaces, and Bristol aiming to reconcile the functions and users of public space by addressing apparent conflicts within the city. Collectively, the range of public space objectives ranged from strategic to operational concerns, and covered both outcome and process-based dimensions of management.

Public space management structures and coordination

The national survey suggested that the majority of local authorities in England do not have fully integrated coordinating structures for managing

Marketgate, Warrington

BOX 6.2 WARRINGTON: ATTRACTING INVESTMENT

Warrington Borough Council saw the need to enhance their town centre public space, initially to reverse the retail competition from neighbouring centres and out-of-town retail schemes, and latterly following the 1993 IRA bomb that had a devastating effect on the vitality of the town centre. Strong political support backed an initiative to improve the public realm in the town centre which become possible when in 1996 the council received unitary status and inherited a windfall tax from the county council. Part of the windfall was put towards the regeneration of the town centre after match funding was received from the Regional Development Agency.

An innovative high-quality scheme was completed by the American artist Howard Ben Tre and the Landscape Design Consultancy in January 2002 within an overall budget of £3.25m. The centre of the town centre is now the focus of a pedestrianised retail quarter with steps, a water feature, and an impressive lighting scheme. Marketgate links to a series of 'commons and garden spaces' set within two other streets, each with its own character, providing a wide variety of visual and sensory experiences.

Strong political support was crucial in seeing the scheme through, not least for ensuring that the quality of the initial vision was reflected in the execution and post-completion management. Initial scepticism from the local press and some residents has been replaced by a recognition that the scheme is unique and greatly enhances the town centre, and that it is beginning to fulfil what it was commissioned for, to attract new investment to the town. Recent research shows increased numbers of users in the town centre and renewed interest from private developers.

public space, with potential knock-on effects on the quality of the public spaces in their areas. The 20 case studies where partly chosen because they demonstrated more integration than their counterparts. Their aspirations and practice seemed to be to try and include responsibility for as many types of public space and management processes in one department as possible, rather than separating public space typologies and management processes into fragmented units. Furthermore they tried to ensure that public space policy and operations are not artificially separated and that different types of public space professions are brought together in one structure. They often promoted a single point of contact for public space issues within the authority (see below).

The intention was to improve coordination between different public space management processes and to overcome largely historic rationales for fragmentation. As an officer at Bristol described it: 'many local authority structures have evolved for non-strategic reasons, which results in no clear vision for public spaces within an authority and problems for the general public in understanding local authority public space responsibilities'. Waltham Forest was amongst those authorities that had attempted to overcome the problems through its Environmental Services Department. This single department has responsibility for highway engineering, highway maintenance, street scene design and construction, street cleaning and refuse collection, parking, and green space development and maintenance.

Despite efforts to coordinate activities through local authority structures, the interviewees still reported a range of problems, the majority of which related to the link between local authority public space policy/strategy and implementation/operations. Moreover, policy is often translated into a large number of disparate initiatives, with different timescales, compounding the problems of coordination. Typically, it seems, this is also aggravated by a lack of clarity concerning where responsibility for each policy area lies within local authorities. Furthermore, coordination problems still emanate from professional boundaries between public space responsibilities, for example between highway engineers, planners, and public space managers, even if within the same department.

Despite the problems, many of the twenty authorities were actively tackling the non-structural issues, again, often inspired by the best value processes. Thus best value appears to have helped in sharpening the focus of corporate policies, improving the links between policy and implementation, and introducing cross-cutting initiatives between different public space responsibilities. East Riding of Yorkshire, for instance, used a best value structural review to achieve clear lines of communication between those responsible for public space policy and service delivery.

In other authorities, a combination of sharper internal management processes and local authority restructuring has helped to improve public space coordination. Newcastle, for example, restructured after the 2000 Local Government Act into a cabinet and leader style council, as many other councils have done. The council now has an executive member for the environment to champion public space issues, and an integrated public space budget. Previous fragmented responsibilities for different space types, management processes, policy and operations, and staff, are now fully coordinated, and focused on delivering a common set of objectives.

Other internal initiatives have been instrumental in bringing about coordination between departments, and the involvement of strategic partners in public space management. Again, these initiatives are often the result of a few dedicated members and officers within an authority who are championing public space causes. An example is Leeds, which through its 'Green Space Implementation Group' (GIG) has linked up work in two separate council departments – Leisure Services and Planning (Box 6.3). This enables a coordinated approach to the creation and subsequent aftercare of public spaces using resources attracted through section 106 planning agreements and national regeneration funds.

In some local authorities, however, the local political context is still creating problems for coordination. In one local authority, a blame culture was cited for the poor working relationships between officers and members, whilst the politicisation of public space issues resulted in a lack of coordination between the different political parties. Another, Great Yarmouth, got round their former problem of changing priorities with successive administrations, by instigating a steering group for major public space projects with multi-party representation. As a result, the recent regeneration of public space along the town's seafront has been free form the short-termism that characterised previous initiatives.

The experiences of the 20 authorities demonstrated that there is no one way to better coordinate public space activities. Fundamentally, whichever approach is taken, and whether this is structural or cross-cutting (i.e. imposed on an existing organisational structure), the key ingredient seems to be a determination amongst individuals and within the different public space services, to work together.

Stakeholder involvement in public space management

The responsibility for managing public spaces does not lie solely with local authorities. Across the 20 local authorities, both the community and the private sector are taking a more active role as local authorities attempt to harness the expertise and knowledge of key stakeholder groups, rather than viewing it as a barrier. Increasingly this is being facilitated through better structures for external coordination.

BOX 6.3 LEEDS: GREEN SPACE IMPLEMENTATION GROUP

Hyde Park, Leeds

Leeds has undergone a widely publicised urban renaissance in recent years. One of its corporate targets was to create more green space for the city. However, in working towards this objective, officers found a mismatch between the resources generated through Section 106 contributions and what was being spent on the ground by the Leisure Department who are responsible for green spaces. Resources were not being channelled efficiently. The Green Space Implementation Group was set up in 1999 and operates at two levels with a strategic group and a site-specific group. The strategic group meets quarterly and is attended by staff from leisure, regeneration and community involvement teams. This group includes high-level decision makers and deals with broader strategic issues such as play space policy. It has demonstrated a high level of effective decision making and is increasingly the focus of external lobbying.

The site-specific group deals with individual schemes coming onto or actually on site. Its membership includes representatives from the Leisure Department, planning staff, landscape design staff, and financial project officers, and, where relevant, representatives from the Regeneration Unit and Community Involvement Teams. The groups aim to improve coordination but have found that overall working relationships have also improved as a result of the regular meetings between previously 'silo'-based officers. They also provide forums for liaising with key stakeholder groups and bodies such as British Waterways.

Having two levels of group allows decision making to remain relevant to those attending. The emphasis at site-specific level is on coordination, but it has also proved important for the lessons shared in the site-specific group to be channelled to higher-level staff on the strategic group, whose members have appropriate decision-making authority.

THE COMMUNITY

Five of the 20 local authorities perceived the community to be a crucial part of their system for managing public space. These authorities tended to involve the local community in public space management in three main ways.

First, by putting systems in place for the local community to channel and report problems, for example through council hotlines; the argument being that one point of contact within the council for all public-space-related issues greatly improves internal and external coordination. In Greenwich, for example, the Cleansweep hotline is now the single point of contact for all public space matters across the borough. A further example was the integrated IT system in East Riding that allows the community to ring, email, fax, use a video kiosk, or personally submit enquiries relating to public space to a specialised team who can track and coordinate problems and their solutions through GIS software (Box 6.4). The same IT system has also been implemented in an urban context across the whole of Newcastle. Here the council has used similar technology to field enquires and solve public space management problems from the private sector and visitors to the city, as well as from the local population. In this regard, a number of authorities showed a heightened awareness of who the users of their public spaces were, in some cases extending well beyond the local population, such as in Westminster where the built heritage is truly international in significance, and so are the users.

The second main method to engage the local community was through active consultation about public space, including through local meetings or liaison officers. An example was the Community Forums in Southwark, which through the Local Strategic Partnership (LSP – see p. 85) are encouraging ward councillors and their local communities to discuss local environmental and community safety issues. It is envisaged that this will extend in the future to devolving powers to the forums for those functions that impact on public space, such as development control and licensing, as well as the monitoring of public space. The national survey suggested that those authorities that give public space issues a lower priority generally consult the community less, choosing to use passive forms of consultation, such as generic annual surveys, and therefore tend to be less responsive to local community needs and aspirations.

The third means through which the authorities were engaging the local community was through initiatives for direct proactive local participation in public space management, such as voluntary park wardens or graffiti cleaning groups. Only in this latter category is there potential for a real shift to a community-centred model of management. Typically, however, the role is one of involvement in, rather than responsibility for, service provision.

BOX 6.4 EAST RIDING: INVESTING IN IT

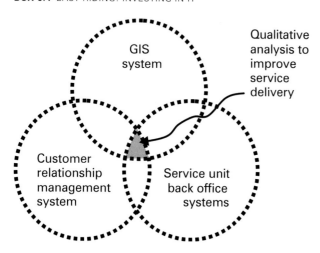

IT system: three linked elements

East Riding has invested in integrated IT systems that have allowed the authority to coordinate resources more efficiency. The system consists of three linked elements – GIS, 'back office' databases and IT systems, and the 'front office' customer relationship management system.

The front office system is used to log enquiries coming through to the customer service team, by phone, email, fax, video kiosk or in person. As the system is linked to the service unit's back office IT systems, it can provide a link to the associated GIS map showing street lamps, maintenance schedules etc., and describe the information required to deal with a problem. The system therefore enables customer service staff to submit maintenance orders directly, and the council is currently working to ensure that the customer service team is able to check if the work has been carried out, thus closing the complaint 'loop'.

A valuable feature is the electronic notice system. As customer service staff have found they are too busy to check emails notifying them of urgent news, a bulletin line is used, continually scrolling across the bottom of computer screens with any urgent information (i.e. winter maintenance delays, critical incidents, etc.). Operators are then quickly aware of relevant news to pass on to callers.

The IT system has been developed incrementally, spreading the investment burden. The IT department is a corporate unit, separate from the larger Operational Services Department (responsible for delivering public space services). Its corporate status eases the process of securing resources as they are funded, in part, through a 'tax' on departmental budgets.

The integrated system also allows for quantitative analysis of thematic information i.e. roads, street lighting, open spaces, 'hot spot' sites for complaints. Qualitative analysis has resulted in improvements in service delivery outcomes, for example by improving refuse collection routes to minimise customer complaints. The use of the IT-based information system is also increasingly building up a corporate memory that can be shared, and is changing the culture of staff members who, previously, defensively protected their own knowledge.

Greenwich, for example, had a graffiti strategy that supplies young people with cleaning materials to help clear up problem areas in local social housing estates, schools, and youth clubs. The authority is also inviting community organisations to contribute to monitoring of graffiti and training in its removal through the 'Adopt a Building' project (Box 6.5). Newcastle has implemented an Environmental Ward Stewardship Scheme that allows communities and other stakeholders to directly influence public space investment, while also acting as an 'umbrella framework' for the city's public space investment. Thus all public space improvements suggested by residents and supported by the council are logged into a database for each of the 26 city wards. Environmental ward stewards then coordinate internal and external funding to resource the schemes.

A minority of authorities went further, encouraging direct public involvement with specific types of public space such as housing estates or parks, and even devolving aspects of management to particular groups, including friends schemes for parks. Newcastle, for example, has also developed an 'Adopt a Plot' scheme, where local individuals or groups can manage any piece of council owned land as long as they can demonstrate they can manage it to a higher standard than the council themselves. Elsewhere, a community-oriented rather than community-driven approach was more common.

THE PRIVATE SECTOR

The 20 local authorities differed in their experience of private-sector involvement in public space management, depending on the general ethos of the individual councils and their particular local contexts. However, most authorities had some experience of working with the private sector.

The private sector as major landowner/leaseholder were involved in public space primarily through the ownership of space to which the public are allowed access. These public/private spaces in the 20 authorities were either managed by the owner, or by the local authority through arrangement, sometimes through the granting of commuted sums for the task. It was noted that typically the private sector have greater resources to spend on managing their public spaces, often achieving higher standards. The interviewed authorities accepted and were grateful for this, often benchmarking their own public space services against these standards in an effort to make the case for more resources to create a seamless transition from council owned and managed spaces to privately owned spaces. The Corporation of London's attempts to create such a seamless public realm have already been discussed. In Bristol, the harbour manager liaises with local landowners to secure public access to all riverside areas, whether through negotiations to transfer the land directly to the local authority, and/or through providing a commuted sum to maintain the space.

BOX 6.5 GREENWICH: ANTI-GRAFFITI INITIATIVES

Operation Clean Sweep leaflet, Greenwich

Greenwich Council operates a 'graffiti strategy' through which it involves the community to tackle this aspect of antisocial behaviour. The council is working with young persons to remove graffiti, especially in areas covered by the Cleansweep initiative that have a high proportion of social housing, and where graffiti is downgrading the environment and increasing the fear of crime. Working through schools and youth clubs, the initiative targets the age group that is responsible for the graffiti. At weekends and during school holidays, teenagers are supplied with materials and receive supervision to tackle the problem.

Officers have observed that not only do the young persons enjoy the work, but often the graffiti does not reappear in the locations that have been cleaned. Thus the initiative not only improves the public realm but also educates those sectors of the population who are likely to exercise peer pressure on the offenders. As part of this pilot, the council is also working with traders to prevent young persons gaining access to materials that can be used for graffiti painting.

The council involves community organisations in areas outside the Cleansweep pilots in the 'Adopt a Building' project, through their 'graffiti monitoring officer', by giving members of these organisations relevant training and offering information packs and the necessary tools to remove graffiti and fly-posters. The Greenwich Society is one of these, and over 18 months, 50 volunteers were recruited and went out once a week or every fortnight, removing 3,000 'marks' over the period.

At the start, their work was limited to private buildings but it has now been extended to street furniture. The key to their success has been a quick response and good monitoring; the sooner graffiti is tackled the easier it is to remove and repetition is discouraged. The society aims in the future to divide their area into zones and encourage volunteers to take responsibility for a zone.

In some cases the private sector manage public spaces in their entirety, for example in Brindleyplace in Birmingham. There, through mutual agreement, a coordinated approach to the maintenance, cleaning and sign posting of these public/private spaces and the surrounding council owned and managed land has been developed. Many such public/private spaces were reported to be the result of Section 106 planning agreements, through which developers have negotiated with local authorities to create or improve the public realm. While Section 106 contributions on the whole tended to be more concerned with new projects, rather than with managing the existing environment, the monies they provided were also used for long-term management. In the City of London, for example, the 'Street Scene Challenge' initiative is half funded by Section 106 contributions, and half funded from the parking surplus fund. The initiative has been used to provide public spaces for the general public, but also provides the only forum through which local authority officers and departments come together to discuss public space issues.

The private sector in the form of local business contributed to public space management in several ways, the most widespread of which was the sponsorship of street furniture, hanging baskets, lampposts, flower displays and verges. Lancaster reported the successful implementation of

such a scheme for its flowerbeds and street signage. However, several of the authorities felt that such schemes can lead to visual clutter.

Increasingly popular was the involvement of local businesses in partnerships covering public spaces, whether this be through representation on a not-for-profit company board with other stakeholders, including the local authority, or through commuted sums or financial contributions to public space services. Coventry City Council had gone one step further by setting up an independent not-for-profit city centre management company – CV One – in an effort to improve the city centre's image, in part through the quality of its public spaces. Under the auspices of CV One, city centre businesses pay an annual membership fee which the company uses to invest in improvements to the city centre. With a flexible budget the company reported that it can respond to public space management issues quickly and efficiently (Box 6.6). The legislation included in the 2003 Local Government Act to allow local authorities to set up business improvement districts (BIDs) and levy extra charges for public space management was eagerly awaited here, and in many of the authorities interviewed.

Businesses that create a high impact on public space, such as licensed premises and fast-food takeaways, were generally singled out for particular criticism. Local authorities thought that most local businesses did not recognise that public space quality affects them, and in the cases where

BOX 6.6 COVENTRY: CV ONE

Coventry city centre

Throughout the 1980s Coventry gradually lost its city centre shopping trade to new out-of-town retail centres. If business was to be attracted to the centre, an initiative was required that would improve the physical character, build up the marketing profile of the central area, and regain the trust of the private sector. In 1996 the council took the step of creating an independent not-for-profit city centre management company – CV One. The move did not involve privatising the council's building assets, but the council did contract out the management of the entire city centre area to the new company. The company was charged with attracting new investment through a dedicated commercial focus on the city centre that the previous 'silo'-based council department had not been able to take.

The company received start-up funding from the council, which demonstrated the council's commitment, and represented the crucial first step in earning the confidence of the private sector. It was given a five-year contract (currently renewed on a year-by-year basis) to provide maintenance services and to use environmental improvements to lever further revenue. Under the strong leadership of CV One's CEO, from 1998–2001 the ten-year decline in footfall was reversed and some £2.4m extra revenue was generated for environmental improvements.

Maintenance has improved, and proactive marketing through the press and events has attracted new interest, but much of the achievement of CV One stems from the relationships established by CV One with business, for example through its Business Membership Scheme. Retailers pay a membership fee to join the scheme which CV One invests in improvements to the city centre. The associated Business Forum provides CV One with a vehicle through which to coordinate the different interests and offer a lobbying route to the council to direct future investment.

The company benefits from both a clear mandate, flexibility, and clear geographical operational boundaries. Provided the company's overall business plan is approved each year by council and it continues to meet its contract, it is able to undertake other activities as it wishes under the direction of its board. As an independent entity, CV One is free from council procurement regulations, allowing it to be flexible in sub-contracting maintenance and managing those contracts to high-performance standards. It is able to generate, and similarly spend, its own revenue, and resources can be easily redirected where there is a problem to be solved.

they did contribute, it was generally only as far as the public spaces that directly interfaced with their businesses.

The private sector, as public space contractor, represented a further relationship highlighted by the case studies. In this regard, the private sector may provide numerous services, but the most common seemed to be street sweeping and cleansing, and waste-collection services. Interviewees suggested that local authorities are generally less antagonistic to the use of private contractors in the post-CCT environment, although perceptions of public space contractors were rarely positive, and generally suggested a concern for the quality of the service delivered. Nevertheless, authorities did recognise the value of private sector contractors for their specialised knowledge or for the services that the council could not always provide, including tree surgery or chewing gum removal from pavements.

A key lesson from the 20 authorities seemed to be the need to recognise where and how best to involve other stakeholders in public space management, be that the community, private sector, or other public bodies. Generally there was little hesitation in using the private sector

where they could do the same quality of job for less. A broad acceptance also existed that the private sector, in a range of guises – from owners of space, to managers, to commercially interested parties – has an important role to play in managing public space, and that this energy, interest and source of recourses and best practice should be harnessed wherever it exists. A wide variety of approaches and models were apparent to achieve this, which not only had the potential to deliver resource savings, but also to raise the profile of public space management services. Despite this, there was no support for a more dramatic move towards a market-centred model of public space management.

Challenges facing local authorities

There was considerable consensus regarding the key problems and challenges associated with managing public space, which can be grouped into three main categories: investment, regulation, and maintenance, whilst the coordination of these issues was an overarching concern. As such, concerns in the public sector mirrored the range of issues identified by the key stakeholder groups discussed in Chapter 5.

INVESTMENT

When asked what the main challenge facing public space management was, the most common answer amongst the 20 authorities was insufficient financial resources. The lack of resources for staff was a key problem, particularly for enforcement activities, hands-on maintenance roles, and to coordinate activities. In Sandwell, for example, grounds staff have been reduced from 220 to 30 in 20 years, and despite increased mechanisation, the service is under extreme strain. In other places, the ongoing management costs associated with physical regeneration had been causing a strain and had not been factored into regeneration activities. Greenwich fell into this category, where much regeneration activity is delivering large areas of new development, but where severe resource constraints on the local authority are preventing it from managing the new public spaces to the standard it would like.

Even the relatively wealthy Corporation of London reported resourcing and staffing problems, and at the time of interview had only one dedicated fulltime enforcement officer for public space in the Square Mile. In relation to enforcement, local authority officers were often well aware of the lack of police support to help in regulation and enforcement activities. In this regard, enforcing fixed penalty notices seems particularly difficult for local authorities. Not only is it time-consuming and staff-intensive to issue notices, but authorities reported that it is difficult to successfully prosecute those who do not pay.

In an innovative move to improve efficiency and to overcome the lack of enforcement staff, Newcastle retrained traffic wardens to issue fixed penalty notices for litter, on top of their normal duties. However, only a small number of authorities argued that there was still scope to operate more efficiently within existing resource levels. Instead, most described doing the best they could with limited resources. This attitude was typified by a Great Yarmouth officer who described his job as 'to decide how best to do things with the finances we have available'.

A number of suggestions were made to improve the resource problems. Local authorities were particularly keen to have greater financial flexibility when managing public space. Kensington and Chelsea, for example, argued that local authority parking reserves should not be ring-fenced for highways improvements, as stipulated by central government, but should be available for the local authority to spend anywhere in the public realm. Similarly, Southwark suggested that monies received under Section 106 planning agreements should be available for spending on projects not directly related to the specific development being considered for planning permission. By contrast, Harlow argued that funds granted by government under the national spending formula for the management of public space should be ring-fenced to prevent them being used to fund other political priorities.

Many of the emerging good practice local authorities reported their concern with the numerous public sector funding streams which authorities rely upon for public space investment, including regeneration and lottery schemes, but also that each has a particular emphasis and are rarely joined up. Authorities faced a number of barriers to access this funding. First, the funding is often delivered in compartments covering particular types of public space or management processes, so that new investment, when it comes, may not match local priorities. Second, public space investment does not cover all aspects of public space management, for example, public conveniences are rated as a high priority by the public, but often fall outside of dedicated funding streams and so tend to be neglected. Third, competitive funding is costly to bid for, putting pressure on scarce local authority staff, and making it difficult for small authorities to access funds. Finally, many public funding streams are focused on capital spend, and tend to ignore longer-term public space management issues altogether.

Local authorities reported that this bias to capital investment over revenue expenditure was a major problem. Thus, although most regeneration projects that deliver new public spaces allow for some limited post-completion maintenance, funding quickly runs out. Projects funded through the National Lottery often do not cover revenue expenditure at all, and tend to assume maintenance procedures and resources that are not realistic for limited local authority budgets. Some interviewees suggested that these problems needed to be overcome through harder bargaining with sponsors, others that Section 106 resources should be used for maintenance.

It was clear, however, that the local authorities themselves also share the blame for under-investment. Many authorities, for example, were ready to admit that they had not developed a coherent strategy through which to frame public space investment. One local authority officer mentioned that development decisions taken at sub-committee level do not always fit with overall council strategies for public space. Others suggested that there is limited knowledge of the different public space funding sources and their requirements by local authority officers. This specific barrier dates from practice that until recently ensured that officers responsible for public space management acted as service providers only, and were not concerned with the more strategic processes of policy-making, resourcing, or making a case for their activities.

Most authorities reported that the changing macro-context within which they operate was negatively impacting on public space. For example, three-quarters of the interviewed local authorities named the deregulation of utilities as a major barrier to achieving good quality public spaces, as utilities companies can effectively ignore any public space strategy established by the local authority. This problem, it was argued, has been aggravated by the reduction in funding for highway schemes which means that too often only the strictly necessary maintenance gets done. The deregulation of bus services outside London has had a similar impact through the increased strain on public spaces around transport nodes.

With regard to the first of these problems, Oxford has commissioned and adopted a public realm strategy that includes a chapter on managing the public realm and managing private utilities. As part of the strategy the council is developing a maintenance manual setting out standards to ensure the correct reinstatement and replacement of materials when roads and pavements are dug up, as well as general principles for private utilities to adhere to. In the meantime the council has set up a 'Utilities Liaison Group' through which private utilities are requested to notify the council when they carry out works, in order to better monitor and coordinate the work (Box 6.7).

Some local authorities also saw the involvement of other stakeholder groups such as the private sector and the local community in public space management as a means to reduce the resources authorities need to commit to managing public spaces (see above). Other authorities are looking to educate users of public space into better behaviour, thereby reducing the need for public space management and enforcement in the first place. This, they admit, is a long-term objective.

REGULATION

The second main challenge for public space management was the regulation of public spaces and their users. Problems include anti-social behaviour and the general maintenance problems associated with increasingly heavily used spaces. Sandwell, for example, noted the problem of managing 24-hour public spaces for the local authority, which is not a 24-hour organisation. Others cited problems with enforcing byelaws to regulate users, and regulating the 'illegal' use of streets by shops and businesses.

Authorities identified a 'shopping list' of regulatory powers they wished to have, covering a broad range of public space activities. Many authorities described the need for byelaws to control activities such as music/busking, skateboarding and alcohol consumption, whilst admitting that the time and resources for enforcement would be limited. Other more commercial public space activities such as leafleters, fly-posting, and unauthorised trading have proved particularly difficult to regulate, although some local authorities are thinking their way around the problem. Newcastle,

BOX 6.7 OXFORD: PUBLIC REALM STRATEGY

Oxford Public Realm Strategy

After traffic was removed from Oxford city centre in 1999, a Public Realm Strategy was produced as a means to improve the centre. Whilst the city's historic college architecture provides a dramatic and distinctive street scene, increased traffic and a lack of investment in the city's public spaces left the streetscape looking tired.

Commissioned and adopted by the city council, the Public Realm Strategy analyses Oxford on a historic, urban design, and policy basis, and suggests design ideas and guidance on a range of improvements to the city centre. The strategy also includes thorough guidance on how to manage public realm improvements following completion, for example the need for maintenance manuals to ensure correct reinstatement and replacement, and principles to be followed by private utility companies

Despite the positive investment in the strategy, implementation has proved problematic. Historic antagonism between the county (who as the highways authority have a large part of the public space powers and resources) and the city council has in the past resulted in limited coordination on public space services. The Public Realm Strategy suffered the same fate and was never adopted by the county council, ultimately limiting its impact

Oxford have nevertheless succeeded in commissioning a high-quality public space design framework that covers important management and maintenance issues, and establishes a clear vision for the city's streets and spaces. Work has now also begun on implementing the new streetscapes envisaged in the document utilising both county and city resources. The experience indicates that despite strong backing within the council at councillor and officer level, without coordinated preparation and joint ownership, improving public space can be very challenging.

BOX 6.8 WALTHAM FOREST: STREET WATCHERS

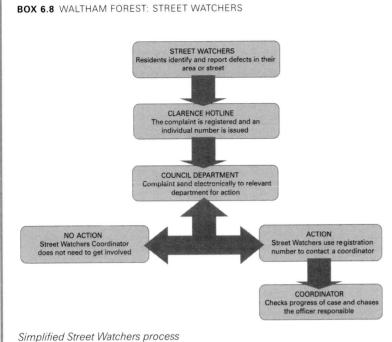

Simplified Street Watchers process

The Street Watchers initiative in Waltham Forest comprises local volunteers who report on problems affecting their immediate environment, but which relies for its success on the enhanced coordination of the follow-up delivery mechanisms. The initiative integrates with mainstream management but remains at arm's length from council operations; the only contact being the Street Watchers coordinator who is responsible for the interface between the Street Watchers and service delivery through routine programmes. The initiative has led to improved services by keeping the pressure on officers to deliver.

As most enforcement powers are within the same directorate, matters reported by the Street Watchers relating to private premises or land (i.e. overhanging vegetation, fly-tipping, abandoned cars) can be dealt with quickly. However, delays in response times can arise when action involves other directorates who need to contact their own contractors before action is taken, or other agencies i.e. (electricity companies for street lighting) who do not necessarily share the council's priorities. Residents, however, do not distinguish between different agencies or departments.

Street Watchers started in 2000 following a suggestion from a 'Citizens' Jury' to involve residents more in council activities. Originally set up within the Highways Maintenance Section, it was subsequently transferred to Customer Services acting as an extension to their free-phone hotline. The post of Street Watchers coordinator is therefore customer-focused and perceived to be independent from service delivery.

Initially, the pilot involved twelve residents and 400 defects were reported in the first six months of operation. The scheme's long-term success was assured by the borough-wide expansion of Street Watchers to 224 volunteers (the objective being eventually to have one for every street). Recruitment of volunteers is through the local press and attendance at events, and on joining they are equipped with information packs and relevant contacts. The initiative has so far failed to engage young persons and ethnic minority volunteers, and to redress the balance the council will be targeting these groups in future. It has nevertheless resulted in an improved environment by bringing problems to the council's attention before they become serious.

for example, has introduced, via a new byelaw, a licensing scheme for businesses which distribute free literature around the city. Birmingham have a 'Street Entertainment Policy' that identifies official 'Busk Stops' in the city, and employ an 'Alternative Giving Strategy' to reduce begging. Finally, Coventry and Great Yarmouth have taken advantage of powers in the 2001 Criminal Justice and Police Act to establish alcohol-free zones in previous problem areas.

Local authorities also identified vehicle abandonment, fly-tipping, littering and dog fouling as further areas in need of greater powers for control. Waltham Forest 'Street Watchers' programme tries to do this, by engaging members of the local community to act as the eyes and ears of the council, picking up problems – both actual and potential – and reporting to officers who are then able to respond more quickly than if they had to wait for scheduled inspection visits (Box 6.8). This has proved to be effective for improving services like street cleansing, refuse collection, abandoned vehicle and graffiti removal, and the control of fly tipping,

but depends on very good relationships between the volunteers and the relevant officers. Westminster has brought together enforcement powers under the Leicester Square Enforcement Initiative, where cross-council departments, the police, a specific Leicester Square Action Team, and the Leicester Square Wardens combine to create an integrated enforcement team to tackle problems.

Authorities tended to express frustration with the tardy response of some government agencies to public space matters, in particular the Environment Agency, Driving and Vehicle Licensing Agency (DVLA), and National Rail, and argued that powers should be devolved from central government and its agencies. Southwark even supplied a specific list of powers that would improve the authority's management of public space, including signing a joint agreement between the borough and the Environment Agency on information sharing, the power to seize those vehicles identified as being involved in fly tipping, which would enable them to interview owners and if necessary prosecute, and powers

BOX 6.9 GREAT YARMOUTH: ENVIRONMENTAL RANGERS

Environmental Rangers publicity

On a mission to improve coordination, education, and enforcement in the public realm, Great Yarmouth Borough Council have trained and deployed Environmental Rangers. Directly relating to the improved management of external public space, the rangers are recognisable council operatives who have the means – through each having a dedicated van and equipment – to quickly respond to and coordinate public space management issues. This includes cleaning, collecting and cleansing anything from broken glass to fly-tipped items, and if the problem cannot be solved immediately, to liaise with other council services, including the Boroughworks depot team.

The ranger will typically inform the public of what the council systems are for dealing with the problem, how long it will take, and any contact numbers if a member of the public wants to follow issues up. The rangers also liaise with Neighbourhood Wardens who patrol the residential areas in the town, and with the Town Centre Wardens. All three sets of employees were trained together.

The second role of the Environmental Rangers job is education, getting out onto the streets, being friendly, meeting the local community, meeting parish councils, and visiting schools. The rangers recruit community voluntary wardens to help educate local people about using and caring for public space, and to help in 'detective work' (i.e. finding out where perpetrators of antisocial activities live). Rangers will also help in getting local environmental initiatives off the ground such as community litter groups.

The final role of the rangers is enforcement under byelaw by issuing of £50 fixed penalty notices for littering or failing to clear up after dogs. While the fixed penalty notices are not easy to enforce, they are effective in educating residents to change bad habits; as one ranger said: 'word gets around'. The council are also considering investing in portable CCTV equipment to help gather evidence for when those issued with fixed penalty notices appeal.

to destroy abandoned vehicles with a value of less than £300 and then recover costs from the last registered keeper.

Collectively, the authorities identified three key challenges. First, the proper coordination of the range of agencies with responsibility for different regulatory and enforcement regimes. Authorities admitted that these problems exist within local authorities as well as between different authorities and other agencies and in part relate to the severance of those responsible for enforcing laws and byelaws from those charged with delivery tasks. The second key challenge has already been mentioned and concerned the lack of resources to employ monitoring and enforcement staff. A police-led scheme in Lancaster involving the retraining of surplus traffic wardens to police anti-social behaviour instead is helping to address this problem.

The final challenge surrounded the difficulties in successfully prosecuting those in breach of public space related regulations (i.e. fly-tipping). In many cases there is the need to establish proof of culpability at the higher level of criminal law rather than at the level of civil law, leading to costly and time-consuming legal cases which local authorities cannot afford and are very difficult to win. Indeed, the time-consuming nature of all enforcement procedures was a point raised by many local authority officers.

In this context, many of the officers interviewed emphasise prevention and education rather than enforcement as the best approach to delivering good quality public spaces. An example was Great Yarmouth who had appointed Environmental Rangers with the power to issue fixed penalty notices for dropping litter or failing to clear up after dogs. The rangers regularly speak to community groups and schools in the area, while enforcement is seen as a last resort (Box 6.9).

MAINTENANCE

Authorities identified a number of major process problems relating to the third major challenge, maintenance. Perhaps most fundamental was the insufficient level of investment in maintenance, for three key reasons, because this activity has historically not been recognised as important by council members, because of an associated squeeze on local authority finances generally, and as a result of CCT contracts in the recent past that have driven costs and service levels right down. The latter problem was gradually being corrected as best value mechanisms encourage a more holistic and integrated approach to service provision. Harlow, for example, has recently completed a best value review and has now set about implementing a 15-year plan to improve street scene maintenance services. Bristol, on the other hand, improved the maintenance of its public realm in one residential ward by integrating the maintenance, cleansing

BOX 6.10 BRISTOL: PROJECT PATHFINDER

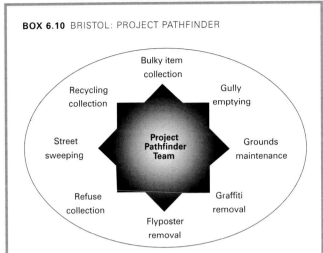

Project Pathfinder team maintenance duties

The national New Deal for Communities (NDC) programme has been used in Bristol to improve the cleanliness of the public realm in a deprived residential area. The impetus to improve the external environment came from the community themselves following consultation. At operational level, Project Pathfinder has managed to integrate the maintenance, cleansing, and waste management of the public realm into one team of nine public realm operatives, consisting of employees from the city's private waste management and cleansing contractor. The key to Project Pathfinder's success is the use of a multi-skilled, flexible, and dedicated team of operatives.

The diversity of skills needed for managing and maintaining the public realm often leads to an uncoordinated approach with segregated responsibilities and regimes that is confusing to all involved. In the case of Project Pathfinder, a single dedicated team carry out eight different duties. The cleaning and maintenance regimes of the operatives are flexible and responsive rather than sticking to a rigid contractual regime. So, for example, if a street is not dirty it will not be cleaned, and something more pressing will be done.

The team members were chosen for their commitment to their work and their willingness to learn new skills, with the NDC having a contractual agreement with the waste contractor that the same operatives do the job everyday, in a specific uniform, with a dedicated vehicle and equipment. A single dedicated team has the benefits that operatives know where problem areas are likely to be, develop a team morale, are accountable, and most importantly, get to know the community. Project Pathfinder is an example of a genuine partnership between the local community, the council, and the city's private waste/cleansing contractor. The council is currently rolling out similar schemes in other residential areas in Bristol.

and waste management of its public spaces into one team of nine public realm operatives. Such a multi-skilled team of operatives, it was argued, can be responsive and flexible in completing maintenance duties, and can get to know the local community (Box 6.10).

A second problem relates to procurement practices and the relationship between client and contractor functions within local authorities, and between the authority and external contractors. The practice of tendering out services, although not viewed by interviewees as a problem in itself, has sometimes led to a lack of ownership of maintenance processes, as there might be several layers of management, and reduced responsiveness due to long lines of communication between council management and those actually doing the work. The Envirocall system in Newcastle helps to overcome this, increasing responsiveness and shortening lines of communication through one point of contact reporting 45 different public space services, all recorded on GIS software through which an operator can locate public space problems and coordinate responses (Box 6.11).

Local authority officers also highlighted barriers to the coordination of maintenance routines and standards in areas where two-tier local government regimes are in place, and between local government and other organisations. West Sussex County Council, for example, tries to coordinate maintenance between itself and its district authorities by promoting the use of shared contracts. District councils can use the shared contracts for their own maintenance work or to enhance the standard of the county service, for example by specifying an increased frequency of routines, for which they pay the county. Warwickshire have pioneered a monitoring system called the 'Streetscape Appearance Index' (SAI) which also helps to coordinate actions and responsibilities. The system relies on the council and local community scoring different elements across different types of public space, with the scores being used to highlight where investment is needed. Complementing the SAI is the 'Streetscape Maintenance Log' which identifies responsibilities for spaces, infrastructure and buildings, and helps to ensure that problems on private land are quickly remedied.

A final set of problems relate to conflicts between design and management objectives, for example designing out crime objectives versus the provision of attractive landscaping, street cleaning versus tree planting, and short-term development costs versus long-term maintenance concerns. These challenges were regarded as structural, and not amenable to easy solutions. Thus interviewees suggested that some public spaces are designed with inadequate landscaping, materials or street furniture, or with features that subsequently prove too costly to maintain to an acceptable standard. The solution was believed to lie in the better involvement of public space managers during the design phases of new or refurbished public spaces.

BOX 6.11 NEWCASTLE-UPON-TYNE: ENVIROCALL

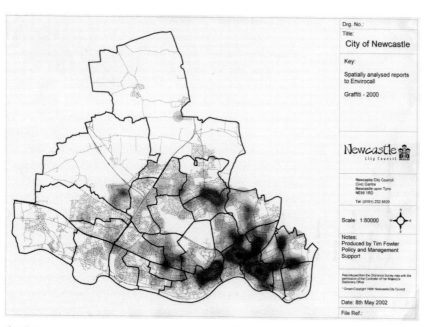

Drg. No.:

Title:
City of Newcastle

Key:

Spatially analysed reports
to Envirocall

Graffiti - 2000

Newcastle
City Council

Newcastle City Council
Civic Centre
Newcastle upon Tyne
NE99 1RD

Tel (0191) 232 8520

Scale 1:80000

Notes:
Produced by Tim Fowler
Policy and Management
Support

Reproduced from the Ordnance Survey map with the
permission of the Controller of Her Majesty's
Stationery Office
© Crown Copyright 1996 Newcastle City Council

Date: 8th May 2002

File Ref.:

Graffiti incidents across Newcastle as mapped by Envirocall

Envirocall is a call centre and one-stop shop that coordinates and monitors resident and business environmental and public space enquiries for 45 public space services. Aware that many residents and businesses were ringing the council and either not getting through or reaching any number of different departments and staff, Newcastle created Envirocall to make public space services quicker, more responsive, and more consistent.

The Envirocall HQ is staffed by up to 25 telephone operatives, six days a week, from 8am–8pm. Residents and businesses can either telephone, email through a dedicated Envirocall website, or report personally in any City Council Customer Service Centre. Operators have the means to answer enquiries (i.e. about waste collection times) and can organise council services such as bulky item collection. Operatives can also arrange to rectify public space problems that are reported.

The coordination and distribution of such a wide range of public space services is made possible through structured case-based reasoning software. Envirocall operatives log public space problems or requests using GIS and Windows software to locate and track complaints, and then coordinate council public space services. GIS software is used so that an operator can locate exactly where a public space problem is, what land the council owns, can track routes of waste collection or cleaning regimes, or even find the number of a broken streetlight.

The software will automatically assign a team, vehicle and depot to handle the problem, with jobs electronically sent to the correct depot. The software will also tell the Envirocall operative if a charge is associated with the service, such as for commercial waste collection. Once a job is completed the depot staff will update the file to a 'done' status.

GQL software is used in combination with the GIS software to print off maps for any part of the city for different instances of the public space management issues covered over any time period. The results of the monitoring are used for a number of important purposes: the identification of outstanding jobs; compilation of maps and statistics to monitor staff working and efficiency; and performance management of set public space standards and council response times at city and ward level.

COORDINATION, THE OVERARCHING FACTOR

Collectively, the three key challenges presented significant but not insurmountable problems to the chosen authorities, and each was related and dependent on the others, for example better regulation depended on adequate resources for enforcement and went hand in hand with every-day maintenance tasks. An overarching solution was frequently found in the better coordination of efforts within the constraints established by local spending priorities. This encompassed the better coordination of investment – funding and human resources – the better coordination of regulatory powers and activities, and the better coordination of maintenance roles and responsibilities.

Conclusions

The case studies established that although authorities had not clearly defined public space, most favoured a broad, inclusive definition, not least because the key problems facing public space managers were re-occurring across space types and contexts. Most fundamental seemed to be the general lack of resources for public space management, and perennial difficulties with coordinating activities.

The interviews confirmed the four key sets of barriers to delivering better managed public space. First, barriers to the better coordination of policies, programmes and actions:

- lack of funding;
- lack of linkage between policy formulation and implementation;
- vaguely formulated policies;
- fragmentation of initiatives;
- the persistence of local authority 'departmentalism'.

Second, barriers to the better regulation of public space:

- lack of coordination between regulatory regimes;
- lack of resources especially for enforcement;
- the patchwork nature of laws and byelaws;
- insufficient powers to prosecute;
- insufficient enforcement powers.

Third, the major investment and resource barriers:

- fragmentation of public funding streams with different requirements;

- the cost and time involved in getting and managing these funds;
- the fact that funds do not cover all aspects of public spaces;
- many authorities do not have a cohesive strategy to frame investment and an ad-hoc approach dominates;
- capital investment funding tends not to cover on-going revenue requirements;
- de-regulation and decreased subsidy to some services makes environmental quality objectives more difficult to achieve.

Finally, barriers impacting on maintenance routines:

- an insufficient level of investment in maintenance;
- problematic relationships between client and contractor functions, reinforced by the impact of now abandoned CCT practices;
- a lack of coordination of maintenance routines and standards between agencies (internal and external to the local authority);
- a mismatch between community expectations in terms of standards and what can be accommodated within the local authority's budget;
- design conflicts and lack of concern with maintenance during design;
- intensive use of some spaces leading to conflict between maintenance routines and some users/uses.

Some authorities argued that there was scope to operate more efficiently within existing resource levels, but most argued they were short of manpower and expertise across management services. In addition, more effective enforcement powers were generally seen as a pre-requisite to the better management of public space.

In this regard, the community was sometimes seen as part of the problem, particularly in hard-to-manage contexts, such as heavily used city centre locations. Nevertheless, engaging the community in public space management was seen as important by most local authorities, for example through basic consultation exercises, friends and community groups for particular open spaces or areas, community planning events and meetings, education initiatives, and formal community councils (area forums).

The role of the private sector in better managing public spaces was also seen as very important and preferably part of a positive two-way relationship, for example through their role as landowners and investors in the public realm; via direct sponsorship schemes; as a result of direct contractual arrangements, bringing specialised knowledge to tasks; as members of partnerships paying for wardens, town centre managers, CCTV systems, and so forth; as contributors of Section 106 planning gain funds; and, in time, through formal town centre management and BIDs schemes.

To sum up these two relationships, between the state and, first, the community and second, the private sector; the former might be seen as a community-oriented rather than community-centred approach, whilst the latter amounted to an acceptance of varied market involvement in a pragmatic manner, but with little stomach for a market-centred approach. In this sense, both the community and the market were seen as partners in, but not drivers of, public space management. This contrasted with how the public sector increasingly saw themselves, as the instigators and arbiters of a more controlled environment, in which the interests of the majority, rather than the activities of any particular minority groups, was the priority.

Overcoming the challenges

Key challenges facing local authorities in managing public space are clearly numerous and diverse. Some challenges are old, such as fragmented local authority organisational structures and outdated working practices that do not foster a holistic approach to public space; other challenges are newer, for example high-density, mixed-use public space contexts that are constantly evolving and changing. What is clear is that public space and public space management are concepts which local authorities have not fully grasped, and as such have suffered from a low political priority. However, some authorities are beginning to change their approach to public space management, and in a piece-meal manner, new approaches to managing public space have been emerging.

Indeed, a wide range of initiatives were in place within the 20 local authorities, many of which deal directly with the problems listed above. These included:

- initiatives that involve the restructuring of the way public spaces were managed, towards more focus on crosscutting approaches and joined-up action – these varied from changes in the local authority structures to temporary street scene working groups, liaison offices, creation of single points of contacts and area-based management teams;
- initiatives aimed at making existing resources go further, for example by changing and integrating procurement practices;
- the creation of forums to involve the community and voluntary sector in deciding on public space strategies and actions;
- initiatives involving partnerships with private sector organisations to fund and implement public spaces improvements;
- initiatives involving the participation of the community in implementing public spaces policies, including neighbourhood and street warden schemes;

- initiatives focusing on safety and crime reduction such as crime reduction partnerships and cleaner and safer environment campaigns.

Many reflect the eight types of management solutions advocated in the range of national research, reports and policy statements on public space. Although few authorities are actively engaged in more than a few of these initiatives, many of the approaches identified in the 20 authorities cut across the different categories.

They suggest in turn eight cross-cutting steps to better practice (Figure 6.1) that represent a somewhat idealised iterative process of public space management that should start and end by monitoring the context in order to devise a plan for action. In this regard, the problems and pressures might equally be viewed as opportunities: opportunities for a radical re-think of priorities and processes; and opportunities to move towards more sustainable models of urban management.

Redefining roles and responsibilities?

Overall, the picture that emerges is a complex one. It is not so much about the retreat of the state and consequent privatisation of public space, but instead reflects a limited transfer of powers and responsibilities for its management to a range of stakeholders, varying in degree from place to place and from one type of public space to another. Although there are instances of a corporate thrust towards control of some high-value public spaces whose quality more directly affects business performance, often this transfer of power also implies the involvement of residents and user groups in management processes through neighbourhood management initiatives. Moreover the finding that some authorities are more concerned to create a seamless public space network, rather than necessarily seamless ownership or management responsibility, was important as it emphasises that with the right public-interest management regime in place, safeguarded by appropriate agreements and/or powers, the actual ownership (and potential privatisation) of space may not matter.

As with similar changes in other spheres of public-sector provision, different stakeholders have assumed a variety of roles in policy design and implementation as a response to changing demands over public spaces that defy the capacities of existing governing arrangements. Through redefining roles and responsibilities in the provision of public space services, stakeholders are seeking a more effective way of producing collectively agreed policy outcomes (in this case, the long-term preservation of a degree of quality in public spaces), although here, as elsewhere, the definition of collectivity is a matter of debate.

Define clearly and early a vision for public space and its management that explicitly prioritises 'quality' as the first and overarching objective

On the basis of the vision, carefully define and integrate all key responsibilities for planning and delivering the better management of external public space – cross-responsibility, cross-departmental, intra-governmental and inter-agency

Actively monitor the success and effectiveness of management processes and initiatives, including the well-resourced enforcement of public space infringements, and continually question, what could be done better?

Be inclusive in developing strategies for the better management of public space, communicating with and actively involving private sector partners and the community wherever possible

Carefully consider the particular requirements of the full range of local contexts, where necessary modifying standard space management approaches, or defining dedicated management strategies to avoid key areas falling through the gaps

Aspire to deliver higher quality services and outcomes (public spaces) by actively challenging existing practices, design thresholds and specifications, and raising standards and expectations

Invest and regulate wisely and for the long-term by thinking of management and maintenance requirements early in the development process and by building processes and places to last

Allocating sufficient core resources to the management of public space to deliver high quality public space, whilst actively seeking additional public and private sector resources to add value over and above established standards

6.1 Eight cross-cutting steps to better practice

A hypothesis suggested by the research findings is that these collaborative arrangements of multiple stakeholders seem to be emerging particularly strongly in this field because the management of public space is in many regards a new area of policy. The absence of a previous codification of roles and responsibilities with a focus on public space quality, of an established policy culture with clear expectations in terms of responsibilities and power, and of clearly defined and widely accepted routines, is likely to have made it easier for local stakeholders to be more receptive to collaborative forms of policy making and delivery. There is certainly evidence of similar processes in the field of environmental policy and new areas of social policy in the UK and elsewhere (see Hajer and Wagenaar 2003, Andersen and van Kempen 2001).

The restructuring of public space management reveals an ongoing process of refocusing separate public services and their respective policies around the locus of their delivery – the public spaces. As already mentioned, if this was true in the past for many parks and green areas, it was certainly not the case for the majority of public spaces. Although this is still a process in its early stages, it already suggests the emergence of a better-defined field of policy, concerned with public space quality, focusing on the processes of management and maintenance, encompassing national policy and local initiatives, and with its own practices, programmes, policy actors and stakeholders.

Borrowing from Marsh's (1998) concept of policy networks, this suggests that restructuring seems to be leading to the definition and consolidation of new networks focusing on public space management issues. Emerging multi-sector public space governance mechanisms, such as town centre management companies, area management partnerships, BIDs and neighbourhood management schemes are the most structured ways of formally arranging roles and responsibilities of stakeholders in a policy field in the process of definition and consolidation.

Finally, although most of the processes described in this chapter are still tentative, there are clear signs that they are already changing the shape of public space management practices. The more defined policy focus on public spaces in their own right revealed in many national and local initiatives, together with the formation of explicit coalitions of interests around public spaces have increased the profile of public space issues within local governance institutions and have, therefore, put public space

services in a better position to compete for policy attention and resources. This increase in profile for public spaces seems to have gone hand-in-hand with a better collective understanding of their roles in achieving a wide range of policy objectives. The recent emphasis on public space quality and its long-term management in the prominent urban regeneration interventions of large English local authorities such as Manchester and Birmingham seems to confirm that.

At the same time, the collaborative arrangements that have emerged for the implementation and long-term management of public space, even if still localised and incipient, are already signalling a weakening of conceptions of management based on narrow, functional views of such spaces. As users, dwellers and others get a say in what happens to the streets and squares they use, it becomes increasingly less possible to see and treat these public spaces as mono-functional containers of facilities, infrastructure or movement corridors.

Therefore, the interplay of national initiatives and local responses and actions, based on a broader understanding of public spaces and cross-sector policy making and delivery, is shaping a public space management policy field that has the potential to be more effective, more responsive to context and thus more relevant to promoting 'liveability' in urban areas. A better understanding of this new policy field and its governance is required to fully understand these new arrangements, their potential and their limitations.

It is yet to be seen whether the increasing interest at the national level will be sustained enough to move practice decisively on from the top-down, or alternatively whether – in time – the bottom-up innovations being introduced by the sorts of local authorities discussed in this chapter will spread and become more widely adopted. Presently, the evidence in England suggests that the top-down initiatives from national government have been important in beginning to inspire a burgeoning range of local initiatives below. Equally, a number of local authorities are beginning to establish a corresponding bottom-up agenda that seems to offer potential for better public space management in the future. Unfortunately, as the national survey demonstrated, the vast majority of local authorities still have a long way to go. The next chapters in this book will show that many of the problems experienced in England, as well as some of the burgeoning solutions, are universal.

Chapter 7

Eleven countries, eleven innovative cities

The context for open space management

The next two chapters look at public space management from an international perspective. The focus, however, is not on public spaces in general but instead on a particular type of public space from the typology in Chapter 3 (see Table 3.1), namely public open (or green) spaces. The research looked at the experience of open space management in a number of cities around the world; cities chosen because of their reputation for high-quality public open space, and/ or for their innovative management practices. The aim was to identify lessons from these experiences that could be applied elsewhere where management practices are less developed. This chapter first discusses the analytical framework against which comparison of the cases was made. Next, the research methodology used for this and the next chapter is discussed. The third and fourth parts of the chapter focus on the first two of six dimensions of open space management identified in the analytical framework: the types of public open space and their needs; and the aspirations for public open space. Finally a set of conclusions are extracted from the analysis.

A basis for comparison

Whilst Chapters 5 and 6 focused on the management of the full gamut of public space types in one country – England – it was important that the research was not over-influenced by the inevitable peculiarities of one country. At the same time, when looking internationally, across a diversity of cultural, political and governance contexts, it was equally important that the focus for investigation was narrowed from public space generally to a more limited typology of public space if meaningful comparisons were to

be made. As perhaps the most developed area of public space management practice, the opportunity was taken to focus on the management of public urban green or open spaces.

In the post-election period following the return to power of Tony Blair in 2001 a considerable body of research on green public spaces in England was launched (Urban Parks Forum 2001; GLA 2001; University of Newcastle-upon-Tyne 2001; Sport England et al. 2003). This work culminated in the setting up of the Urban Green Spaces Task Force whose own work included a comprehensive review of the available literature and an in-depth discussion of potential policy solutions to the perceived decline in the quality of urban green space across England (DTLR 2002a; DTLR 2002b). The research provided an invaluable basis for comparing the urban open space management systems of eleven cities in different countries around the world.

From this basis, issues and challenges could be distilled into six themes, bringing together and juxtaposing the key challenges for open space managers. In doing so it reflected the 'process nature' of open green space management moving through understanding context, to defining a vision, to combining and coordinating actions to deliver change on the ground. The first two themes cover the context for open space management. They refer to the understanding of the set of open spaces to be managed and their needs, and to the aspirations that inform management objectives.

Management context

UNDERSTANDING THE TYPES AND NEEDS OF PUBLIC OPEN SPACE

The first set of issues related to the ability to understand the nature and purpose of public open space and the needs and values that are attached to them by different stakeholders. This ability can be negatively affected by the lack of information about different types of public open space and about the different problems and opportunities they present for open space managers. It depends upon clarity about where responsibilities lie, but also upon more fundamental concerns about what spaces exist, how large are they, what they are used for, what qualities they have (including ecological), what needs different spaces have, and how they should be cared for.

SETTING ASPIRATIONS FOR PUBLIC OPEN SPACE

A particular issue in England had been the lack of local political support for public open spaces which, as a consequence, became a low priority in local government. This has led to poorly formulated policy frameworks for open space, which have not provided strategic guidance, vision and leadership, and clear relationships to other related public policy frameworks. Related to these concerns is the issue of stakeholder involvement in setting aspirations for open space policy. The concerns here are with the degree of civic pride and engagement from local communities, local interest groups and from local businesses, and how well open space management systems are grappling with the changing demands from an increasingly diverse urban population, particularly from the range of 'excluded' social groups.

The remaining themes address the delivery of open space management, and cover the four key management dimensions of coordination, regulation, maintenance and investment discussed in previous chapters

Management priorities

COORDINATING PUBLIC OPEN SPACE MANAGEMENT ACTIVITIES: ROLES, RESPONSIBILITIES AND COORDINATION MECHANISMS

Different degrees of political priority will affect the status of public open space management services vis-à-vis other public services. A key issue was where powers for public open spaces lie in the urban governance hierarchy, what roles are played by different stakeholders within and outside formal governance structures, and how powers, decisions and implementation actions are coordinated among stakeholders, across levels of governance, and with other policy areas. In England, there has been a history of local government splitting up the responsibility for managing open spaces between different departments and contracting out implementation, resulting in confused and poorly integrated organisational structures and a lack of coordination of activities, services and responsibilities (see Chapter 5).

REGULATING PUBLIC OPEN SPACE: REGULATORY INSTRUMENTS AND MONITORING SYSTEMS

The key issue regarding the instruments available for managing open spaces was whether adequate powers exist, whether they are utilised adequately and what the drawbacks of their use are. As analysis in the previous chapter demonstrated, the perception amongst public space managers in England has been that greater use of regulatory powers is required. The connection with broader policy and regulatory frameworks (socio-economic, health and well-being, education, environmental quality, urban regeneration, and so forth) and the sensitivity to local contexts is also important, and this requires monitoring. The issue here was whether data collection systems are adequately developed, including systems for monitoring/auditing open space such as its biodiversity.

MAINTAINING OPEN SPACE: MAINTENANCE DELIVERY AND REINVESTMENT

The setting, delivery and monitoring of maintenance routines are as important as the initial design in determining the long-term quality of public open spaces. The English experience discussed in Chapters 5 and 6 demonstrates how under-funding, lack of prioritisation and unimaginative planning of maintenance led to a decline in the overall quality of public spaces. Key issues were how maintenance routines are designed and delivered and who they involve. How maintenance routines deal with variations in context arising from local circumstances, new demands and expectations were also critical. In particular should maintenance be run on the basis of generalised or specialist teams, and should it be devolved to local areas or centralised for maximum efficiency?

INVESTING IN OPEN SPACE MANAGEMENT: FUNDING AND SKILLS

Adequate funding of urban open space management has been an issue in England, both in terms of the quantity of funding, but also as regards the ability to explore alternative sources, and the emphasis on capital funding for new projects. This raises questions about how to maximise the potential of existing funding streams and to exploit alternative sources of funding through partnerships, sponsorship, trusts, local charges/taxes, grants, how capital investment in new spaces is matched by revenue investment in maintenance, and about how reinvestment in renewing existing spaces factors long-term maintenance into the process. However, investment is not just about money. The quality of public open space management is directly related to the investment in people through the recruitment and retention of staff with adequate skills, both at management and

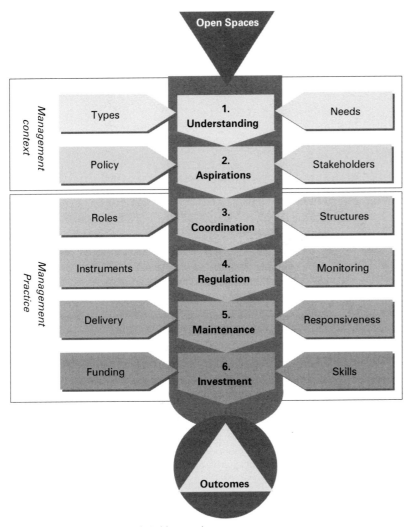

7.1 Open space management analytical framework

operational levels. In England the falling status and budgets of open space management have resulted in an increasingly poorly motivated staff, and increased reliance on out-sourcing of open space services, with long-term implications for management standards. Key issues related to means of creating and maintaining the skills base required to face the challenges of managing open spaces.

Research methodology

The research aimed to identify good practice in parks and urban open space management around the world using the experience in England as a comparator. As such, an early phase of the work involved the formulation of an analytical framework based around the key issues introduced above relating to the context for, and practice of, open space management in England. This is summarised in Figure 7.1 and formed the basis from which practices in different urban governance contexts, cultures and public space traditions could be compared and lessons drawn.

The analytical framework was used as a basis to commission a series of reports from international experts on the management of urban parks and open spaces. Each expert was formally commissioned to prepare a report on their local practices following and addressing the structure and issues specified in the analytical framework. Time and resources did not allow original research in each country, although experts were encouraged to elicit the views of other key stakeholders and stakeholder groups within their country before compiling their reports. Reports were therefore based upon key stakeholder views, available existing case study data, secondary published sources and expert opinion.

Although the approach carried risks, these were heavily outweighed by the benefits. Foremost amongst among which were:

- the collaborative nature of the research and ability to tap into existing networks of expertise
- taking advantage of local in-depth knowledge about open space management and its institutional, political and financial context

- the ongoing assistance of international partners throughout the research project in interpreting the results and undertaking the comparative analysis
- the different perspectives brought to the research by the international partners.

A key task was the identification of cities that could provide examples of good practice in the management of urban open space, and the identification of expert partners who were knowledgeable about urban open space management in the selected cities. An expert steering group was put together to guide the research, and from this, city and international expert nominations were gathered. From these initial nominations, a rolling and expanding network of international experts was soon created as initial contacts suggested other cities and other experts. Using this rolling and expanding network, eleven cities were eventually chosen for the research as those that most consistently came up in discussions amongst the network. They were:

1 Århus , Denmark
2 Curitiba, Brazil
3 Groningen, Netherlands
4 Hannover, Germany
5 Malmö, Sweden
6 Melbourne, Australia
7 Minneapolis, USA
8 Paris, France
9 Tokyo, Japan
10 Wellington, New Zealand
11 Zürich, Switzerland.

Suitable experts were eventually identified and commissioned for each city.[1] The remaining sections in this chapter examine how the eleven cities deal with the first two themes of the analytical framework regarding the context for open space management in the cities. They look at how open spaces are understood and how policy aspirations are formulated. Chapter 8 then goes on to look at the remaining four, focusing on questions of management practice. Both chapters conclude with an overview of the general lessons that come from the eleven cities. Interspersed with the text are insets that focus on the individual cases and on aspects of open space management practice in the cities.

Understanding the types of public open space and their needs

The process of public open space management should logically begin by understanding the nature of that space (i.e. what spaces exist, of what types, what conditions they are in, what pressures and opportunities they are subject to, and how open space is currently used and managed). Different types of public open space will inevitably be subject to different pressures, and to different aspirations and management regimes. Therefore it is important to know what types of public open space exist in what places and to be able to categorise them.

Open space typologies

The types of public open space for which city managers are responsible varies considerably from large expanses of open land, to the smallest green squares. Many of the cities were actively involved in managing large areas of natural or semi-natural landscape that have been incorporated into the city because of topographical constraints, by historical accident, or sometimes by design. These are now highly prized and valued parts of the cityscape.

Nearly all the cities use public space typologies as part of their approach to public open space management, most often classifying spaces by size and function, but variously also by:

- location according to their position in the city (i.e. Wellington's city open spaces, suburban open spaces, inner green belt, the bays, outer green belt);
- environmental criteria and natural value/protection;
- potential uses as well as existing uses;
- ownership;
- relative protection from development;
- heritage value;
- management responsibility;
- professional responsibility (i.e. gardeners or foresters);
- required maintenance approaches and tasks;
- special equipment requirements.

In Malmö and Tokyo, the classifications also have a long-term planning function, as a tool to try and ensure an even distribution of open spaces by function across the two cities.

The exceptions were Paris and Minneapolis. In the former, there is no official typology of urban open space for management purposes, and although there are clearly differences between the city's spaces in terms of their management needs, apart from the large urban forests, all open spaces are classified as gardens. In Minneapolis, almost all urban open spaces are classified as local parks, and rather than a hierarchy of open spaces, the Minneapolis park system is centred around a system of trails, paths and roadways incorporating several lakes, parks and both banks of the Mississippi.

Most typologies represent non-statutory, locally derived systems inspired by local contexts and open space types and often by management convenience. Occasionally, however, systems are based on nationally created typologies. In Curitiba, for example, the local open spaces classification was revised through municipal legislation in 2000 in line with federal legislation of the same year in order to better control the development of unsuitable land and protect existing open space; particular problems in a city subject to squatter settlements. In Japan the public open space typology is defined nationally on the basis of size, location and function as part of a national policy to provide various kinds of open space within walking distance of residential areas. In New Zealand, the Reserves Act of 1972 requires all reserves to be classified according to purpose (recreation, historic, scenic, nature, scientific, government, or local). However, under this broad classification, most councils have their own more detailed breakdown of types, determined mainly for operational management purposes.

Open space ownership

With relatively few exceptions, most public open space in the eleven cities is owned and managed by the state and, in the main, this ownership is exercised through local government in various guises. The exceptions to local government ownership and management include open space controlled by national or regional government because of its present or past strategic nature. Open space along major roads, riverbanks, canals, and other waterways often fall into this category. Similarly, culturally and/ or historically important parks and gardens are often owned and managed by the national government, frequently by historic accident, as is the case with a number of key Parisian parks. Public open space within post-war housing estates is also an exception, as in many places it is owned and managed by housing corporations and not local government. This is the case in Malmö and also in Groningen, where local housing corporations also manage the neighbourhood parks. In some of the cities there are also spaces managed directly by user communities themselves. In Minneapolis, for example, a number of community gardens are owned and managed by a coalition of not-for-profit organisations, whereas in Tokyo, the management of small public green spaces have recently been taken on board by voluntary organisations.

A number of innovative practices with regard to ownership and legal responsibility can be highlighted. First, the dissociation of ownership and management, for example in Hannover, where the banks of the Mittellandkanal are owned by the state but managed by the city, as are a number of privately owned forests with public access. This arrangement brings with it distinct benefits by allowing the management of these spaces to be coordinated by the city and with that of other local open spaces. In Groningen, all nationally owned space is managed locally by the municipality, and offers similar benefits.

Second, is the practice of temporary ownership for park use. In Tokyo, the Urban Park Act of 2003 allowed temporary open spaces to be created on unused private land and even on private structures, for example in the form of roof gardens. In essence, the legislation establishes a right of use separate from ownership, and the resulting spaces are managed by local government on the basis of flexible contracts established for specified periods of time between the local authority and the owner (see Box 7.1).

Third is the specific case of Minneapolis, where the management of urban parks along with the larger regional parks, parkways, boulevards and trails falls under the authority of the Minneapolis Park and Recreation Board (MPRB). This is an independent elected board with law-making and tax-raising powers, which manages 30 regional and 140 neighbourhood parks, plus 49 recreation centres and 43 miles of bike trails in Minnesota. Some smaller open spaces along rights of way and adjacent to buildings are owned and managed by the City of Minneapolis, but in essence the board represents an independent form of local government dedicated to the provision and management of public open space.

Finally, there are examples of focused arm's-length local government agencies, set up specifically to manage public open space on behalf of local government. In Tokyo, the majority of parks are managed by the Tokyo Park Association, a public corporation with a dedicated remit. In Melbourne, all open space is crown land, but Parks Victoria manages much of it, amounting to a network of 37 metropolitan parks, the recreational aspects of Melbourne's major waterways and the trails network throughout the city. By contrast, the City of Melbourne is responsible for a much smaller amount of open space in and around the city centre. Parks Victoria was created in 1966 from the amalgamation of state and municipal agencies, and given legal status as a statutory authority providing services to the state and its agencies for the management of parks, reserves and waterways on public land. In addition to urban parkland, Parks Victoria manages national and state parks around the metropolitan fringe, and, like the Tokyo Park Association and Minneapolis' MPRB, is able to focus on this task alone.

BOX 7.1 TOKYO

Tree planting by volunteers

Tokyo (population 12 million within Tokyo Prefecture, 8 million in inner Tokyo) suffers from an endemic shortfall of green open space (6.1m2/person, compared to 8.5m2 /person in Japan) and it has been a long-standing goal to increase this figure.

Meeting national aims through private-sector involvement
There has been a shift in the approach to urban open space management since the Japanese economy went into recession in the early 1990s from the historic interest to increase provision, to current goals concerned with achieving better quality. This resulted in attempts to develop parks as public amenities suited to diverse social needs rather than to provide standardised urban open spaces.

Traditionally, political power has been highly centralised in policy and budgetary terms but recently greater independence of local government has been encouraged. After 1998 Japan started to look at partnerships with the private sector, including open space provision in the context of limited land resources. This has involved:

- agreements for the rooftops of privately owned buildings to accommodate open space in addition to the established land use;
- relaxing of regulations to allow for agreements with private owners in order to establish urban open spaces for limited periods on unused land;
- extension of the PFI approach to urban open space management;
- introduction of more competitive practices by contracting out maintenance work.

Community participation
Revisions to the Urban Park Act 1956 have fostered greater community participation in the management of open spaces by enabling NGOs and community groups to establish and manage facilities in public open spaces. Local government retains overall responsibility, but can entrust management to other organisations such as the Park Preservation Society in the case of 'city-wide' parks, or much more local organisations in the case of smaller community parks, which tend to be used by the elderly. The Inquiry Commission of City and Regional Planning, established by central government, has proposed three kinds of action to improve the involvement of active local groups:

- better support for local groups already engaged in management;
- a comprehensive system for training volunteers to maintain skill levels and establish standards;
- relaxation of restrictions to allow the construction of park centres as bases for these local volunteers.

Contextual needs in urban open space

The eleven cities also illustrated a number of problems associated with particular types of public open space. Many of these relate to the intensity of management responses required in highly dense urban areas where open spaces and green features are coming under pressure for a variety of reasons:

- conflicts between green and the built structures in urban areas;
- difficulties controlling development pressures in areas of high land values in order to keep existing open space and provide new spaces where none exist;
- the intensity of use of city-centre parks, requiring intensive management regimes, often exacerbated by the original (often highly particular) design solutions adopted;
- conflicts between occasional events and everyday leisure use, the former bringing with them problems of littering, noise, drug use and vandalism;
- the challenges associated with the replacement of ageing street trees and green landscape features without undermining visual qualities in often sensitive areas;
- differences in management and maintenance expectations and therefore between the quality expected by different organisations responsible for public open space, for example between the municipality and housing corporations in Groningen;
- standardised and insensitive legal duties towards traffic safety which tend to shape the management systems for the spaces to which these duties are applied.

Another set of problems relates to the diversity of open space needs and the existence or otherwise of management systems that explicitly acknowledge diversity. This was relevant in relation to the control of introduced pest plants and animals in natural or semi-natural areas in Wellington, related to the need to conserve the sensitive ecology of New Zealand.

In some places, this diversity was inspiring innovations with regard to some particular space types that were then transferred to others. In Hannover, for example, the cemetery sector was the first to adopt more innovative and effective management systems tied to legislation specific to its needs. Legislation in the 1970s determined that cemeteries should be financially self-sustainable, and that cost should be covered by income. This led to new, decentralised management practices which were later adopted in other parts of the open space management service. In a number of the case studies, new opportunities were being seized around a water theme, with recent developments or collaborations in Århus,

Groningen and Malmö leading to the creation of new water-based spaces with specific management systems. In Malmö, for example, this has led to the integration of drainage ponds and canals into the park system.

Formulating aspirations for public open space

Defining a clear set of aspirations for the different types of public open space is an important stage in developing and implementing an open space management strategy. For individual public spaces these are likely to be quite specific, but should also reflect the different forms of value added by public open space. Just as the problems associated with particular public open spaces vary, so aspirations are also likely to vary, depending on who is defining them, the nature of the space being considered and the functions that a space needs to cater for. It is therefore important to understand who defines the aspirations for public space, who is involved and through what mechanisms. In this regard two key sets of aspirations are of particular importance: the aspirations for public open space defined through the political process at different spatial scales (national, strategic, local) and the aspirations of the wider community (residents, businesses, users of public open space, particular interest groups, children, etc.).

The policy context

Three types of policy were apparent across the eleven cases: national policy, spatial planning policy, and local open space policy. The extent to which urban open space represented a national interest varied between cities, from no explicit national interest – in the USA and Australia – to open space policy being almost entirely established at the national level. Tokyo was the clearest example of the latter approach, where an aspiration to increase the area of open space per capita has been a longstanding national goal for urban areas. Thus since the 1920s, national policy has viewed open space as a refuge from the effects of natural disasters such as earthquakes. More recently, the policy has been viewed as a countermeasure to the heat island phenomenon; as a boost to the tourism potential of Japanese cities; and as part of the effort to provide for the leisure needs of children and the increasing numbers of elderly.

Sometimes, however, particular forms of open space are subject to their own legislation over and above general open space policy provided elsewhere. In Denmark, for example, allotment gardens were recently preserved by special legislation, and can only be removed for national purposes.

In cities with a strong national policy context, the development of open space policy usually links back to spatial planning policies established through national statutory planning regimes, such as those in Sweden. This link can bring with it distinct advantages. One is long-term certainty: in Groningen, the city's municipal structure plan has included policies on hard and soft landscaping from the late 1980s onwards, and now outlines ambitions for open space development and management in the city for the next ten years. Another is a higher level of protection for open spaces: the Danish Planning Act 1970 makes open space a formal land use category and as a result there are few changes from open space to developable land use categories, because changing land use designation is a time-consuming process that includes public consultation.

Elsewhere, other broader environmental legislation establishes a similar framework. In New Zealand, there are three main statutory mechanisms defining the aspirations for open spaces and their transformation into policy. The Reserves Act 1977 sets out powers and responsibilities for creating and managing specific reserves and imposes management obligations on the reserve administrator. The Resource Management Act 1991, which aims to achieve the sustainable development of New Zealand's land and physical resources, is applied at local level through regional and district land use plans (tree listing and protection is part of this process). Finally, the Local Government Act 2002 empowers local government to use various statutory and non-statutory tools for fund raising, spending, managing the environment and providing services and facilities for the community.

For a number of the cities, an open space policy hierarchy began at the national or state level but cascaded down to lower tiers of government, and sometimes vice versa in a two-way process. In Melbourne, Parks Victoria works within a number of state government policies and strategies, from the overall vision for Victoria ('fair, sustainable and prosperous'), through environmental policies, and to a public sector management reform programme addressing resources management practices to achieve improvements in service delivery. In 2002, after two years of public consultation, Parks Victoria produced its own strategy for Melbourne's open space network, focused on six principles: equity, sustainability, diversity, flexibility, responsiveness and partnerships. These principles were later incorporated into the state's overall metropolitan strategy around the vision statement of 'a linked network of open space for all to enjoy as part of everyday life, preserved and enhanced into the future'. Parks Victoria is also required by statute to produce a corporate plan each year that includes a ten-year vision, three-year strategies for progressively achieving the vision, and a one-year business plan detailing programmes and activities.

OPEN SPACE PLANS

Significantly, many of the study cities have open space plans of some form to articulate their open space policies; plans that varied in their spatial scale and level of detail. Collectively, the eleven cities demonstrate the function and value of open space planning. It has been instrumental in securing an adequate provision and protection of open space in urban areas, establishing coherent approaches to balance recreational, ecological, and heritage concerns, and in setting guidelines for day-to-day open space management. It has also helped to facilitate indirect but equally important outcomes such as a shared strategic vision for open spaces between city government departments, residents and politicians, or more visible connections to other policy frameworks and responsibilities.

In Denmark, municipal green structure plans are requested under the Planning Act, and are used in Århus as tools for planning and to enable public debate on their strategic urban open spaces policy. In Groningen, the municipal structure plan serves as a framework for sectoral plans and the zoning plan (the only physical planning instrument directly binding on citizens). The former include the 1990 policy plan 'Giving Colour to Green', which formulated a vision for each park. Three municipal structure plans dealing respectively with trees, ecology and the linkage between the overall open spaces vision and the Groningen public spaces management system have also been prepared. To add to this already comprehensive policy framework, the municipal council has been working on a green spaces structure plan, linking the various instruments to the structure plan and thereby creating greater coherence.

Green plans have also helped in assisting in decision-making priorities about land acquisition for open space and disposal for other purposes. This was particularly important in the development of Zürich's 1999 'Open Space Concept', which established the broad aims for open space planning and urban development within the city, following discussions between different municipal departments and external experts. The document establishes a range of quantitative standards which should guide the implementation of new open spaces, from targets for overall amounts of open space per person and per workplace, to catchment areas for different types of open spaces, to amounts of undeveloped land that should be acquired for new open space per square metre of new development.

Only in Minneapolis was there no comprehensive planning with which to marshal resources and provide a vision for the future of the city's open spaces. Short-term overarching goals for the park system are instead set by the nine elected commissioners of the MPRB (Box 7.2). Their ideas are subsequently distilled into four or five narrowly focused goals that are used for evaluating the board's performance. Recently, however, the lack of planning has impaired the agency's ability to react to changing user

BOX 7.2 MINNEAPOLIS

Pedestrian concourse along the Mississippi river

Minneapolis (population 380,000, 2.9 million in metropolitan area) in the north-central USA, began as a mill town on the Mississippi surrounded by lakes and tall-grass prairie. The lakes are still there and the city's natural parklands are today part of an interconnected system of parks encircling the city.

The Minneapolis Park and Recreation Board (MPRB)

The MPRB is a much admired model for other park management agencies for its blend of public accountability, financial independence and the expertise of its long-serving staff. It was created in 1883 by the Minnesota state legislature and consists of nine democratically elected park commissioners. It can hold legal title to property and develop and administer land for use as parks or parkways.

The MPRB is now responsible for local and regional parks, forming a system of well-planned and interconnected parks, lakes and greenways; the 'Grand Rounds' that almost encircle the city in a 50-mile loop. Unusually for the United States, Minneapolis has a regional authority, the metropolitan council, responsible for planning functions and for establishing guidelines for the regional parks system. The board makes sure that Minneapolis's park priorities are enshrined in the regional plan.

Authority to levy taxes

The MPRB enacts its own laws and has statutory authority to define and regulate the use of all its landholdings. Regulations are enforced by the resident park-keepers and by the Minneapolis Park Police Department; a law-enforcement agency created to protect park users and park property.

To pay for its parks, the Minneapolis city charter gives the MPRB the authority to levy a tax on residential property. The Board of Estimate and Taxation sets the tax rate, allocating about 9 per cent to the parks system. The rest of the MPRB's budget comes from the state of Minnesota and a small amount from revenue income (user fees, facilities rental). In addition, the MPRB supplements its tax-based income through the Minneapolis Parks Foundation, a non-profit group, and through revenue-generating public–private partnerships.

Staff expertise

Part of the success of Minneapolis's parks comes from its cadre of long-serving employees with extraordinary seniority and expertise and near encyclopaedic knowledge of the board's historic practices. Staff loyalty is largely due to a strong union that guarantees good working conditions, good benefits, job security and competitive salaries. The job specialisation imposed by union rules allows staff to develop knowledge and experience that no contractor or part-time employee could match. However, high pay results in high operational costs per resident, the second highest in the USA.

needs or to adopt the sorts of innovative park practices that are common elsewhere such as separate pet-friendly areas or teenage skating areas. Given the long-term success of the parks system in Minneapolis, the laissez-faire attitude of MPRB towards planning is surprising. Being an elected board with one responsibility only – to manage the parks system – and with guaranteed income through its own tax raising powers probably explains how the MPRB has been able to achieve such levels of success, with single-mindedness substituting for long-term planning.

POLITICAL WILL

These advantages are not shared by the other ten cities included in the study, although in different ways each confirms the importance of political will and vision to delivering well managed urban open spaces. The experience across the eleven cities has generally been that the commitment and performance of individual local administrations is more important for the quality and quantity of urban open spaces than the national legislation, reflecting the largely devolved nature of powers and responsibilities in these areas.

Perhaps the most obvious demonstration of this local political dimension is provided by Paris, where policy for open spaces is defined exclusively by the Mayor of Paris and under him by the Deputy Mayor for Green Spaces, subject to approval by the city council. Because there are no other stakeholders statutorily involved in deciding on open space policy, lines of political accountability and responsibility are very clear and helped by the fact that open spaces together with public transportation have consistently been the main priorities of the municipality. This commitment to open space was demonstrated by the elected mayor's pledge in 2001 that at the end of his term there would be 100,000 trees along the streets of Paris and that new open spaces would be created wherever possible, so that no one would live more than 500 metres from one.

Linking local open space agendas to broader national policies and priorities can also be important in raising the profile of open space management. In Århus, politicians have long given priority to environmental issues, reflecting such concerns in the plans and practices of the municipality, as demonstrated through the adoption of eco-accounting. This emphasis on sustainability has influenced open spaces management in the city and has led to a strong emphasis on open space management issues in the Agenda 21 strategy for 2002–2005. It has also meant that open space management issues themselves have become a political priority, with the Århus Green Structure Plan benefiting from a wide cross-political consensus (Box 7.3).

In Curitiba, the open space vision dates back to the 1940s. Since that time, open spaces in the city have been conceived as places not only for leisure, but for the protection of native forest, waterways and for flood control, and have become a major political priority. Consequently there has been a continued effort by the city administration to convince citizens in general and businesses in particular of the importance of investing in open spaces. This open space consciousness has now become a part of the city's self image.

Likewise, in Hannover open space policies rank high amongst city council policies even if they are not included amongst the statutory duties of local government. The main vision for open spaces is summarised in the slogan 'Hannover – City of Gardens' which underpins the political vision and physical strategies of the council. All the political parties see open spaces and their management as important to the image of Hannover, and therefore a political consensus on this issue has emerged.

In Malmö and Melbourne the open space managers themselves have successfully taken the initiative to raise open space issues up the local political agenda. Malmö Streets and Parks Department, for example, has been very successful in marketing the benefits of parks and open spaces to their local politicians by ensuring that every opportunity is taken for securing positive headlines for their work, and by inviting politicians to launch events arranged to mark the opening of new or refurbished local spaces. In this way, they argue, public open space is not simply seen as a drain on resources, but instead as a way of actively improving the city's quality of life. In Melbourne, Parks Victoria has tried to demonstrate and quantify the wider benefits that accrue from parks, from environmental, cultural, economic and health benefits, to benefits in community cohesiveness, as a means to influence government funding priorities and increase community support. Their report *Healthy Parks/Healthy People* was commissioned and launched as part of a marketing campaign to demonstrate the health benefits of interacting with nature and which successfully partnered the agency with the National Heart Foundation, Asthma Victoria, Arthritis Victoria and the Royal Australian College of General Practitioners.

Stakeholder involvement

Together with political commitment, the issue of user involvement in the management of public open spaces was taken seriously by most of the eleven cities, not least as a means to garner public support for open space and thereby raise the issue up the local political agenda. Issues vary from place to place, but amongst recent concerns have been social issues such as safety and security in Malmö, a demand for more and better play spaces in intensely populated Paris, and the issue of improving accessibility to the widest possible section of the public in Curitiba.

A wide range of mechanisms are being used across the eleven cities to encourage involvement, and range from one-off initiatives or tokenistic

BOX 7.3 ÅRHUS

Århus' riverbank

Århus (population 292,000) is the second largest city in Denmark and dense for Danish standards. However, with the exception of the old city centre, the rest is a garden city.

The Green Plan's vision and opportunities

The priority given by politicians over many years to environmental issues and the ability of management to seize every opportunity have turned Århus into a 'green city'. Open space provision and the quality of management are considered close to optimal.

The green structure plan was prepared as part of the planning reforms of the 1970s and since then the political vision of 'Århus surrounded by forest', has acquired strong public support. The physical consequences of the widespread acceptance of the plan's green policies are best appreciated in the public support for various projects, such as the transformation of the Århus river valley from a sewage and waste outlet into a major recreational amenity through the creation of a continuous, publicly accessible path. Today, the well-designed urban space along the banks of the re-opened river is one of the most popular inner city recreational areas.

Environmental concern is one of the main driving forces of the plan and extends to maintenance practices; a more natural appearance for parks has been adopted and has become very popular, achieving greater visual variety and maintenance savings. An innovative environment-friendly, though costly, approach is the irrigation of sports grounds with water collected from their own drainage systems, reducing water pollution from excess fertilisers.

Staff performance

A recent analysis concluded that the skill of some long-standing staff to manoeuvre in the political environment and in identifying/following-up opportunities proved crucial to successful management of open spaces. By not using seasonal workers, operational staff are encouraged to train during the low workload periods. A profit-sharing scheme has also been adopted as an incentive for operational staff coping with difficult maintenance tasks and savings are shared according to hours worked or invested in the district's equipment.

Complaints are a good indication of the success of open space management and a 'balanced scorecard' is used to measure staff performance. Management is assessed by the local council every year and by external experts every three years, the aim being a score of at least 1.4 on a scale from 1 (for best practice) to 4. This regular feedback allows the administration to react swiftly to changes in residents' attitudes and needs, to their own operational practices, and to new knowledge about open space management.

consultation (i.e. on open space related spatial planning policy), to the direct involvement of communities in the management process or indeed across the range of open space related activities. In Melbourne, the community is consulted with regard to the development of the metropolitan open space strategy, the organisation of recreational activities, and in specific park planning processes; the approach being to involve as early as possible and to encourage the airing and discussion of all views. In Hannover, the city council also has a statutory duty to ensure the participation of the community in the planning process and has to respond to formal complaints. As a result, whenever a new park is planned, an existing one is refurbished or even a task such as tree cutting undertaken, the council seeks to involve the community. In Wellington, several key statutes affecting open spaces require formal public consultation. These include preliminary input of ideas to help policy formulation, followed by formal written submission and hearings on draft plans with recourse to higher levels. The benefits are thought to be the gathering of community support for processes, and the consequential reduction of adverse criticism, although only when the council uses a range of methods to communicate effectively.

Beyond statutory consultation, proactive initiatives to involve and communicate with open space users have been used in the eleven cities. They include Voluntary Neighbourhood Boards which have been introduced in Århus, and are made up of local residents and businesses. There is now an obligation to involve these local boards in all matters concerning local areas. In a similar fashion, Minneapolis utilises Park Activity Councils which bring together park users, local residents and MPRB staff to develop and run recreation and sports programmes and other park services. They also include local partnerships, such as the three-way partnership which forms the basis of a new 'collaborative model' being introduced in Curitiba to involve the city government, the community and the private sector in open space planning. The initiative represents an attempt to overcome a decision-making processes which, previously, was highly centralised.

Another initiative is the involvement of users in open space appraisal in Groningen though the BORG management system which gives residents a role in assessing open space quality as a means to raise awareness of their surroundings. Participation through design operates in Malmö on an ad-hoc basis when parks are being renovated, and in Zürich where the former industrial areas of the city are being converted into parks, with the direct participation of local residents, business and key local organisations. Similarly, in Minneapolis, Park Planning Citizen Advisory Committees are utilised for new capital improvement projects and consist of volunteers or citizens appointed by the Commissioners.

Volunteer rangers are a particularly successful initiative in Wellington, assisting with patrolling and inspection of open spaces, especially the larger areas, becoming the councils eyes and ears. There is now a paid volunteer coordinator working for the council to coordinate the activities of the rangers which has helped to establish better lines of communication between the council and the community.

In Paris and Hannover, involvement is encouraged though the strategic use of lower-tier district councils as conduits for organised community participation on open space matters, and as a means to disseminate plans and policies to the community. In Hannover, this latter role is supplemented by the large number of publications produced and events run especially for that purpose. In Zürich this role is played by local open space administrators such as those employed in each district of the city. Because they are usually well known to the local population, they act as a direct conduit through which residents can engage with the city council on open space issues.

PROBLEMS WITH INVOLVEMENT

Overall, two types of problems have been encountered by the cities, problems broadly associated with too little participation, or at least an unwillingness of groups to get involved; and conversely, less frequent problems associated with too much involvement.

Too little involvement was not usually associated with a lack of effort on behalf of particular cities to involve their citizens, but more often with a lack of response to their effort. In Århus, for example, despite provision for public participation in municipal planning, the actual levels of participation mean that most decisions on the strategic management of open spaces are taken on an administrative or political basis. Local citizens are far more concerned with influencing the quality of their own local open spaces, and seem content to leave more strategic decisions to those with direct responsibility for such matters. The city has found it particularly difficult to involve the business community and minority ethnic groups in decision making.

However the opposite – too much involvement – has been a problem in Groningen and Minneapolis for different reasons. In Groningen, the Dutch tradition of high levels of public participation has led to a situation where public works (including open space management) has sometimes been demand-led rather than planned. The implication has been ad-hoc city management approaches and a tendency for those who shout the loudest to get the most out of the system. To resolve the problem, project development (including major repair works) is now the only aspect of open spaces management in which there is direct public participation. In Minneapolis, although there are around 20 'adopt-a-park' agreements in place between the MPRB and community groups, unsolved disputes between the MPRB and its highly unionised workforce around the nature of citizens' possible roles in the day-to-day management of parks have restricted the extent to which communities can become directly involved.

Community participation is, however, a vague term, which can mask different degrees of involvement from different groups of the population. Where the cities encountered difficulties in involving minority groups in open space management, this issue was being tackled in a number of ways. In Århus, recent immigrants often occupy the less desirable 1950s housing estates with their poorly defined public/private space relationships, making their lack of engagement in open space issues a particular problem. In an attempt to reverse the situation, the city is trying new approaches through an EU-funded URBAN initiative which aims to enable excluded and deprived communities to influence changes in their own environments. The approach is aiming to involve these groups directly in the ambitious Hasle Hills project, not least through the direct employment of these groups in the operational staff.

Although there is little specifically done to address other minority ethic groups in Wellington, consultation with the Maori is obligatory when formulating open space management policy. The Treaty of Waitangi is based on the principle of autonomy for the Maori and of mutual consultation, and it forms part of the original constitutional settlement between the indigenous Maori peoples and the Crown. Iwi (Maori) Management Plans are now produced as a vehicle for local Maori to articulate their aspirations, including the protection of Maori heritage sites.

In Hannover, the council works with identified representatives of the disabled, migrant communities, the elderly and women's groups, who are informed about any proposal that might affect them before any decision is made and who have the opportunity to thereafter help to shape the proposals. Similarly, in Tokyo, residents are increasingly being directly involved in various stages of open space management, from local to large scale parks, from planning to operation (see Box 7.1). The initiative is particularly focused on the increasing numbers of elderly residents, as a means to tap into their knowledge and skills. Some 30 community groups are now directly involved in restoration and beautification projects.

Conclusions

Understanding the types of public open space and their needs

The experience of the eleven cities confirms that the good management of open spaces depends upon a correct understanding of the nature and needs of different types of public open spaces, and that one-size-fits-all standardised approaches are rarely appropriate. Therefore, a typology to differentiate amongst open spaces can be a useful management tool to establish common management regimes within categories of public open space.

Those experiences also suggest, however, that this should not be primarily a matter for standardised national classifications to which local open space managers have to conform, but the result of locally generated criteria, shaped by history, geography and ecology, as well as by national standards where they exist. In the international cases where formally defined typologies have been particularly beneficial (e.g. Wellington, Curitiba, Groningen and Malmö), clear linkages are also found between open space typologies and active management strategies, explicitly connected to clear, but differentiated, public space quality aspirations.

Typologies also offer the opportunity in several cases to explicitly establish a link between the open space classification and broader local government policy objectives, especially as regards issues of sustainability. Taking this broader policy context on board has not only helped to deliver overarching policy objectives, but also reinforced the position of open space management and its needs and priorities within other areas of the local government remit.

As regards the ownership of open space, the ideal scenario seems to be one where one organisation both owns and manages all key open spaces across a city, from the large to the small. Minneapolis is perhaps the closest to this ideal, with MPRB being almost the sole agency in charge of deciding on management policies for the city's public open spaces; a set of responsibilities aided greatly by the conflation in one organisation of the financial and legal means to implement its own policies. However, the case is unique, and most of the other cities have had to operate within a historic legacy of different types of open spaces being owned by different agencies and levels of government.

A key lesson that emerges from their experiences is therefore the need to establish a coherent management strategy to cope with the diversity of open spaces, integrating and unifying management regimes, preferably under the auspices of one organisation. The dissociation between ownership and management responsibility seen in many of the cases seems to be the key to achieving that unification, with, for example, open spaces owned by multiple organisations, but managed collectively by one. How this has been done, to what extent open space owners have transferred power and control to management agencies, whether this has involved setting up new organisations or using existing ones, and so forth, is a function of the institutional, legal and political context of each of the cases, and no single 'right approach' is apparent. The benefits of a dedicated public open space agency/authority are nevertheless readily apparent. Removing ownership rights – even if temporarily and by negotiation, as in Tokyo – from under-utilised spaces in areas with open space deficiencies so that they can be utilised as public open space – also carries obvious benefits.

As seen in the international case studies, the nature of, and the pressures on, open spaces can vary either as a function of their location in the urban fabric, the uses they have, or the expectations of the different agencies with a say in their management. The natural dynamics of changes in society adds to the variation, as exemplified by the new demand for play areas in the historic parks of Paris.

Very often these pressures lead to real threats to the quality of the open spaces concerned. However, the international cases suggest that whereas these problems cannot be avoided, they can be dealt with quite successfully if they are openly acknowledged by management strategies. Thus, in many of the eleven cities, special management regimes have been set up to tackle types of open space where particular problems are more acute (e.g. Zürich's lakeside parks, the neighbourhood parks in Tokyo, or city centre parks in Groningen). In some cases this has meant more intense maintenance routines, in others a closer involvement of park users in management decisions, in others still, the introduction of more sophisticated monitoring tools. The key message is therefore that diversity in problems as well as opportunities needs to be acknowledged and dealt with.

Formulating aspirations for public open space

The international cases cover examples where there is a strong national policy framework shaping open space aspirations and examples where open space policy and strategy are entirely a local affair. No matter how different these contexts might be, a common thread is the ability to link closely their visions for open spaces to broader national, regional or local economic, social and environmental aspirations through effective use of the available policy instruments.

In many of the cases, the spatial planning system has provided the instruments for that linkage, with particular success where open space is at risk from development pressures and/or there is a pressing need for an expanded network of open spaces. Thus often the simple inclusion of open space issues within powerful statutory spatial planning documents – even when this is not a legal requirement – has helped to raise the profile of those issues. In some cases this linkage with the spatial planning policy has come together with an equally effective connection with environmental sustainability policy instruments such as Local Agenda 21 initiatives.

A key lesson is therefore that open space aspirations need to be considered within the broader context of other relevant policy areas if they are to have resonance beyond specific open spaces interests. An important

means to achieve this has been the positioning of open space policy in a hierarchy of policy instruments ranging from the national to the local, and incorporating detailed open space plans reflecting both a spatial vision and day-to-day management policies. The example of Denmark, where the requirement for municipalities to prepare 'green plans' is established in national legislation, has potentially important lessons to offer.

Significantly, in most of the cities, the commitment and performance of local administrations seems to be a much greater determinant of the quality of open spaces and their management than the national and regional legislative framework. This is not only a reflection of the devolved nature of most responsibilities and powers for the management of open spaces, but also because no matter how decisive national open space policy frameworks are, most of the concerns that define the quality of open spaces and their management can only be effectively tackled at the local level. This seems to be equally the case where the formal power is concentrated locally such as in Paris (with no national role), or where state or federal authorities have delegated their formal powers to the local level as in Melbourne or Hannover.

Strong local leadership is therefore a key determinant of success. Another key lesson emerging from most of the cases is that successful open space management depends upon a long-term commitment to a vision for open spaces that by its nature cannot be restricted to a single party agenda. All the cases have achieved results only through a sustained commitment to open spaces over many years, often through changing political administrations and priorities, and through different economic and social contexts. Only a level of consensus on the relevance of open spaces and the importance of adequate management across the political spectrum can secure that commitment.

Experiences in a number of the cities (e.g. Curitiba, Hannover, Århus and Groningen) also suggest that shared aspirations for open spaces need to go beyond the political spectrum to be incorporated by the citizenry in the image they have of their own city. The cities where this has been the case suggest that this collective 'green' image of the city contributes to convince politicians to maintain a high level of support for public open space management.

In some cases this commitment by politicians and citizens has been the result of the efforts of technical staff in the relevant open space agency, in others, of a few visionary politicians. Rarely, however, has it simply been a result of formal policy-making procedures. In this regard, marketing open spaces, both internally and externally, appears to be an important task of the open space management agencies. Indeed, agencies across the eleven cities have devoted considerable effort to persuading local politicians and citizens of the importance of well-maintained open spaces in social, economic and environmental terms.

All the international cases illustrate a proactive attitude towards the involvement of the community in open space management. Although there is no one common approach to how this should be done, or to what extent communities should participate, a key dimension of successful open space management seems to be a willingness to engage local communities in the task, and to use creative means to make this happen.

The challenges faced by each of the eleven cities to create a framework for community involvement where none exists already vary considerably, from the complete restructuring of management systems, so that they are not simply reactive, to developing better direct channels of communication with local communities. In some places resulting participation have been mostly at the level of statutory consultation about new capital investment in the neighbourhood (e.g. Paris), whereas in others, an actual transfer of management responsibilities to volunteers and neighbourhoods has been achieved (e.g. Tokyo).

Despite this variety, some common themes have emerged, and those can provide the basis for useful lessons. First, in all cases, there have been clear benefits from sharing with the community the responsibility for managing open spaces. The most obvious benefit has been the harnessing of active support for open space issues that is vital if those issues are to remain on the top of local, regional and national political agendas. The power of neighbourhood-level organisations to influence higher-level resource allocation decisions in Århus was a clear example of this. In some places, technical staff in the municipal parks and open space department have been very skilled in using this pressure from below to shape decisions from above.

Second, in cases where community involvement is well established, even if just on a consultation basis, it provides a ready means of assessing changes in the needs and preferences of users of open spaces. These can subsequently be factored into open space management systems and either provided for, or their impacts ameliorated.

A further key lesson is that whatever its form, effective community participation needs an information system to facilitate the dialogue between open space managers and the community. The BORG system in Groningen with its visualised scenarios is perhaps the most sophisticated example of this (see Box 8.4), but much simpler processes of discussion and exchange of information between municipal staff and the community seem to work equally well.

Lastly, the cases suggest that whereas increasing community involvement in open space management adds to the quality of both management processes and the open spaces themselves, this is not without its problems. Active communities can skew priorities towards their immediate concerns and leave other equally important issues and sectors of the community without the necessary resources. In this context open space management can too easily become primarily reactive, whilst long-term or strategic objectives can be neglected. Paris and Groningen provide examples where this occurred. The lesson is that community participation in open space management is immensely beneficial, but needs to happen within a framework that gives weight to different voices within the community and takes into account immediate and localised demands as well as long-term aspirations and city and region-wide objectives.

Notes

1   The experts were: Århus: Karen Atwell, Danish Building and Urban Research; Curitiba: Eng. Carlos Eduardo Curi Gallego, Cobrape Curitiba; Groningen: Gerrit Jan Van't Veen, Kirsten Mingelers and Iefje Soetens, STAD BV; Hannover: Kaspar Klaffke and Andrea Koenecke, Deutshe Gesellschaft für Gartenkunst und Landschaftskultur; Malmö: Tim Delshammar, Swedish University of Agricultural Sciences; Melbourne: John Senior, Parks Victoria; Minneapolis: Peter Harnik, Trust for Public Land; Paris: Michel Carmona, Le Sorbonne; Tokyo: Aya Sakai, Royal Holloway, University of London; Wellington: Shona McCahon, Boffa Miskell Limited; Zürich: Professor Peter Petschek, HSR Hochschule für Technik. See (CABE Space 2004) for a more detailed discussion of the case studies.

BOX 8.1 CURITIBA

Polish Immigration Memorial Park

Curitiba (population 1.7 million, 3 million in the metropolitan area) is one of Brazil's large regional cities. It is recognised in Brazil and internationally as a model for urban management, largely based on environmental achievements including its integrated transport system. Its success is attributed to political will, leadership and efficient marketing and environmental education which has ensured widespread support for the creation and protection of the city's open spaces.

Achieving environmental objectives

In 1966 the Curitiba Master Plan's designation of Environmental Protection Areas created a framework for the creation of large parks along its main rivers as places for recreation, reserves for native vegetation, protection of water resources and watercourses, and flood control. These areas have been successfully reclaimed as open space, reaching 51.5 m^2/person in the late 1990s.

In 1986 responsibility for all environmental matters was handed to the newly created Municipal Secretariat of the Environment (SMMA), including functions previously within the remit of the state and the federal government. It became Curitiba's most influential local government agency thanks to its legislative and financial autonomy and its broad remit.

SMMA continues to identify locations to be transformed into open areas and has the powers to appropriate privately owned land or negotiate land exchange with the owners. In 2002 the exchange of conservation guarantees for the right to build outside the protected areas led to the preservation of around 9500m^2 of green open space. Fiscal incentives and other mechanisms also encourage conservation. With the city's continued expansion, the future challenge will be maintaining the current ratio of open area per person. Moreover, in 2000, 70 per cent of squatter settlements were located on the banks of watercourses and nature protection areas.

Effective communication and environmental education

Since the 1970s the city authorities have been engaged in selling their open space policies to the community. As a result of growing popular support for them, politicians of all parties subscribe to their broad thrust, endorsing the work of the administration and guaranteeing long-term continuity for Curitiba's open space initiatives.

Environmental education was also used to win support for the city's policies, having been introduced in 1989 into the curriculum of all municipal schools. It has now extended to the community, reaching people of all ages and social backgrounds, comprising a variety of educational activities and fostering community participation in the protection of natural resources. Environmental education was bolstered in 1991 with the creation of the Free University of the Environment in Curitiba to receive and disseminate the most recent ideas on urban environmental management.

Chapter 8

Eleven innovative cities, many ways forward

The practice of open space management

This chapter continues the comparative study of public open space management and follows the analytical framework sketched out at the beginning of Chapter 7. Whereas that chapter looked at the context for public space management, this chapter looks at management processes and practices through the four key dimensions of coordination, regulation, maintenance and investment. It examines how the eleven international case study cities have dealt with those four topics and it concludes with key general lessons that can be taken from their experience. As before, boxes throughout the chapter looking at individual cities and detailing relevant aspects of their open space management practices provide the empirical background for the discussion. The four dimensions are each discussed in turn in the first four sections of the chapter. A final section draws out conclusions from this work.

Coordination of public open space management activities

A wide range of stakeholders play a part in public open space management. A key objective is therefore to understand their roles and responsibilities in the different international contexts and to examine how they are defined and coordinated.

Roles: the key stakeholders

Both within and outside local government a wide range of stakeholders have an interest in public open space management or are directly involved in its delivery. These ranged considerably amongst the eleven cities and sometimes revealed a fragmented network of responsibilities. In Groningen, for example:

- housing corporations manage spaces around their housing estates;
- the Water Board manages the banks of canals and waterways;
- an independent trust owns and manages nature reserves in and around the city;
- green spaces around public facilities (e.g. schools and hospitals) are managed by their respective departments;
- the national government manages the open space along the national trunk road network.

Yet despite this seeming fragmentation, the city has managed to maintain high-quality open spaces, suggesting that the mechanisms for, and coordination of, management responsibilities may be more important than the particular structure of responsibilities.

Universally, it seems, local government carries primary responsibility for managing public open space, although management operation may involve a broader range of actors who may also be responsible for certain discrete categories of space. Århus is typical: the municipality is responsible for managing the landscape, forest areas, parks and other public open spaces; the royal grounds are also managed by the municipality in return for public use; whilst garden allotments and golf courses are managed by user organisations. In Curitiba, the majority of public open space is council property, and the Municipal Secretariat of the Environment (SMMA) has overall responsibility for its management. SMMA is directly involved in planning and maintaining public open spaces, licensing land uses, and land division of protected private land, as well as for the felling of trees on public and private land (Box 8.1).

In both cities, open space management responsibilities are largely focused on local government. By contrast, in cities such as Tokyo this activity takes place within a comprehensive national policy framework, effectively creating a dual responsibility involving central and local levels of government. Hannover also falls into this camp. Its local authority is responsible for open spaces and their management, but important exceptions are found in nature conservation and the protection of garden monuments which are duties of the state government.

A clear distinction emerged in the case studies between day-to-day management and long-term development responsibilities, usually through the division of responsibilities within one overarching department. In Groningen, the management of open spaces owned by the municipality falls under the responsibility of various divisions of a single department, the Department of Physical Planning and Economic Affairs (ROEZ). Within this, open space management is the responsibility of the Urban Management Division, whereas open space development is undertaken by the Physical Development Division. The former is responsible for open space upkeep and replacement, the latter for expansion of the park system, reconstruction, and other large-scale changes. The arrangements are complicated by the facts that cleaning responsibilities for open spaces (including litter disposal) are carried out by Environmental Services, whilst the city architect plays a pivotal role in the relationship between new development and subsequent management.

Even where the majority of responsibilities for open space management were coordinated through one local government department, other departments also retained an involvement to a greater or lesser extent, such as those in charge of spatial planning, highways, sports and leisure, health, and real estate. In Zürich, for example, Grün Stadt Zürich (GSZ) is part of the city council's Infrastructure Department and has separate planning and maintenance units. As the parks/environment agency for the city of Zürich, GSZ is legally responsible for managing all urban open spaces, however, these responsibilities cease when open spaces have either large areas of hard surfaces or significant levels of traffic, in which case they are managed by the Traffic and Civil Engineering Office. Public sports grounds and swimming pools, by contrast, are owned by GSZ but managed by the Environment and Health Agency.

In some cities, a further local tier of government has a role to play. In Germany, large cities have had district councils for the last 20 years, and Hannover has 13 of them. The arrangement has the advantage that even if the city council does not regard open spaces as a priority, the district councils certainly do, and although their formal power is limited, their political influence is considerable. Political decisions regarding open spaces are first debated in the district councils before a political committee advises the city council on priorities.

Some cities are influenced at the more strategic level by regional policy. At this level, the Metropolitan Parks and Open Space Commission sets general open space strategies for the metropolitan area of Minneapolis-Saint Paul through its Regional Parks Master Plan. The influence of strategic parks panning is also felt particularly strongly in the Melbourne area through the work of Parks Victoria which directly manages the urban open space network around Melbourne, whereas the City of Melbourne manages a much smaller area of open spaces in and around the city centre.

Local politicians played a decisive role in all but one of the cities examined. In Paris, for example, power resides in the hands of the elected city mayor who has ultimate decision-making responsibility for the Department of Gardens and Green Spaces. In Hannover, the mayor and the directors of the municipal administration are all politicians, and one of the latter is directly responsible for open spaces as director of the Environment and Green Spaces Division (FUS).

The exception to this general rule was Minneapolis, which is unique amongst large US cities in having an independent park board, separate from the mayor or the city council. Within the board, management responsibility lies with nine elected park commissioners, six of whom represent the six geographical districts of the city, whilst the other three represent citywide interests. Although the commissioners are elected, they are not politicians in the conventional sense because their remit is highly focused on developing general park policies and delivering open space management.

THE USE OF PRIVATE AND COMMUNITY STAKEHOLDERS

As well as the multiplicity of public-sector roles and responsibilities apparent in the eleven cases, a range of private and third (community) sector stakeholders were also involved in open space management.

The extent of private sector involvement varies considerably. At one end sits Minneapolis, with almost no private involvement in public open space management. There, widespread contracting out has been avoided through a strategy of in-house job specialisation that cannot be matched by external contractors. Groningen, on the other end, has 80 per cent of maintenance work carried out by external contractors. In between these cases, the general approach seems to be one of using the private sector in various forms of partnership.

In Hannover, most new construction work within public open space is undertaken by private contractors. However, only 10 per cent of maintenance work is contracted out. More recently, city-owned sports fields have also been transferred to private sports clubs who receive a grant from the council to fund maintenance work. In Malmö, the Streets and Parks Department is responsible for managing all public open spaces, but it contracts out much construction and maintenance work to a mix of public and private contractors. In Paris, all works are undertaken by private contractors under the system of public bidding, and private architects and landscape architects are used to design major new parks, through the same system.

In Wellington, the private sector has been involved in the management of open spaces in a more comprehensive manner. It provides contracted services such as design and management consultancy, weed spraying, and so forth. It is also involved in sponsorship. Council-controlled trusts and companies have been set up to manage certain facilities or areas suitable to be run as business enterprises such as the Regional Stadium. The private sector was also involved in negotiating incentive development rights in the city centre in the 1980s and 1990s when extra building height was allowed in exchange for open space provision at ground floor level. The outcomes of this practice were not good and, as a result, it has been discontinued. Collectively, however, few problems were reported by the eleven cities concerning their use of private contractors, as long as work is carefully specified, properly integrated with other operations, and carefully monitored.

In contrast to this widespread use of private contractors, involvement of the voluntary sector in public open space management was not common in the cities, although a number of initiatives existed to improve the situation:

- In Melbourne 50 voluntary Friends Groups contribute to regular programmes and projects.
- In Århus voluntary neighbourhood boards are given direct support by the municipality, and are involved in decisions about open space management in their areas.; in these areas, some smaller open space projects have only been implemented and maintained by mobilising local voluntary labour, delivering viable open space management on a shoestring and creating a long-term sense of responsibility within communities.
- In Zürich the management of playground areas and open spaces close to residential buildings have sometimes been contracted out to voluntary parents groups.
- Community gardens have been created in Minneapolis which are managed by a coalition of not-for-profit organisations.
- In Curitiba a schools' initiative in deprived neighbourhoods has helped to train young people in gardening and related activities through an extra-curricular programme that also offers the opportunity of long-term employment in open space maintenance.

Tokyo had perhaps the most developed system for involving community organisations in managing their public open spaces (see Box 7.1). There, local government sometimes manages open spaces directly and in other instances management is contracted out to external organisations, either in the form of voluntary groups made up of local residents to manage community parks, or as private contractors for larger parks. This reflects a concerted effort being made in Japan, where national legislation was changed in 2003 in order to promote greater involvement of the community and voluntary organisations in the management of open spaces. This idea is being translated into the production of model contracts and the setting up of information exchange networks between voluntary and community organisations.

Elsewhere, feedback from open space users and other municipal staff has been used to define maintenance priorities, although the community remains a still largely untapped source of enthusiasm, labour and expertise.

Management structures

A number of the cities had recently been engaged in management reforms as a means to improve the delivery of public services in general. These were often inspired by 'new public management' approaches (see Chapter 5), including the streamlining of responsibilities, the introduction of cross-service community planning mechanisms, and a focus on outcomes as well as processes. In Hannover, for example, during the 1990s a national initiative to reformulate local government emphasised the decentralisation of responsibilities, considering citizens as customers and understanding local authority services as products. Open space management was chosen as a pilot sector for several of the new management initiatives, including the better coordination of responsibilities through a dedicated division of the city administration (Box 8.2). The state of Victoria was also a few years into a management reform programme for public services focusing on outputs from service delivery activities rather than on service processes. Departments are now accountable to the state government for their outputs, and key output groups are identified for each service against which performance is measured.

Debates concerning methods of managing open spaces have been widespread since the mid-1990s in Japan through the auspices of the Parks and Green Spaces Committee, set up by central government. The outcomes from this work were reflected in revisions to the Urban Park Act 1956 which included legal mechanisms to create new open spaces in built-up areas, promotion of community involvement in open space management, and the better enforcement of open space regulations. In Tokyo, the recent metropolitan Inquiry Committee for Urban Green Spaces

BOX 8.2 HANNOVER

Former gravel pits, now recreational spaces

Hannover (population 500,000), the capital of Lower Saxony, benefits from a rich heritage of urban open spaces ranging from the historic gardens of stately homes to the extended meadows of the river Leine and the Eilenreide forest. These have earned the city the label 'City of Gardens'.

Improving maintenance through decentralisation

Hannover, like other German cities, has experienced a decrease in population. To address this problem, the authorities focused on making urban living more attractive including efforts to maintain the quality of its open spaces.

Budget cuts and outmoded management practices led to new initiatives to re-focus the maintenance of open areas. In the 1980s the KGST, an institution under the German Cities Federation, engaged consultants to improve efficiency in local government. Hannover, like most German cities influenced by the KGST initiative, adopted a more business-oriented management style based on the new principles of citizens being treated as 'customers'.

When the open space sector became one of the pilots, the previous management approach was reversed and outcomes and user satisfaction rather than cost became the measure of efficiency. The need to improve communication with the public on open space matters also helped to redirect the focus of the administration's work.

All employees of the Environment and Green Space Division (FUS), responsible for managing Hannover's open spaces, receive ongoing training in innovative management practices through the city's association with the Faculty of Landscape Architecture (University of Hannover). Recently the KGST added another initiative, its IKO Network, a forum to compare management efficiency in different cities using key indicators. Hannover also participates in the Standing Conference for Green Space (GALK), which brings together local urban parks and open space administrations throughout Germany for the exchange of information and experiences and the discussion of management problems.

FUS (part of the Environmental Services Department) coordinates all management tasks (planning, construction work, maintenance) and is responsible for the overall financial coordination although each section has its own budget and can reinvest any income from charges. Maintenance tasks are fixed in working plans with responsibilities clearly defined geographically so that each group feels responsible for an individual site or group of sites. About 90 per cent of maintenance work is carried out by the city's own workforce.

Departing from previous standardised maintenance practice, decentralisation has permitted maintenance activities to be determined individually taking into account the special character and function of open spaces. Different regimes can now be applied to historic gardens, to the highly used 'promenades' or to the wildlife parks.

supported the need for reform of open space planning, maintenance and operations, incorporating business-inspired management practices.

In New Zealand and Brazil, there has been a growing emphasis on cross-service planning. The 2002 Local Government Act in New Zealand requires every local authority to prepare long-term plans that describe key strategies and policies for funding, financing, investment and spending. One aim of this is the better coordination of strategic and regulatory policy. In Curitiba, the establishment of the Municipal Institute for Public Administration (IMAP) has allowed a similar focus on cross-departmental planning. The body formulates and oversees management strategies throughout the municipal administration to ensure that departments coordinate their actions. Since 2000 IMAP has been in charge of the municipal Management Plan, which is now used as a reference for planning, running and evaluating the management of public organisations at the city level.

Key amongst the organisational objectives stressed by the eleven cities were the importance of good day-to-day personal working relationships, the value of inter-departmental cooperation and the benefits of integrating public open space responsibilities. The emphasis on personal working relationships could be seen most directly in Århus, where the continuity provided by long-serving senior staff has made an important contribution to successful open space management. In particular, the close personal contact between four senior officers made for smooth cooperation between the Natural Environment Division (NED), the City Architect's Office, the Road's Office, and the office of the mayor.

INTRA- AND INTER-ORGANISATIONAL COORDINATION

Beyond personal working relationships, the cities demonstrated a commitment to overcome organisational barriers thrown up by the different departmental/organisational responsibilities for different dimensions of the open space management remit. A number of approaches were adopted to achieve this. The first is coordination through higher government tiers such as through the offices of the metropolitan council (regional government) focusing on planning and development activity in the Minneapolis metropolitan area. As such, it is both a planning agency and service provider (transport, housing, sewage) and is in charge of managing the regional park system. In so doing it operates primarily as a planning agency for the regional parks system, helping to coordinate across jurisdictions whilst leaving most of the implementation and day-to-day management to the local parks agencies (including the Minneapolis Park and Recreation Board – MPRB).

Although there is a clear structure of local, regional and national government in New Zealand with distinct jurisdictions, 'grey' areas inevitably emerge between open space jurisdictions leading to funding

tensions between regional and local councils. Typically these are solved by adopting memoranda of understanding or partnership agreements between authorities. At city level, open space management is organised in Wellington into several management teams involving various aspects of policy and operation, all under the Built and Natural Environment Committee. A key difficulty has been in-house communication within Wellington City Council where responsibilities still overlap and conflicts arise (e.g. conflict between the needs of roads, cabling and drainage and those of open infrastructure in the city centre where space is limited).

In Zürich, the GSZ routinely works together with other departments in the city administration such as the Civil Engineering, City Planning, City Development, Transport Planning, and the Health and Environment departments. External links are also prioritised, including at the operational level where weekly contact meetings between the maintenance crew of parks and local police are now commonplace. The initiative builds on a project called 'Security and Cleanliness', which, in order to raise the image of the city and its open space has put together a team with representatives of GSZ, the police, PR professionals and council members.

Not all attempts at intra- and inter-organisational coordination have been successful. In Wellington, for example, recent restructuring of the council has improved clarity in the division of responsibilities and funding, including the separation of regular maintenance responsibilities from one-off capital projects. In the short-term, however, it has negatively affected open space management through the loss of institutional knowledge as a result of staff transfers and changed lines of communication within the council and with external stakeholders. In both Groningen and Malmö, attempts to combine the maintenance of public open spaces with those belonging to public housing providers have proved unsuccessful. In both cities, housing corporations work to much higher standards and to a more intensive management programme than the municipalities can hope to meet.

INTEGRATED STRUCTURES

Significantly, the good practice exhibited by the majority of the eleven cities was built upon a move towards unifying responsibilities for public open space in more integrated open space management structures. In Malmö, for example, park management is part of the Streets and Parks Department and is coordinated with the management of streets, bridges and squares. Planning of new parks and management and maintenance of existing ones is coordinated with the same functions for all types of public spaces.

In Groningen, the development and management of open spaces has been the preserve of a single organisation – Municipal Services.

This was organised by sector, was internally orientated, and tended to emphasise technical specialisation, with little understanding of priorities outside its remit. Changes in the 1990s led to a devolved neighbourhood management model that was also more outward looking and focused on results. Today, the division of responsibilities within the various divisions of the Department of Physical Planning and Economic Affairs (ROEZ) aims to ensure that interventions in open spaces should be approached in a more integrated fashion. This example suggests that coordination through a single management structure is not by itself enough to deliver integrated management. What is more important is the integration of planning, expertise, and day-to-day operations at the local level and the degree to which the organisational structure allows this to happen, or, conversely, militates against it.

Regulation of public open space

Open spaces are subject to a variety of pressures coming from the different functions they perform and the varying nature and intensity of the uses they accommodate. Two key factors affect whether and to what extent these issues impact on the quality of those spaces over the short and the long term. These are the instruments regulating such spaces, particularly how different activities are sanctioned or discouraged, and the monitoring systems that aid regulation and which feedback into the policy-making and implementation processes.

Regulatory powers and instruments

The eleven cities rely on a range of powers for public open space management, but these powers are rarely neatly packaged from one source. Nevertheless most cites have a clear statutory basis for at least some of their open space management activities.

In Hannover, the federal construction law and the federal nature protection law form the legal framework for open space management, together with the federal and state planning legislation – which define open space as an important land-use category. City councils in Germany also have a statutory responsibility to ensure safety in public open spaces. Similarly, the Japanese Urban Park Act of 1956 established that local governments have specific statutory responsibilities over the development and maintenance of open spaces within their boundaries. It sets basic rules which local governments should follow, including the types, sizes and functions for new open spaces. In New Zealand, under the 1991 Resource Management Act, city councils can require developers to set aside land as a reserve, or pay a levy towards a reserve acquisition fund. Land designated as a reserve is vested in an appointed administrator (usually the city council) who has to prepare a management plan.

Sometimes powers apply specifically to organisations set up with the specific purpose of managing public open space, as in the case of the Minneapolis MPRB. Parks Victoria was also created in such a fashion, by legislation to manage the state's national, state, regional and recreational parks, providing services to the state or its agencies for the management of parks, reserves and waterways on public land.

Conservation powers offered some of the most robust powers available to the cities. In Århus, for example, 33 per cent of the municipality is affected by landscape protection legislation that imposes specific obligations and restrictions on open space management and requires the preparation and implementation of management plans. In Paris, if open space is classified under national heritage legislation, or is attached to, or indirectly connected with, a protected building or landscape, then the Architect of Historical Monuments in the Ministry of Cultural Affairs has a regulatory role.

Spatial planning powers were available in all cities. In Zürich, the key legal instruments framing the management of open spaces are the city's zoning plans. In Japan, the 1956 Urban Park Act also establishes what uses and activities are and are not allowed in open spaces and which of these require local government permission. In Wellington, under the planning legislation, the city council is responsible for providing consent even for tree removal and substantial pruning.

DEVOLVED POWERS

A further set of powers related to those that local management agencies were able to establish themselves, through powers devolved down to them from state or national governments. The key pieces of legislation for open spaces management in Curitiba are of this nature. Local legislation includes the city's Zoning and Land Use Plan (open spaces are defined as a specific land use), the transfer of development rights, and law giving protection to open areas through a conservation unit system, which includes public and private land. In the latter case, it might involve the transfer of development rights, agreed through negotiations with landowners and facilitated by fiscal incentives. Similarly, in Minneapolis, as an independent law-making authority, the Minneapolis Parks Board can enact ordinances addressing the use of parks, planting policies, standards for construction, and so forth, provided that they comply with US and state laws and with city ordinances.

In many countries, however, open space management has largely remained a non-statutory activity for local authorities. In France, for example, every local community has all the powers they require to manage or change public open space, but no statutory duty to do so. The city of Malmö is only required by law to ensure that parks do not pose a health and safety risk to the public. The fact that these cities still maintain their public open spaces to a high standard is testimony to the political priority given to public open space in each city.

At the level of localised, day-to-day management of open spaces, a range of powers exist in the eleven cities in addition to the broader powers described above. The responsibilities for enacting these powers vary between cities, as do the range of problems and their solutions.

The prime responsibility for detailed regulation of public open spaces in all the cities falls on the municipal authorities. Typically local byelaws form the basis for regulations dealing with such matters as litter and control on dogs, often as a complement to national legislation. Thus in Wellington, operational regulation of activities within open spaces is governed by reserve management plans prepared under national legislation to regulate public uses in each reserve, whilst the Wellington Consolidated Byelaw contains standard rules and provisions for all the city's open spaces.

ENFORCEMENT

Enforcement of local byelaws is also often a municipal responsibility. In Hannover, for example, the city's Environment and Green Spaces Department (FUS) is responsible for enforcing open spaces regulations. These are initiated variously by the city council and district councils, or are the result of higher level legislation. In Curitiba, the regulatory basis for the management of public spaces is almost exclusively municipal, and the responsibility for enforcing it falls with the Municipal Secretariat of the Environment (SMMA) and the municipal guard.

In Zürich, the city has clearly defined park and open spaces regulations for its territory and conducts a communication campaign to explain to park users what is and is not allowed, helped by a permanent, visible presence of maintenance staff in all key public open spaces. However, enforcement of regulations is the responsibility of the police. Indeed, the police have an important role to play in most of the eleven cities, and generally the relationship between city authority and police is viewed as an important partnership, with clearly prescribed roles for each party.

Common across the eleven cities was the use of parks keepers or managers in an enforcement role. The Department of Parks and Gardens in Paris is responsible for enforcing open space regulations throughout the city. Every park has at least one park keeper whose daily reports form the basis for the department's actions to tackle vandalism, safety issues, or, in the worst cases, to make structural changes in park layout (Box 8.3).

In Hannover, park managers within FUS are also responsible for ensuring that regulations are complied with. However, their role is more to observe and advise than to punish, and they operate closely with the police, social services and the youth services (particularly relevant in the case of anti-social behaviour).

In Wellington, a safe city programme for the city centre has included uniformed officers providing a visible and approachable patrolling presence in all public spaces. These services are contracted out to a local security firm. Volunteer rangers also assist fully paid rangers with patrolling and inspecting open spaces in the larger 'natural' areas, whilst in the most visible open spaces in Paris, park keepers are helped by municipal security. Although in many respects akin to the police, municipal security officers do not bear arms and are limited to patrolling the city's open spaces.

In Melbourne, Parks Victoria is responsible for administration and enforcement of a wide range of legislation. There, only authorised officers who are properly trained, including on how to use their discretion on whether to inform, educate, issue a warning, a penalty notice or pursue prosecution can conduct enforcement activities. Education and interpretation programmes are also used as an initial approach to achieve compliance with the regulations.

Only Minneapolis had the advantage of a dedicated force to police the city's parks. Parks regulations are enforced by the resident park keepers and by the city's Park Police Department whose role is to protect park users and park property. Park police officers are professionally trained police officers of the State of Minnesota and are responsible for visitor and resource protection, emergency services, maintenance of good order in parks, law enforcement, and information and public service.

RECURRING PROBLEMS

Three issues seemed to create the greatest range of enforcement problems across the eleven cities: anti-social behaviour, vandalism and dog-related problems. Significantly, however, they were never described as major problems, and instead were usually kept under control by efficient enforcement mechanisms and/or programmes of repair. Such problems are nevertheless most apparent in central areas because of the intensity of their use, corresponding with the fact that these areas are also the highest maintenance priority.

Anti-social behaviour is considered a problem particularly in Paris, Malmö and Zürich. In Zürich, however, negotiation rather than outright enforcement has been adopted by the city's Social Services as a means to resolve conflicts between different social groups and their use of parks. The approach has led to the 'Sip züri' initiative, a programme to encourage the coexistence of different groups in public space that relies on regular

BOX 8.3 PARIS

André Citröen Park

Paris (population 2 million, 10 million in the Île de France region) has a long tradition of open spaces, including forests, major parks and squares.

Centralised powers

Paris is perhaps unique for the historic importance of city government and the fact that all powers over open space are vested in the municipality. Since it gained the right to self-government in 1977, all decisions rest with the Mayor who also has fund-raising powers. Responsibility for open spaces is delegated to the Department of Parks and Gardens, under the Deputy Mayor for Green Spaces. All the resources come from the city's budget. Advertising is banned in the city's open spaces and income from leasing facilities goes back to the city treasury since French law forbids hypothecation of revenue.

Open space provision

Although 20 per cent of the city's area is open space, its distribution throughout Paris is uneven. In 1973, the Paris Region adopted a standard of 300 metres as the maximum distance to the nearest park, but at the time this excluded 75 per cent of the population. Successive administrations seized every opportunity to create new open spaces within deficient areas, almost doubling the overall open space area. Despite clear progress, the difficulties in achieving the 1973 target prompted the incoming mayor to adopt the more pragmatic 500 metres goal.

This has led to all opportunities being systematically considered, including the creation of new open spaces in all major urban renewal projects, but also the creation of micro open spaces by acquiring and demolishing derelict housing. Major urban renewal sites tend to be located in the peripheral districts of Paris whilst in the high-density areas most new provision relies on the micro open spaces. In such cases the municipality either purchases derelict housing with the specific purpose of creating open space (*dents creuses*) or alternatively they landscape the backyards of existing houses.

Antisocial behaviour

Traditional management has kept antisocial behaviour under control with very low levels of vandalism in the city's open spaces, despite the absence of community involvement in the decision-making process. This is a tribute to the traditional doctrine of discouraging these acts by repairing any damage without delay. Where vandalism or antisocial behaviour persists, the layout and design of open spaces are modified to address the problem.

Park keepers based in each garden are responsible for supervision, enforcement of regulations and locking the gates. In the last ten years the department has created its own parks police who are trained to intervene in conflict situations. This force is mainly deployed in tourist or problem areas.

meetings between the various conflicting parties and partnerships between the city authorities and key stakeholders.

Waste disposal and vandalism have been a problem in Zürich and Århus. In Zürich the solution has been a much more intensive programme of maintenance and cleaning in the heavily used lakeside parks where the problem is more intense. In Århus, solutions have included the employment of a gardener to travel around on a full-time basis to report problems and, if possible, to identify culprits who are then reported to the police. Theft of expensive plants has been a particular problem and is being solved by tagging plants with GPS chips in order to track their movement and arrest the culprits. Although vandalism is not a major issue in the Parisian parks, where it does occur, the solution has been to redesign the affected area in order to discourage or prevent it from happening again.

Dog fouling and other dog-related problems were reported in a number of cities. In Zürich, for example, efforts to regulate dog access to parks have failed. The alternative has been to discuss with representatives of all affected parties a set of measures that will have broad acceptance, emphasising the need to involve key interest groups in decision-making if regulation is to be effective. In Malmö, there are no special programmes to deal with the issue, but better information and facilities have helped to alleviate the problems it causes. In Wellington, a council policy document – the Dog Control Policy – sets out the responsibilities of dog owners and establishes the areas that dogs are allowed to use.

Monitoring open space

In all the cities, monitoring was both a citywide and site-specific activity. The former focused on the effectiveness of the urban management systems and public opinion, and the latter on the success or otherwise of managing specific open spaces.

A number of the cities employed GIS systems as a continually updated record of the condition of their open space resources. In Århus, for example, management systems allow for the continuous electronic updating of plans, programmes and budgets. In Malmö, all areas managed by the Streets and Parks Department are logged into a GIS system containing data on the location, the characteristics of the area itself, and maintenance routines. This is used to inform maintenance plans and budgets.

Inspection regimes are used in Paris as an additional layer of monitoring conducted by a special body – the Inspectors – within the Department of Gardens and Green Spaces. In Minneapolis, parks are monitored daily by their resident park keepers for hazards and maintenance problems, whilst periodic inspections by crew leaders and the district foreman are intended to keep park keepers motivated. More complete and rigorous

inspections of all parks are conducted semi-annually by the Director of Park Operations and the Maintenance Supervisors.

The most sophisticated systems employ a range of measurement systems to carefully monitor and record the conditions of public open space. In Groningen, the Beheer Openbare Ruimte Groningen (BORG) system of management information for open spaces links management options directly to visualised target scenarios (Box 8.4). It also allows the condition of open spaces to be regularly recorded or the success of management policies and processes to be assessed on the basis of clearly specified and visualised quality thresholds. In Melbourne, Parks Victoria uses an asset management system to record the condition of their parks. The system is based on a comprehensive database covering the value, condition, life-expectancy and future maintenance requirements of each park, information which is then used to compare maintenance levels with industry standards and to calculate asset replacement costs.

The asset management system used in Wellington is also effective at evaluating the durability and physical condition of the city's parks, particularly their furniture, paving and planting. The system has therefore proved to be a useful tool to recognise trends such as consistent damage to particular types of equipment or consistent failures of particular aspects of maintenance.

A further important category of monitoring occurs through the various methods used to gauge citizens' opinion on open space quality and its management. Two basic approaches were found. The first were dedicated complaints management systems, with direct accountability to complainants, as well as inputs to internal management practices. User complaints in Curitiba are dealt with by a 24-hour helpline that manages complaints and queries related to a broad range of municipal services, not just open space. Complainants and municipal staff can follow progress of the complaint through the various levels of the administration. Similarly, complaints by the public in Hannover are managed through a citywide complaints management system that includes prescribed times for complaints to be answered. Complainants are routinely kept informed of progress.

The second are internal feedback systems, in which users' views were used primarily as a way of reorienting internal management processes. In Malmö, the Customer Services Division within the city's Streets and Parks Department deals with all complaints and comments from residents. This information feeds into a three-yearly performance evaluation of all private contractors. Good feedback triggers automatic extension clauses to come into play, thereby extending contracts for a further two years. In Melbourne, Parks Victoria relies on regular surveys of visitor opinions and telephone interviews to gauge the awareness of, and satisfaction with, the services provided. These surveys are also used to develop predictive models to access the likely impact of changes in management strategies

BOX 8.4 GRONINGEN

Scarce inner-city green space

Groningen (population 177,000) is known throughout the Netherlands for its progressive policy on public spaces. It became the first city in the mid-1970s to give priority to people over cars and to subscribe to the concept of neighbourhood-based services.

The BORG management system

At the end of the 1980s various developments at the national level led to a shift from a centrally managed, sector-based approach, to an emphasis on consultation and civic participation, an approach that became known as neighbourhood-based services.

However, experience showed that focusing services on the local area had been taken too far and the introduction of a complaints hotline in particular had lead to ad-hoc, reactive problem solving, at the expense of regular, planned work.

Moreover, devolved management and neighbourhood-based services were resulting in inefficient management and to wide discrepancies in the state of repair of public open space in the various neighbourhoods throughout the city.

A policy reversal was adopted and management programmes were once more determined centrally. This resulted in a new system of management, known as 'Groningen Public Space Management' (Beheer Openbare Ruimte Groningen – BORG) through which the condition of all public open space can be regularly evaluated, tailored management programmes can be developed, results can be verified and, more importantly, opened up for discussion.

Under BORG management, desired outcomes are informed by visualised target scenarios. The town is also divided into structural elements such as the city centre, parks, trading estates, and so on, whilst the management quality to be attained in each structural element is established on the basis of both photographs and predetermined criteria. This enables all stakeholders, experts and lay people, to agree on the desired quality of public open space management and to determine precisely the results that need to be achieved. The system can also assess the effect of damage and pollution on the visual quality of public space.

Experience has shown that the link between outputs and projected costs established through BORG has given open space managers a greater level of trust in budget applications. This has to some extent also reduced threats of cuts in the budget for open space as it becomes easier to demonstrate the benefits for the city as a whole of open space management expenditure.

and practices. Wellington also conducts regular public satisfaction surveys on its various services, whilst the Parks and Gardens Business Unit has started to conduct its own visitor surveys to provide direct feedback on its problems and successes. Similarly, the Department of Park Development in Tokyo has recently adopted Internet surveys backed by site surveys on specific initiatives as a means to monitor public opinion on public open space management.

Open space maintenance

Different policy aspirations and responsibilities for open space management eventually make themselves felt on the ground through day-to-day maintenance and periodic reinvestment. Maintenance processes relate to the ongoing care of public open spaces to maintain their quality. Reinvestment processes relate to the far less frequent decisions to totally or partially renew public open space infrastructure. In this area, the exact nature of the delivery processes are inevitably shaped by the specific nature of each open space, whilst their effectiveness depends on how well they adapt to each local situation.

Maintenance delivery

The large majority of time, resources and expertise of the eleven cities' public open space managers was spent on maintenance work, which, because of its widespread impact, has potentially a much greater contribution to make to environmental quality than comparatively rare reinvestment activities. To guide this process a number of the cities prepare specific maintenance plans to guide the operational delivery of open space management. Such plans allow long-term maintenance priorities to be established and properly resourced, and for key policy priorities to be interpreted in the context of everyday responsibilities. In Århus, maintenance is undertaken on the basis of four maintenance districts, and a general park maintenance plan, in combination with detailed maps of each locality. This provides the basis for operational work. In Paris, maintenance plans are prepared based on the natural agendas of gardens and plants and on reports by caretakers and park security staff. However, the most sophisticated maintenance planning approaches are found in Melbourne, Wellington and Groningen.

Parks Victoria's levels of service (LOS) framework is a key management tool used to establish the 'optimum' quantity and mix of visitor services, given forecasts of user demand and availability of resources. It uses data on visitors, on the park assets and on available resources to define service standards across the different park settings, ensure that resourcing decisions match visitors' demands and to balance those against the capacity of Parks Victoria to meet them. Through the process, the system defines the kinds of maintenance services applicable to each park (see Box 8.7).

The Asset Management section of the Wellington Parks and Gardens Unit uses asset management software to programme maintenance, inspections, replacement and funding under a number of asset management plans (Box 8.5). These plans have improved the ability to recognise trends in the performance of open space facilities and equipment. Links between the council's GIS database and the asset management database has proved particularly useful in helping to locate and check overlapping areas of responsibility.

In Groningen, the BORG system uses visualised maintenance scenarios in the form of actual images of how an open space should look, depending on the level of quality and intensity of the maintenance regime selected (see Box 8.4). Intended results of management action can then be assessed and discussed by experts and lay people. The system allows different types of open spaces to be managed to suit their particular requirements.

Each city organised the routine delivery of open space maintenance in their own way. In Paris, approaches to maintenance are decided at a more strategic level by staff managers in the Department of Parks and Gardens. Much of the work is based on routine patterns, but the strategic approach means that the department is also able to react promptly to emergencies and can quickly re-design routines and practices and re-deploy staff.

In Århus, operational staff of the Natural Environment Directorate (NED) are subdivided into four district groups. Within each district, smaller groups are responsible for specific geographic areas. Annual meetings between district staff help to link site-level action to overall citywide plans and policies. Similarly, Minneapolis is divided into four districts to facilitate maintenance in a large and diverse park system, and maintenance at the level of individual parks is carried out by park keepers assigned to specific geographical areas.

Although the Parks and Environment Secretariat (SMMA) is directly responsible for open space maintenance in Curitiba, the task is shared with the Public Works Secretariat (SMOP) and the service units of eight district administrations. These agencies have specialised teams to look after streets and squares and are contracted by SMMA to do so. SMMA directly maintains the larger parks and woods. Responsibility for day-to-day management programmes in Groningen lies with the Public Green Space Team. Their job is to ensure that aspirations laid down in the management quality plan are fulfilled through proper specifications, monitoring contractors work, and supervising jobs. Target specifications are formulated locally on the basis of BORG parameters and the expertise of municipal staff.

BOX 8.5 WELLINGTON

Wellington's Park Rangers

Wellington (population 175,000), New Zealand's capital, was covered in forest until the Europeans settled in 1840. Although large tracts of open space remain, partly due to the Town Belt, they are under development pressure.

Asset management

During the early 1990s, the condition of many of the city's open spaces was deteriorating due to long-term deferred maintenance, but long-term budgeting and asset management plans introduced in the late 1990s improved the situation. Wellington's Parks and Gardens Unit use asset management software to programme maintenance, inspections, replacements and funding. Standard life expectancies cannot be applied due to unpredictable factors (differing site conditions, political demands and public complaints) therefore all assets are inspected, their condition assessed, asset management plans are prepared and priorities set.

These asset-management plans have proved useful in providing documented justification for securing funding. Another useful outcome has been an improved ability to recognise trends in terms of depreciation and maintenance needs. The advent of ten-year financial planning also allows for commitment to long-term works.

Inner city greening

Although Wellington has 200m²/person of open space, its historic development resulted in a serious deficiency within the city centre. However, the council can increase provision in the inner city through powers under the 1991 Resource Management Act, which requires developers to set aside land as reserve contributions or pay into an acquisition fund.

During the 1980s to early 1990s when high-rise development replaced older buildings, the council negotiated open space provision through development control, allowing increased building height in return for on-site open space. However, public tenure of the open space was not secured at the time of negotiation and some of the sites were not suitable. Consequently, several of these spaces have been built over and negotiated rights are out of favour. Given the high cost of city centre land, Wellington has had to revisit mechanisms to improve the distribution and quality of inner city open spaces.

More recent projects such as the waterfront development are leading the way towards the next phase. Two decades ago Wellington's port activities relocated, and the waterfront adjacent to the city centre opened up to the public. When it came up for redevelopment, the controversy surrounding the proposed plans raised the issue of balance between buildings and public open spaces. In the late 1990s, the original proposal was replaced by a Waterfront Development Framework. The lessons were that the amount of development needed for the operation to be self-funding was unacceptable to the community and that only additional public funding could ensure open space provision and reassure the community about its continued vested interest in the area.

OUTSOURCED APPROACHES

In the context of the ubiquitous pressures on resources, each of the cities are striving to deliver more efficient maintenance services (i.e. more service for less resources). This has often meant the contracting out of previously in-house maintenance operations. Curitiba, Groningen, Malmö and Tokyo have all gone down this path.

Many of the operational activities of the SMMA in Curitiba are outsourced, including the maintenance of open spaces. In the case of larger parks and woods, maintenance procedures and standards are defined on an annual or bi-annual basis by SMMA, and this becomes part of the contract put out to tender. The maintenance packages take into account seasonal variations and include an inspection regime by SMMA staff.

For the last few years Groningen has worked with target specifications. In the BORG system the contractor is free to choose the inputs and the kinds of expertise to be deployed, but has to meet carefully prescribed visual outcomes. The system relies on the selection of experienced contractors and works well only when the contractor is familiar with the area and can estimate the nature of the tasks correctly. Malmö Streets and Parks department employs both private and public contractors working to maintenance standards defined by the Streets and Parks Department. The system relies on close cooperation between the city and contractors, with contractors expected to take the initiative in innovating and improving their practices (Box 8.6).

The Tokyo Parks Association – the large public corporation in charge of maintaining 64 out of the 76 large parks in Tokyo – has also adopted competitive practices for contracting-out maintenance work. In this instance, the approach follows recent policy directives emanating from central government, with new directives requiring local government to adopt more business-like approaches coming into force.

The experience of the cities suggests that of critical importance is the need to view contracting out as a mutually supporting and long-term partnership and not as simply a way of driving down costs in the short term. Also important is the need to specify quality expectations as carefully as price on the basis of outcomes rather than inputs, and to monitor delivery.

IN-HOUSE APPROACHES

Not all of the cities had gone down the path of contracting out maintenance responsibilities. In Minneapolis, for example, most maintenance work is still conducted by in-house teams. The benefit has been the generation of a high and enduring sense of responsibility for the city's parks amongst the in-house staff. The downside has been the relatively high cost to operate the system.

Other cities have retained their maintenance work in-house but have been innovative in the way they pursue efficiency gains. Wellington is quite unusual in New Zealand for having retained in-house operational parks functions. One of the major benefits has been the flexibility to respond to unexpected needs without the need to renegotiate external contracts. To achieve adequate standards of efficiency with in-house services, the council manages much of its operational responsibilities as 'business units' whose standards of service can be compared against benchmarks and who are run along self-contained business management lines. In addition, maintenance programmes in Wellington are carried out by mobile specialised park and gardens crews operating throughout the city from a central depot. Centralising staff into functionally-specialised teams has proved more efficient, with less idle time for equipment and a general improvement of skills and knowledge through specialisation.

In Århus, a new profit-sharing approach has been implemented in the NED in order to encourage in-house operational staff to cope with difficult maintenance tasks. Savings from an accepted contract describing aims, budget, timeframe and so forth may be either shared among staff or used for investment in the district's equipment. This is part of a series of measures taken to restructure open space management services in a context in which outright outsourcing is not viewed as desirable. The approach has been to involve private contractors more widely whilst giving the internal units the chance to bid for work. The municipal open space maintenance unit has therefore been reorganised as a contractor arm of NED and has to tender against private contractors for maintenance work.

In Hannover, although the majority of work is undertaken by public sector employees, seasonal working peaks are often met by using private contractors. There is also a flexible system of working time within the city's Environment and Green Space Division (FUS) that has been accepted by employees and which assumes longer hours at certain times of year in exchange for shorter hours elsewhere.

Responsiveness, reflecting local needs

To a greater or lesser extent, all the cities have attempted to be responsive to the needs of different types of public open space, recognising that they present different maintenance problems and require different solutions. The problems associated with small open spaces in inner urban locations, for example, have presented particular challenges in a number of the cities. This reflects the intensity of maintenance required, which is of a different order altogether to that required in outer areas, or in larger parks, and to which centralised management systems seem to have difficulty in adjusting.

BOX 8.6 MALMÖ

Malmö parks, designed for reduced maintenance

Malmö (population 250,000) is Sweden's third largest city. An industrial centre until the 1990s, the city has experienced an increase in immigration over the last decade, resulting, for the first time, in a housing shortage that places new development pressures on the city.

The Green Plan

Malmö's Streets and Parks Department, one of the most successful in Sweden, has received a number of national awards. As part of the process of preparing the General (land use) Plan, Malmö produced a Green Plan providing guidelines for future requirements for all open areas within the city. The Green Plan aims to ensure adequate provision and distribution of urban parks and to protect existing open spaces from development. The Green Plan is not legally binding but serves as guide in decision making. Its impact relies on its acceptance by the key stakeholders and the relevant departments' commitment to its implementation. Communicating effectively is therefore an important aspect of the department's work and it has been very successful in marketing the value of parks as a way of improving the quality of life in the city.

The city has powers to acquire land and to negotiate and agree with other landowners to develop areas for public recreation. In Sweden, all natural areas are publicly accessible irrespective of ownership, but agreements can improve accessibility and provide facilities for visitors. One such agreement with the Water Authority allowed the integration of ponds and canals in parks, increasing water-based recreation and biological diversity, with costs borne by the Water Authority who benefited from a less expensive option for managing storm water.

Contracting out maintenance

Maintenance operations are financed entirely by the city. The Streets and Parks Department employs both private and municipal contractors; over the years it has progressively increased the demands on the contractors' expertise and as a consequence, they have shouldered increasing responsibility for delivering quality.

The department sets the standards and the contractors are responsible for their implementation and the coordination of operations. For the last decade there has been a gradual move from the issuing of specific instructions to contractors towards a more flexible system based on them achieving the department's broadly defined key goals. Contractors are encouraged to take initiatives to deliver continuous improvement, which should improve their chances to be awarded future contracts. Although each maintenance area of the city has a manager who is the contact for the area's contractors and acts as a supervisor, it is the responsibility of the contractor to oversee his own activities and to report any problems to the city council. This demands skilled contractors and good communications between the commissioning body and the contractor and should lead to mutual learning.

Maintenance routines are thus closely related to local context. In Hannover, standardised approaches are not used, and instead regimes are determined by the special character and function of individual open spaces. Therefore, more complex approaches are used in the iconic Herrenhausen gardens, and more intense daily routines are implemented in the summer along the city's lakes and canals. Location-specific maintenance is also part of general practice in Minneapolis. The lawn-mowing programme, for example, is divided into different categories of open spaces depending on the required intensity and frequency of mowing, taking into account dominant uses and the nature of each open space, cultural features, ecological conditions and the regional and historic context.

A logical progression of these more locally responsive approaches has been the devolution of responsibilities to levels below the citywide scale. In Hannover, maintenance groups are responsible for individual sites or small groups of sites, and carry out all the maintenance work in them. In Minneapolis, strategic decisions on park services are made at a regional or district level with the coordination of contractors or internal staff on the ground being carried out by the respective park managers. In Paris, operational staff are attached to geographical areas of the city, and are responsible for day-to-day maintenance in those areas. In addition, each park has at least one dedicated park keeper responsible for a range of day-to-day management functions.

Despite the benefits that such approaches bring through the greater tailoring of management regimes to local circumstances and the greater responsibility felt by local staff, they have not been without their problems. In Groningen, the emphasis on devolved management led to wide discrepancies in the state of repair of open spaces throughout the city and so in the mid-1990s greater centralisation was adopted. Management programmes are now determined centrally, following local consultation. Within the maintenance unit of Zürich's Green Planning Office (GSZ) there are still open space managers for every city district who are in charge of the day-to-day maintenance of open spaces in their areas. However, there has increasingly been a drive towards citywide specialist teams and away from geographically-based teams in order to drive up efficiency through optimising the use of specialist machinery, and through raising the skill levels of specialist staff.

Investing in open space management

The quality of open spaces is related to the size of budgets for management and maintenance and to the efficiency with which financial resources are utilised. In a general context of reduction in public expenditure for parks and open spaces, the issue of alternative sources of funding becomes a priority. However, money is not the only part of the equation. The quality of public open space management also depends on the recruitment and retention of staff with adequate skills, both at management and operational levels.

Funding open space management

Two basic forms of funding open spaces management were available to the cities: core funding, more often than not biased towards revenue expenditure; and supplementary funding, often with a capital expenditure bias. Most cities utilise both.

CORE FUNDING

The primary sources of core funding are local tax revenues and recurrent central/state government grants. Although core funding levels have not fallen dramatically anywhere, few of the eleven cities could achieve all they wished through core funding only, with investment and reinvestment in capital works often the chief casualty. In addition, the general state of public finances across the world seems to have placed a squeeze on recurrent maintenance activities.

Two basic approaches to core funding were found in the eleven cases, the first of which is by far the most common and takes the form of an allocation from the general municipal budget, for which the management of public open space has to make its case alongside a multitude of other calls on that same budget. As an example, in Århus the management of open spaces is funded through municipal tax revenue, with allocation decided by the city council (Municipal Board). Funding has so far been adequate for the maintenance of existing open spaces, but funding for capital projects (renovation and new parks) is more difficult to come by, has to be especially applied for by NED to the city council, and is not always forthcoming. In Curitiba, most of the resources for public open space management come from the municipal budget made up of municipal taxes and federal and state transfers. Within SMMA, allocation to the different divisions is undertaken by the head of the agency according to the administration's priorities, although usually there is not enough for all priorities. A similar situation was found in most of the cities.

Dependence on the general municipal budget often brings with it the threat of funding cuts as more pressing needs make themselves felt. In Groningen, the high annual expenditure on open space management has tended to make it a popular target for cuts, and in recent years there has been very little scope for new investment in existing or new open spaces.

BOX 8.7 MELBOURNE

Albert Park Lane

Melbourne (population 3.5 million), Victoria's capital, is Australia's second largest city with an extensive integrated network of open spaces. Virtually all the open space is Crown Land but managed by various tiers of government.

Parks Victoria

Parks Victoria is statutorily responsible for managing 40 per cent (6,200 hectares) of the network of open space within metropolitan Melbourne (the rest falling under the jurisdiction of local councils) as well as national and state parks around the metropolitan fringe. The agency was created in 1996 from the amalgamation of the Victoria National Parks Service and Melbourne Parks and Waterways, to manage most of Victoria's national, state, regional and recreational parks. Through the merger, declining funding levels could be maximised by directing resource priorities across the whole system and eliminating duplicated services between government organisations.

Key output focused groups have been identified to describe Park Victoria's service delivery obligations to government, namely Natural Values Management, Cultural Values Management, Wildfire and Other Emergencies Management, and Visitor Services Management. The latter is directly responsible for the management of open spaces.

Funding the metropolitan park network

The primary source of funding for Parks Victoria's metropolitan parks is revenue from a 'parks charge' levied on all domestic, commercial and industrial properties within metropolitan Melbourne, and collected and administered by the state government. Parks Victoria receives about two-thirds to spend on its corporate governance and the management of open spaces. Even with this discrete funding, Parks Victoria continually needs to present its case to government for additional funds to meet increasing costs and the growing scale of its asset maintenance/replacement liability.

The 'levels of service' (LOS) framework

The delivery of sustainable visitor services and facilities with limited resources requires a strategic context for the management and creation of built assets. Parks Victoria has developed the LOS framework to establish the 'optimum' quantity and mix of visitor services, given forecast user demand and the level of resources available.

The LOS framework uses a comprehensive, regularly updated, database of visitors, assets and resources, which includes the value, condition, life expectancy and future maintenance requirements of the built assets to develop optimum approaches for each park according to its relative priority in a park-wide context. This process quantifies the gap between model levels of service and actual levels for each park, generating appropriate service level scenarios. When applied to determine future asset replacement costs, it indicated that Parks Victoria is significantly under-spending on maintenance, and is facing major replacement costs in the next ten years.

In Malmö, severe budget cuts across Swedish municipalities over the last two decades have impacted strongly on all non-statutory municipal services such as parks management. Similarly, new tax laws affecting German local governments have made the financial situation for urban open space management in Hannover increasingly tough.

The second and much less frequent approach to core funding involves monies gathered specifically for the management of public open space, and hypothecated for that purpose. Although this approach does not provide a guarantee that adequate funding will be forthcoming, it nevertheless secures a more transparent collection and expenditure process, and decisions about funding are not played off to the same extent against other calls on the public purse.

The Minneapolis city charter gives the Park Board the authority to levy a tax on residential property and this hypothecated tax revenue is supplemented by state allocations under the Local Government Aid programme. In Melbourne, the primary source of funding for Parks Victoria is a Parks Charge levied on all residential, commercial and industrial property in the metropolitan area. The charge is collected and administered by the state government, which distributes the revenue among all relevant organisations (Box 8.7).

SUPPLEMENTARY FUNDING

Supplementary funding, the second form of open space management funding, comes from a wide variety of sources. Although generally much smaller in quantity than core funding, these resources were particularly welcomed for the ability they provided to enhance the level of the general service, to fund capital investments, and to help establish better connections to the community of open space beneficiaries – including the business community. Århus has been particularly successful in supplementing its budget through utilising the local and national interest in the protection of water resources and the environment in general to lever EU funds for forestation schemes. In Curitiba, pollution-related fines administered by SMMA have been used to fund capital projects, while subsidies from federal and state governments and tax incentives have also been used to attract private money.

In its various guises planning gain has also been a supplementary source of funding in Groningen, Wellington and Zürich. In Groningen, all infrastructure associated with new residential developments, including open spaces, must be paid for from revenues generated from the sale of the houses. In Wellington, land development levies are used in a similar fashion, whilst in Zürich developers donate the land and pay for the implementation of new open spaces in exchange for zoning bonuses. In this case, the system supplements core funding which has been increasingly

squeezed to make it go as far as possible (Box 8.8). In all these cases, the city administration subsequently takes over the management of the new open spaces.

In Minneapolis, in the search for reliable, long-term, non-tax streams to supplement its income MPRB has been looking at private fundraising and fees and charges for services as means to raise income. For private fundraising the Board has worked with the Minneapolis Park Foundation, a charity whose aim is to solicit private funds for the development and maintenance of Minneapolis's parks. Revenue-generating public-private partnerships are also being explored. In Tokyo, following a general reduction in resources available from central government for the management of open spaces, a private finance initiative-type scheme has been introduced as a means to fill the gap. The monies generally only relate to new capital projects and their subsequent management.

Basic versions of partnerships, through private sponsorship of space are found in Hannover, Malmö and Curitiba, usually for special projects. Although total contributions are not financially significant, they are often politically important and help to strengthen connections with communities. In Melbourne, Wellington and Curitiba, income is also derived from rents and licenses to private operators, but the total is small, and further limited in the case of Melbourne by government policy restricting the introduction of market rates. In Paris, some income is generated through granting licenses to private businesses to run sports facilities, restaurants, cafés and events in the city's public open space. However, French law forbids the ring-fencing of revenue streams and, as a consequence, this revenue goes into the municipal budget as a whole, and not back into open spaces.

Finally, a number of the cities have been able to fund capital works through urban regeneration schemes involving regional, national and even supra-national funding. These include capital investment in the open spaces of older urban areas in Groningen, or the significant government resources used for improving living conditions in Malmö, including the renovation of parks in deprived areas around high-rise estates.

REINVESTMENT PROCESSES

Both core and supplementary funding fed into capital reinvestment projects, as well as ongoing maintenance. Indeed, processes of reinvestment were not always seen as distinct from day-to-day maintenance processes, but rather as degrees along a continuum of caring for public open space. Thus some tasks need daily attention, others are on much longer time frames up to many years, as and when reinvestments need to be made. The latter are nevertheless generally funded through different mechanisms, which many of the cities were finding it increasingly difficult to secure.

BOX 8.8 ZÜRICH

MFO Park on a former industrial brownfield site

Zürich (population 361,000) is Switzerland's largest city, situated at the northern end of Lake Zürich. It is surrounded by wooded hills that have been protected since the nineteenth century.

Grün Stadt Zürich (GSZ)

Compared with other cities, Zürich is always rated very highly for the quality of life it offers, enabling the business community to attract employees. Surveys indicate that open areas and parks rate (together with public transport) as the main reasons for this high quality. This success is attributed to the combined efforts of the administration and the politicians, and has been made possible through the workings of GSZ. This organisation is part of the city's Infrastructure Department and has been in existence for over a hundred years. It is responsible, in cooperation with other agencies, for the planning and management of open spaces.

Provision of open space

The main problem facing the administration is the unequal distribution of open spaces within Zürich. In 1999 it developed the Open Space Concept, which proposed public open spaces in all districts at no more than 400 metres from every household.

GSZ's planning activities focus on the redevelopment of former industrial sites or problem estates. Zürich North is an example of former industrial land for which guidelines for mixed-use development were prepared in 1991 preserving an existing park and creating new open spaces. By agreement with the city, local landowners provide these new urban parks as they are seen as positive identification factors that add value to their developments. Ownership and future management are subsequently transferred to the city. Turbinenplatz and Oerlikerpark are two of the completed new parks in Zürich North.

Cost transparency calculation

GSZ recently introduced cost transparency calculation as a management tool to determine the costs and effects for every 'product'. GSZ services are divided into five product groups (open spaces, nature areas, nature enhancement, management, services) each with a manager. Individual district managers have to 'sell' their workforce to the product manager to achieve the high-level maintenance to which they aspire. This system seeks to achieve internal competition and transparency and is expected to lead to cost efficiency.

As GSZ relies on a fixed budget, any efficiency savings are split between the city of Zürich and GSZ. Reports are prepared twice a year for each 'product' group to give a financial overview. Besides assisting management to stay within the limits of the overall budget, the system is designed to make the administration more service-oriented.

The cities exhibit a range of approaches for assessing reinvestment needs, although none had systems in place to automatically track the depreciation of open space assets in order that long-term investment needs can run in parallel with day-to-day maintenance requirements. Instead, the standard approach is for open space units in their various guises to make annual bids for capital expenditure. In Århus, for example, the need for reinvestment in open spaces is initially decided on the basis of agreement between the leaders of the different units within NED. Bids are next cleared with other municipal departments and accepted by the relevant city councillor before being presented for approval to the City Board. Hannover operates a similar process. Each section of FUS is responsible for planning the necessary reinvestment. Their requirements are sent to a central analysis group in the Finance Department, which advises the municipal cabinet in their final decision about budgetary allocation. However, neither system guarantees that requests for funding will be met.

Malmö takes a more systematic approach to reinvestment in their parks and major new investments are usually preceded by a thematic review, for example focusing on city playgrounds. These reviews are dictated by local political agendas, but they enable systematic consideration to be given to the investment needs in a particular area. In Melbourne and Zürich the new management tools reveal the need for reinvestments just as they reveal maintenance needs. In Melbourne, the LOTS framework identifies the need for immediate and long-term decisions to be made on asset maintenance and renewal that reflect both workforce and organisational objectives. In Zürich, decisions on new investment are based on the classification of open space services under product groups, where maintenance and reinvestment priorities can be prioritised.

In Wellington, changes have separated regular maintenance regimes from one-off capital projects. Reinvestment is now managed under the Asset Management Section of the Parks and Gardens Unit. Previously, funding for major open space projects was vulnerable to emerging political priorities and to funding allocations made on a year-to-year basis. Now, with the advent of long-term (ten-year) financial planning, managers' ability to forward plan has been greatly improved and should result in more consistent investment in new and refurbished public open space (see Box 8.5).

A significant trend was the greater consideration to lifetime approaches in investment decisions, with ongoing maintenance costs becoming an increasingly important concern when allocating funding. The experience in Groningen provides a case in point, with the recognition of a general lack of coordination between annually-set maintenance budgets and the maintenance tasks derived from one-off capital investments funded through urban regeneration and housing sales. Open space managers now routinely participate in the development process and are able to project the long-term consequences of different design options, consequences that will eventually make themselves felt on their budgets.

Other cities exhibit a similar concern. In Århus, cooperation between departments of the city authority over new open spaces starts at the project level, ensuring that there is a maintenance input from the very beginning. In Malmö, new projects have aimed to improve quality and reduce maintenance costs at the same time, and both those planning new investments and those responsible for overseeing day-to-day maintenance participate in the formulation of new projects. In Hannover, because divisions within FUS are responsible for both investment and day-to-day maintenance, long-term management issues are considered for all capital investment proposals.

The skills required

A strong theme running though the international cases was the emphasis placed on skills and skills development; both at management and operational levels. The Natural Environment Division of Århus City Council has a strong body of professionals ranging from landscape architects to foresters, botanists and trained gardeners who work in both the administrative and operational sections of the division. These skills are supplemented by those of architects and engineers who work in other parts of the organisation. A focus on ecology in the 1980s and 1990s led to the appointment of biologists and a change in the skills profile, with a consequent change in management practices. Many members of the council also have professional backgrounds and their skills are used in initiating, carrying out and managing projects.

In Hannover, most of the leading staff at the Environment and Green Space Division have a professional background in horticulture or landscape architecture and managers in the division are trained in new management methods. At lower levels, most managers have gone through technical colleges, and specific skills are also sought at the operational levels: cemetery gardeners, perennial gardeners, foresters, and so forth. In Malmö, the overall planning of parks is carried out by architects and landscape architects at the Streets and Parks Department and the City Planning Office. At the operational level, many park keepers have gone through horticultural sciences courses at further education level. Similarly, in Zürich, trained landscape architects are employed by the Green Planning Office of the City Council for planning and management, whilst at the operational level trained gardeners and other specialists are used, many having graduated from the Council's own apprenticeship scheme. In Paris, 'The Grid of Jobs' carefully defines all administrative positions and the qualifications and practice-based experience required for each.

TRAINING NEEDS

The ongoing training of employees was a priority for most of the cities. In Tokyo, the approach has been that workers pick up their skills in park management through doing the job. Nevertheless, because differential skill levels have been a problem, the government plans to provide a comprehensive training system that will ensure similar skill standards across the park system.

In Melbourne, operational staff already undergo a training regime covering core competencies, plus education skills and personal effectiveness. Middle management, by contrast, participate in a 'focused manager' programme, whilst Parks Victoria has initiated a degree course in park management at Deakin University, and actively supports the programme through curriculum input, lecturing and a scholarship scheme.

In Curitiba, the municipality has gone furthest, creating the Free University of the Environment (Unilivre). Its Reference Centre for the Management of the Urban Environment has helped to improve the knowledge of municipal professionals and acts as a reference point for the exchange of experience and research. However, in spite of the initiative, a lack of clearly defined policies on skills has meant that there are few incentives for lower-level staff to upgrade their skills.

In Malmö, there is no shortage of essential skills and training programmes at management levels. The main problem is that for a long time manual labour in parks maintenance in Sweden had a very low status. This led, over time, to low expectations on parks workers and to low performance. Although the Parks Department has been actively investing in a range of dedicated courses for their staff, municipal parks organisations across Sweden still suffer from the effects of the earlier approach.

By contrast, the benefit of a positive approach to skills and training was visible in the number of long-serving staff in some of the cities. In Århus, a recent study on the skills of long-serving staff showed that a major reason for the success of NED has been the acquired skills of its employees to manoeuvre in the political environment. Detailed knowledge of the key people, places and funding possibilities has helped to ensure that the right decisions are made at the right times. Similarly, one of the reasons given for the success of the Parks Board in Minneapolis is the cadre of longstanding senior employees who, between them, have a vast knowledge of the board's historic practices. There are now efforts to record and systematise the knowledge of long-serving staff so that it will not disappear when these individuals retire.

Conclusions

Coordination of public open space management activities

The first lesson that emerges from the international cases is that open space management remains primarily a local government responsibility along the state-centred model (see Chapter 4), and more often than not, local decision-makers and especially local politicians hold the ultimate responsibility. However, fragmentation is a common phenomenon. Indeed, with the notable exception of Minneapolis, the evolution of open space-related services in the different contexts has been marked in the past by increasing fragmentation.

In a few of the cities this has now been substantially reversed through relatively recent amalgamation of responsibilities, leading to organisations in charge of all aspects of open space management. In the majority of cases, formal responsibilities for open space management remain fragmented and dispersed among divisions within a municipal department, between different levels of government and between mainstream public services and special purpose agencies.

However, the fact that they have managed to achieve good results in complex institutional environments suggests that the way different management responsibilities are coordinated is probably more important for the quality of management and open space, than the formal distribution of those responsibilities. The many examples of effective delegation arrangements, multi-divisional strategic plans, service agreements between departments, and so forth, corroborate the point. The key message is therefore that although it would be ideal to have a management structure that replicated the integration and independence found in Minneapolis, Paris and to some extent Melbourne, it is the other cases that suggest more widely applicable lessons.

In the majority of the eleven cities, open spaces management is carried out by a municipal parks/open spaces department, often as part of a larger directorate, which is responsible for most but not all management tasks and has to liaise with other bodies within and outside the municipal administration. In this regard, two points are of particular relevance:

- First, it is the quality of the working relationships between those with responsibility for open space management that is the most important variable in influencing the better coordination of separate open space responsibilities and interventions. Having all key players under the same organisational structure does help, but good coordination can be achieved where this is not the case, as in Århus, Zürich, Malmö and Wellington.

• Second, in nearly all of the cities there have been conscious efforts to remove organisational barriers to inter-departmental cooperation. This ranges from merging departments to delegating responsibilities, setting up fora, or using higher-level authorities to smooth out conflicts and secure coordinated actions. How this has been done depends on historical accidents, the adaptability of existing structures, political will and the advantages and disadvantages of different courses of action.

The key lesson is that there are a number of alternative paths to better integration and coordination of open space management, each with its own costs and benefits that need to be weighed up in light of local circumstances.

Beyond the public sector, the cities indicate no single approach to coordinating the involvement of private-sector players in the management of open spaces. The eleven cities have adopted different attitudes towards how much private-sector involvement they allow and this has been a function of political preferences, employment practices and cost minimisation/service rationalisation policies. The general trend, however, has been towards the contracting out of at least some management tasks. Where this has been most effective in enhancing the quality of open space management it is because there are clear structures to manage the relationship between public bodies and private contractors, as in Århus, Malmö and Groningen. It seems that an explicit concern to strike a balance between quality outputs and a competitive environment is important for success, together with adequate monitoring of standards and vetting of contractors.

It is also clear from the cases that there are advantages and disadvantages in involving the private sector. The cities were conscious of the cost benefits of contracting out management tasks and some have explored them extensively. However, others have acted more cautiously in order to retain the benefits of an experienced in-house service. Melbourne has tried to separate contexts in which private contractors should be used from those where in-house services are more appropriate. In a few cases the solution has been to transform in-house service providers themselves into competitive contractors, with good results.

Similarly, the participation of voluntary-sector stakeholders in open space management across the cases is highly variable, although again the general trend, even if patchy, has been towards the transfer to them of some management responsibilities. More often than not this has been in relation to small neighbourhood spaces, as in Tokyo, but also in more remote regional parks such as those in Melbourne. What is impressive is the variety of arrangements with third-sector parties found among the cities, although this is still clearly an underused management resource.

Regulation of public open spaces

Rather than neatly packaged legal instruments, the eleven cities use a wide range of powers to build up the legal framework for the management of open spaces. At a strategic level, powers come from national, regional and local laws and are often linked to other areas of policy, most commonly land-use planning, environment protection and heritage conservation. At an operational level they are part of criminal law and local byelaws and regulations.

In this regard two more general points emerge. First, from the experience of places like Wellington, Zürich and Århus, it can be concluded that the availability of a coherent, open space-friendly regulatory framework at the strategic level can help (e.g. national legislation on the protection of environmentally sensitive areas, or statutory city- or region-wide open space management plans). However, when such a framework is absent, it can be substituted by political will, as in Malmö or Curitiba.

The second point concerns the capacity to skilfully combine the available powers to their most effective use. The use of the planning system to put open space issues onto a statutory footing in Malmö or Groningen, or the use of health and environmental legislation in Melbourne are good examples of this. The overarching lesson is therefore that although a clear statutory basis for open space management is desirable, what is more important is the political will to use available powers, or to find other means to deliver effective open space management.

The cases also demonstrate that, at the operational end of management, regulating open spaces is primarily a municipal affair. At this level, anti-social behaviour, littering, vandalism and dog-related issues are problems that affect to some degree all of the cities. Regulations are generally in place to deal with these problems, but a key issue is how they can be enforced, a matter highly dependent upon the characteristics of the cities' particular legal, institutional and cultural environments. Thus, whereas Minneapolis and Melbourne have enforcement built into their management system, others depend on collaboration, especially with the police. Given the generally low levels of misuse problems reported by the cities, it seems that all those approaches can be successful, although they are likely to have very different resource implications. Considerable success was reported when enforcement activities were backed up by information, education and consensus building about the relative importance of proper behaviour norms. Zürich, for example, has been particularly successful in solving conflicts between the demands of different user groups that would not have been eliminated by simple enforcement.

A further key lesson is that enforcement action should feed back into open space management systems and into park design so that it is less susceptible to vandalism or inappropriate use. The challenge here is to keep the balance between offering a good-quality inspiring environment, and designing a robust environment that resists misuse.

Several of the eleven cities have developed mechanisms for monitoring the performance of their management systems, the needs of individual parks, and the interaction between the municipality and open space users. Some of these systems are internal to the municipal administration whereas others serve as tools to involve stakeholders in management decisions. Such systems have been put in place to fulfil a number of purposes, chief amongst which is the desire to secure effective cost management. Nevertheless there has been a general trend to move from an exclusive focus on financial aspects, to a progressive concern for open space quality.

The first and quite obvious lesson coming from the experiences is that effective monitoring systems are essential to securing good-quality open spaces. The second lesson is that effective and comprehensive monitoring requires a considerable effort in developing the parameters and the criteria to feed into the system. This is not an easy task as systems have to be generated locally to be appropriate to local contexts, and there are clear cost, time and manpower implications that probably explain why the majority of the cities examined have not yet arrived at this stage.

A final lesson concerns the importance of monitoring users' interactions with open spaces and their management. All eleven cities have well developed complaints management systems, whether or not dedicated to open space issues. The first key point here is the need to link those systems to management and maintenance decision-making, as achieved in Minneapolis, Malmö, Melbourne and Århus. This is not just a matter of securing users' support, but also of making good use of an invaluable source of first hand information on open space performance.

The further key point is the need to carefully consider the equilibrium between understanding and recognising the importance of users' views and responding promptly to these views without losing sight of strategic and long-term objectives of open spaces management. Examples from Groningen illustrate the tensions that might emerge, and the need for public open space managers to maintain an appropriate balance between satisfying local demands and maintaining a strategic perspective.

Open space maintenance

A common trend across most of the eleven cases has been the effort to restructure public services provision and open space maintenance within it. Public-sector agencies in the chosen cities have been experimenting with ways of delivering services that are more integrated and outcome-focused, that decentralise responsibilities and are less bureaucratic.

The degree to which these changes have been implemented varies considerably. How these changes have been implemented also varies, with some cities radically restructuring open space maintenance organisations and others incrementally changing practices without significantly altering organisational structures.

A first important lesson is the importance of clearly defined and properly resourced maintenance plans as tools for structuring, coordinating and delivering maintenance routines. As the experience of Hannover, Groningen and other cities demonstrates, such plans allow for clear linkages between daily maintenance routines and long-term management programmes and policy priorities. Some cities have invested considerable effort in increasingly sophisticated maintenance planning tools. Results so far are encouraging in terms of better use of resources, the quality of maintenance being achieved, the ability to secure funding on the basis of accurate and demonstrable information, and the ability to identify trends in the performance of open space designs, facilities and equipment, and thus prevent costly remediation work.

A second lesson is that there is no single best way of organising maintenance routines. The majority of the cities examined opted for some form of geographical basis, with maintenance teams allocated to areas or districts within the city to benefit from the detailed knowledge of, and sense of responsibility for, individual parks or areas that are fostered by this approach. By contrast, Zürich organises maintenance by task specialisation, with specialist teams covering the whole city and benefiting from the optimum use of specialised skills and machinery. Therefore although there seems to be a case overall for some form of geographical reference to maintenance routines, equally important is the consistent application of whichever approach to maintenance is adopted, so that specialist/geographically bound knowledge can be developed and put into practice.

On the issue of contracting out the management of open spaces, in general the evidence confirms that contracting out should be viewed as an outcomes-focused, mutually supportive partnership between the parties, rather than as a cost-cutting exercise. The experiences in the eleven cities demonstrated that both in-house and contracted-out maintenance services can be organised efficiently, as long as the strengths and weaknesses of each approach are recognised. It is important to emphasise the setting and monitoring of clear standards of delivery, with due consideration to cost/quality ratios, whether the key relationships are between municipal organisations and private contractors, as in Malmö or Curitiba, whether one public body delegates maintenance responsibility to another, as in Melbourne or Hannover, or whether a voluntary sector organisation is the partner, as in Tokyo.

A further lesson is that the delegation of some responsibilities to the operational level is desirable if maintenance routines are going to be flexible enough to incorporate the varied and changing demands of users and the multiplicity of individual open space contexts. The roles of park keepers in Paris and Minneapolis and area maintenance team leaders in Zürich and Groningen seem good examples of this, and suggest that where local flexibility is required, public rather than private employees are likely to be more adaptable, unencumbered as they are by the often highly prescriptive contractual arrangements that define the responsibilities of private contractors.

In each of the eleven cities, maintenance approaches have been adapted to respond to the individual needs of different types of open spaces, even when there is no formal provision for dealing with those. Some cities have developed quite sophisticated mechanisms to cope with a variety of geographical, seasonal and cultural contexts by shaping maintenance approaches accordingly. The key lesson is therefore that individual open spaces have different needs, and the more successful cities seem to be those that openly acknowledge and understand those differences and actively plan for them. In nearly all cases, locally responsive maintenance implies some degree of devolution of responsibility to local areas, together with good communications between management and operational teams and users and a responsive city-wide system. Individual park maintenance plans, dedicated park keepers, area-based managers and user participation can all play an important role here. Thus even where there is a larger degree of centralisation of management decisions, such as in Paris, there is still room for local adaptation of maintenance routines.

Investing in open space management

Although for most of the eleven cities current levels of funding are still satisfactory, all have faced budgetary constraints over recent years, with capital expenditure budgets suffering the most. Nine out of eleven cities depend on allocation from a general municipal budget for their open space core funding. Only Minneapolis and Melbourne benefited from dedicated funding, making resource allocation relatively free from the bargaining and uncertainty typical of the other cases. The latter approach is more likely to secure adequate levels of resources, but the fact that most cases do not have such a system suggests that political and legal obstacles to such a solution should not be underestimated.

For most cases, the key message that emerges is that adequate funding for open spaces is likely to remain dependent on the skills and political clout of open space managers and committed politicians to make the case

for open space investment, and to bargain with providers of other services for a larger slice of a limited cake. In this regard, accounting methods which link more explicitly open space expenditure to other environmental benefits, as in Århus, or that are more transparent in the relationship between the costs and the benefits they provide, as in Zürich, can be powerful tools to promote the cause of open spaces.

A further lesson is that there is much potential in exploring supplementary sources of funding. Particularly promising was the use of planning gain for capital expenditure on open spaces in Zürich, Groningen, Wellington and Curitiba, revenue-generating public–private partnerships and PFIs in Minneapolis and Tokyo, as well as the use of voluntary sector and community resources in Melbourne and Tokyo. An important prerequisite, however, is that resources raised in this way should be returned in full to the departments responsible for their generation as 'additional' funding.

As for reinvestment in open spaces, the constraints in nearly all the cities were discussed above. The main lesson to come out of the cases relates to the potential benefits of planning reinvestment activities in the context of thematic reviews, as in Malmö; asset management systems, as in Zürich, Melbourne and Groningen; or on the basis of long-term financial planning, as in Wellington. This is based on the need to place reinvestment priorities in the context of other open space management needs, thereby providing clear cause/effect links between day-to-day maintenance activities and longer-term reinvestment. Although this process was still in an evolutionary phase in most of the cities, its potential is quite considerable. The aim should be the automatic tracking of depreciation over time, and the factoring in of reinvestment as part of the continuum of maintenance activities, from minor and regular works, to major and periodic work.

Another key lesson concerns the increasing consideration of lifetime issues in investment decisions. Many of the cities provided good examples of efforts to consider the potential future costs of ongoing maintenance in investment decision-making. This has meant a closer participation of maintenance staff in development and investment decisions, including the analysis of development and investment plans by operational managers. A parallel lesson in this regard is the need to reshape monitoring and feedback systems to provide enough information to allow for the long-term maintenance consequences of new investment to be assessed.

Across the eleven cities, there is an explicit concern with the skills necessary for open space management and their development. However, the nature of the skills relevant for each case, as well as their distribution within management structures varies widely. History, organisational arrangements and styles of service, as well as the nature of the main open space aspirations explain that variation.

It is therefore difficult to generalise in terms of the appropriate nature and distribution of technical skills. However, one common trend that comes across strongly is the importance of experience on the job, from the strategic to the operational level. This is given considerable emphasis in cases as diverse as Minneapolis and Zürich and brings to the fore the issues of personnel turnover. The cases did not suggest ready answers for the problem of how to retain skilled personnel, but they do indicate that this is a key element in any skills policy, which is likely to require an emphasis upon ongoing training across all management and operational levels, and a continuous investment in staff.

Chapter 9

One iconic civic space

Managing Times Square, New York

This chapter and the one that follows take a detailed look at well-known and iconic civic spaces to explore how a management regime impacts on users' experience of those spaces. The discussion centres on the greater involvement of private interests in the management of public space and its associated trends towards commodification, control and exclusion. The first part of this chapter outlines the recent history of Times Square in New York and puts its transformation into a business improvement district (BID) into context. It summarises what this has meant for the management of that space. The second part looks at the physical and symbolic characteristics of the space, the way it is used, and how these aspects are affected by the management regime. This part first explores the shape of the place, its legibility, land uses and signage as the system of codes that structure the visual and sensorial experience of Times Square. A micro-analysis of the public space is then undertaken including of the uses and activities it fosters, and how management affects the traditional roles of fostering a sense of civility and community. The chapter highlights the complexity of the relationship between private-led management and the use of space, and the varied and to some extent unpredictable outcomes that result.

A focus on iconic civic space

In this and the following chapter the discussion moves from the models and practices of public space management to a more detailed examination of how emerging design and management practices interact with the characteristics of particular public spaces: physical, sensorial and functional. The two previous chapters focused on a particular type of public space from the typology in Chapter 3 – public open space. To balance the evidence, these chapters switch the focus to a further type of public space – urban civic spaces (see Table 3.1). This type encompasses the great civic spaces of our cities as well as the everyday streets and spaces that make up so much of the public realm. In these chapters, extreme, iconic, examples of the type are deliberately chosen as a means to focus on the types of public space trends that preoccupy so much of the literature discussed in Part One of the book. The choice further balances the evidence by contrasting the management of these spaces with that of the everyday spaces discussed so far in Part Two of the book.

As discussion in Chapter 3 demonstrated, most of the literature on contemporary public space suggests that economic and social change in the later part of the twentieth century has fundamentally altered the nature and character of that space. Indeed, many of the critiques of contemporary public space are rather pessimistic, contending that it no longer plays the role of an open and inclusive space for social, political and cultural exchange, and that instead it has been taken over by exclusionary, commodified, delocalised interests of a predominantly private and corporate nature.

Chapter 3 discussed at some length the key elements of that critique of contemporary public space, from the dominance of private transportation, to the privatisation, commodification and homogenisation of key public spaces. As was revealed there, the emergence of what characterises contemporary public space is linked to changes in the economy and society – the globalisation of late capitalism, mass consumption and the advent of the 'risk society' (Beck 1992). As a result, public spaces and especially the high-profile ones previously associated with civic or communitarian functions have become a valuable commercial commodity. Moreover, the creation of easily recognisable, safe and visually and commercially attractive spaces has become an instrument in the competition among

159

cities for investment. This process is led by corporate interests in alliance with city governments, with an overarching concern for managing public spaces and their image so that they are perceived as conducive to the types of activities and users that can reinforce and increase their value. In the effort to create 'safe' public spaces, the multicultural and pluralistic nature of public space with its perceived risk has had to be controlled, managed and policed. This has often meant banning unconventional behaviour and those who do not fit the purposes of this new space.

There are many elements in this interpretation of what is happening to contemporary public spaces that are less than consensual. However, it is generally agreed that whatever the real meaning of ongoing changes to the nature of public space, management and management regimes play a fundamental part in them, and none more so than the takeover of the management of many important public spaces, particularly in the US – and now of British cities – by corporate organisations and property owners through business improvement districts (BIDs).

However, the economic and social dynamics of public spaces is far too complex to fit neatly into the simplified view sketched above. In order to explore what actually happens to public spaces exposed to a globalised and consumerist society, and managed to some degree by corporate interests, three internationally iconic civic spaces whose histories were briefly examined in Chapter 2 – Times Square in New York and Leicester Square and Piccadilly Circus in London – are returned to here.

Research methodology

Times Square and Leicester Square/Piccadilly Circus have a great deal in common. They are iconic public spaces, at the core of global metropolises, subject to all the pressures described above. They are also high-profile tourist attractions associated with commercial entertainment and leisure, rather than any significant civil functions. They originate mainly from localised private development, and their fortunes have had historical links with theatre, and by the 1930s, cinema, and hence have attracted both the respectable and the 'dissolute'. They suffered decline in the 1960s and 1970s and have undergone repeated attempts at regeneration. Most recently their central location and iconic status has created the conditions for the adoption of new management regimes, most recently BIDs. This chapter focuses on Times Square, while Chapter 11 deals with Leicester Square and Piccadilly Circus.

Fieldwork tested on the ground how far each of the case studies incorporated some or all of the characteristics expected of contemporary spaces, especially issues of exclusion–inclusion, commercialisation, surveillance and control. A wide range of fieldwork techniques were used as a means to deconstruct the cases into a semiotic environment that could be analysed in terms of its symbolism and meaning.

The approach focused on the experience of place, their legibility, land uses, signage and advertising, as the system of codes that structure the visual and sensorial experience of those places. This was followed by a micro-analysis of each space and their compartments, describing what they contain and which uses and activities they foster. Issues of control and surveillance were also examined, and an analysis of the observed behaviour of users of the public spaces and how this relates to physical and management constraints was undertaken. The observation of users was partly done with a camera, using methods borrowed from Zeisel (1984) and Whyte (1988). The period of observation was one week (for each) in the Spring of 2002, including a Friday and Saturday night.

The production and management of public space in Times Square

During the last quarter of the twentieth century, New York City government experienced persistent financial difficulties which in their worst moments brought the city to the brink of bankruptcy. Some of the reasons for this were local, whereas others were part of a more general reduction in state and city funds which came as a consequence of a wider programme of financial cuts by the federal government. Metropolitan areas suffered further loss of finances by being forced into tax cuts by law and the need to stop the flight of residents and businesses to the suburbs outside the city limits. This process reached its climax in the 1980s with the Reagan administration and its strong commitment to 'neoconservatism', a political ideology that saw government intervention as a hindrance to economic efficiency and individual liberty (Loukaitou-Sideris and Banerjee 1998).

As a result, city funding for all public services, including the provision and management of public spaces, was drastically reduced. In the context of a strong neoconservative approach, the alternative to public provision meant privatisation, or 'the introduction and extension of market principles into public service production and provision...[and] the disengagement of the public sector from specific responsibilities under the assumption that the private sector would take care of them' (Loukaitou-Sideris and Banerjee 1998: 76).

The evolution of BIDs in New York and Times Square

As briefly discussed in Chapter 2, private-sector involvement in the provision and management of public space in New York has taken different forms. In commercial areas, especially those with a high profile, the most widespread mechanisms replacing publicly-funded public space services have been private partnerships in the form of BIDs. This mechanism, created in Canada and successfully embraced by US cities, allows for a partnership of local business and property owners to impose a levy on all property and businesses in an area, on top of normal local taxes. These funds are then used to pay for a range of public services within the boundaries of the BID. To establish a BID, a majority of property owners in a designated area must vote in favour of the scheme. Once a BID is formed, all property owners must pay the agreed levy. The compulsory character of the scheme, once it is approved, required specific legislation, which, in the case of New York City was passed by the state legislature in 1983. In the US, services funded by BIDs typically range from street cleaning, to private security, public works, place marketing and the provision and management of public space.

In New York alone there are now over 130 BIDs, covering most commercial districts, with more than 1,500 in the whole of the US (Lloyd and Auld 2003). This enthusiasm for BIDs has created a form of fragmented municipal government where funds for different districts vary vastly depending on district borders and the nature of business within them. Nearly every shopping street in New York City now has a BID in some form, varying from the Grand Central Partnership, a Manhattan BID that contains 53 affluent blocks commanding high property prices and including many multinational corporations, to much smaller local high street BIDs in less affluent residential neighbourhoods. The former can issue its own bonds to pay for ambitious large-scale infrastructure and service improvement programmes; the latter might raise just enough funds to pay for street cleaning services. This disparity raises questions about the control of urban space, and has led to concerns about the corporate take-over of public space by the larger BIDs (Zukin 1995).

It is in this context of neoconservative policies and private provision of urban services that the redevelopment of Times Square has taken place. The emergence of the Times Square BID, for example, was the result of pressures by the business élite with vested interests in the area to improve its image and thus reverse its economic fortunes. In particular, this has meant finding ways to fight high levels of street crime and the dominance of the area's retail sector by the sex industry (Sagalyn 2001). The perception of the consequences of that dominance for the economic fortunes of the area is illustrated by reports on the concentration of adult entertainment establishments in and around Times Square produced by the BID early in its life. These reports explicitly link the agglomeration of pornography outlets to high crime rates, a fall in property values and to negative impacts on other businesses. Together with other similar studies produced at the time, it helped to create favourable conditions for zoning changes banning sex-related business from many of their traditional locations in New York, including Times Square (Papayanis 2000).

A Draft Environmental Impact Statement (DEIS) prepared by the New York State Urban Development Corporation (UDC) in the late 1970s (quoted in Reichl 1999: 61), described those who hung around the area where 42nd Street meets Times Square as 'hustlers and loiterers', who had stopped 'office workers and other positive users having a territorial stake' there. The document goes on to say that 'in a real sense, 42nd Street is their territory [the hustlers and loiterers], and others venturing through it perceive they do so at their own risk'. Racial tensions in American society also played a part in shaping the dominant views of what was going on in Times Square. As Sagalyn (2001: 20) puts it, for the white middle class, 'the loiterers on the street seemed alien, unrestrained by conventional social codes'. The process reinforced racial stereotypes 'as innocent Blacks and Hispanics on 42nd Street were given a wide berth by wary whites'.

Times Square under BID management

Whether or not the reality of Times Square did actually match this picture is a different matter. A contemporary study by the City University of New York (CUNY), also quoted in Reichl (1999), suggested that most of the population in the area were merely 'hanging out' rather than hustling. The study also found that the area featured 'one of the most racially integrated streets [42nd Street] in the city', and that Whites were the dominant racial group at most times of day and night.

That the dominant perception of Times Square amongst suburbanites and planners stigmatised the area as a 'ghetto street' has been explained as the product of anxieties about a minority takeover, deeply rooted in a society that continues to be characterised by racial segregation and inequality (Reichl 1999: 62). This is not to say that Times Square was not dangerous in the 1970s and 1980s, but so was the rest of New York. It was the perception of 'the otherness' of the space and its users that become more important than the reality (Goldsteen and Elliot 1994).

Underpinning the case for cleaning up the Times Square district and 42nd Street was the need for a westward spread of office space in midtown Manhattan (see Reichl 1999). In the context of a city trying to retain its status as a global financial capital, Times Square was almost a default location for office space for large multinational headquarters and financial institutions given the lack of other alternatives within Manhattan and the area's excellent transport links. Initiatives to redefine Times Square's character as a 'cleaner' entertainment and tourism destination have also served as a catalyst for high-rise office space and increased real estate values. Indeed, the potential role of Times Square as an office location is highlighted by the recent completion of a number of major office skyscrapers in the immediate Times Square bow-tie, all with 'respectable' retail and entertainment uses at ground-floor level (Figure 9.1).

As a result of the real and perceived decline and degeneration of the area, in 1976 the 42nd Street Development Corporation was formed, a non-profit public–private partnership aimed at promoting economic development in the Times Square district. In 1992 this partnership took a more structured form, with the foundation of the Times Square BID, an important step in facilitating the private production and management of the district. The jurisdiction of the Times Square BID (renamed Times Square Alliance in 2003) now stretches from 40th to 53rd Street. The BID board members consist of 23 property owners, 13 commercial tenants, 3 residential tenants, 4 New York City government representatives, and 2 'community boards', in addition to 14 administrative staff (Times Square BID 2001). The Times Square BID enables over 400 commercial property owners and about 5000 businesses to pay into an annual fund of around $6m. This fund has been used to finance private sanitation workers, increased security, a visitor information centre, social service providers, and advertisement campaigns promoting the area (Times Square BID 2000). This means that much of the control, management and design of a world-famous public space has been handed over to private interests.

RE-IMAGING THE PLACE

Whether the emphasis is on the clean-up initiatives that have characterised most of the BID's work, or on the corporate appropriation of Times Square, a key part of the story centres on image management as integral to the management of the space itself. This relates to the specific history of Times Square and its perceived problems (see Chapter 2), but it is also part of a more widespread process of place marketing which has characterised the economic development strategies of urban locations in a globalised economy. Zukin refers to this process as the creation of an 'abstract symbolic economy devised by place entrepreneurs' (Zukin 1995: 7).

In the case of Times Square, place making was not about creating an image where none existed. The iconic character of the location meant that there were already many layers of historical symbolism as over the one hundred years of its history the square has come to symbolise different forms of emotional attachments between citizen and civic space. This is nicely captured by Berman's (1999) phasing of the perceptions of Times Square in the American mind, from the period 1900–1945 when the Square and its large signs and civic gatherings symbolised American

9.1 Times Square study area

civic and industrial might, business success and salesmanship, to 1950–1980 when the Square represented New York's increasing disconnection from the country and its stigmatisation as a place of danger and decline, to the post-1980s, with the Square and New York finding a new role as the locus for globalised entertainment and real estate-led corporate power.

However, part of what gave Times Square its iconic appeal was also its long-standing status as a multicultural and socially diverse space of indulgence, containing a mix of the seedy and the flashy, together with the middle-class theatres, restaurants and hotels. The recent efforts at image management through the BID and its initiatives have tried to control this diversity while still maintaining the appeal it brings. Chesluk (2000) suggests this has been done primarily by 'zoning-out' through design and space management those uses and users perceived as more undesirable, thus creating a perceivably safe and sanitised but still exciting area for shoppers and office workers.

THE SECURITY AGENDA

As a result of the emphasis on cleaning up and sanitising the space, active space management and sophisticated surveillance systems have been a large part of the BIDs work. In November 1993, early on in the life of the BID, a $1.4m sidewalk lighting project was completed as a way of addressing and countering the perception of Times Square as a crime-ridden area. By the end of its first five years of operation, the BID claimed significant improvements in safety indicators, such as a 58 per cent drop in crime, over 80 per cent drop in illegal peddling, and closure of over 40 per cent of pornography outlets (Times Square BID 1998: 20–1).

More recently various new management regimes relating to security have been introduced by the BID in the wake of the terrorist attacks on New York in September 2001. The Times Square Security Council was created in 2005, and comprises the security directors from all the major financial, hotel, media and entertainment organisations in the district. Another security-related management regime is a twice-weekly canine patrol with an explosive-detecting dog throughout designated areas. Finally, BID staff are now routinely trained in 'observation skills relating to suspicious behavior in today's world climate' (Times Square Alliance 2005).

This drive to tackle perceived and real safety problems has been accompanied by concerted action to create a visual image that reinforces the sense of a cared-for and thus safe place. The Times Square Alliance now employs around 70 'sanitation workers' in red jumpsuits, to carry out jobs such as vacuuming and disinfecting the sidewalk, emptying litter baskets, removing graffiti, and painting street furniture. In addition the BID

employs an equal number of public-safety officers. Though unarmed, they are trained and patrol the district on foot and by car, and have a radio link to the armed NYPD. More recently, methods to alter the image of Times Square have been both explicit – changes in the design of the district, from the painting of street furniture, to the creation of new signage, and the removal of spaces for loitering such as seating and low walls – and the implicit – through the commissioning of public art and the removal of graffiti.

However, the re-imaging of Times Square has not been done exclusively by the BID. McNeill links it to more general efforts to transform many of New York's more emblematic spaces, and with them, the image of the whole city (McNeill 2003). For example, the revamped image of Times Square owes something to urban design regulations brought in by the New York City Planning Commission in 1987 which tried to preserve the unique qualities of the place. These regulations stipulate minimum sizes of signage, brightness, position (generally going around the corners of the blocks and, exaggerating the triangular plots of the land as Broadway meets 7th Avenue – Figure 9.2), and the percentage of land use that must be dedicated to entertainment uses.

UNDERSTANDING THE IMPACT

These efforts at rebranding have been largely successful in their own terms. Indeed, businesses and advertisers who have located in Times Square during and after redevelopment recognise it as what Sagalyn terms a 'place brand':

> More than just an address in midtown Manhattan, Broadway between 52nd and 50th Streets was a marketable place. It was the new so-called 100-percent location, but for a different reason than what real estate professionals typically mean by that designation: the location could travel across space and culture to consumers worldwide, through communication broadcasts of every imaginable medium, for one simple reason – Times Square is an instantly recognisable 'place brand'.
>
> (Sagalyn 2001: 309)

Today the image of Times Square is of safe consumerism. As a place, it offers a combination of gentrified working and entertainment district and historic, civic and playground space. This has helped the BID to court large multinational developers, and large financial and entertainment tenants, in a process that further reinforces that image (Starr and Hayman 1998: 254).

This raises the question of how different users of this public space react to such a commercialised, surveilled and actively managed space. Research

9.2 Times Square today (Broadway x 7th Avenue)

elsewhere suggests that the reaction to management and surveillance in the minds of the users of public space varies between different groups. Jackson found that generally white middle-class users found 'surveillance cameras and other security measures' reassuring, while others, particularly ethnic minority teenagers and working-class users found it a 'threat to their security and an invasion of their privacy' (Jackson 1998: 184–5). Reichl claims that the new Times Square serves to segregate the ethnic minority and poorer user, mainly through the cultural symbolism of the area (Reichl 1999: 170–1). Whether or not this is really the case can only be investigated through a detailed analysis of the space, which is the subject of the next section.

The public space and its components

This section starts the process of deconstructing Times Square from the viewpoint of the user. First the public space is discussed as a whole, analysing its key elements and constituent parts.

Experience of place

The concept of experience of place describes the simultaneous perceptions users have of space. One is the practical level of sensory perception when in a place, and a second involves experiencing place on a more subconscious level, through an extra or 'sixth sense' (Hiss 1991). Experience of place observation is used to introduce, from the point of view of the user, the first impressions of a particular place. In combination with a description of other non-visual senses, a consequential ambience or 'feel' of a space can be described. Hiss's (1991) work looked at the complexity of contemporary spaces, and his own reading of the experience of Times Square (although published one year before the formation of the BID) raises a number of important points about the space.

Hiss identified the most notable Times Square experience as the 'bowl of light' created by the meeting of Broadway, 7th Avenue, and the low surrounding buildings, which he feared was being lost in the shadow of an increasing number of tall buildings. Second, he noted the lack of places to sit or stand still. Today, the bowl of light is gone, overcome by the shadow of more tall buildings, and there are still few places to rest. Moreover, because the approach from the north or south, along Broadway or 7th

Avenue, contains buildings that generally do not rise above 20 storeys, Times Square's open feeling is now nullified by the 'canyonisation' effect of its flanking tall buildings (Figure 9.3). Times Square today feels more like one of several downtown Manhattan intersections, dominated by international-style tall buildings and traffic. The lack of social space to sit and stand still is certainly to the detriment of any civic or communal feeling the space might have.

Different senses combine to give an overall impression of the space. Ignoring visual aspects, the sounds of Times Square are overwhelmingly from traffic. As tourists build up during the day, and especially in the evenings and during the weekend, street performers, buskers, and the many different languages and accents of users merge into indistinguishable cacophony. The sense of smell is also dominated by traffic fumes, although with a characteristic New York overtone of hotdog vendors' frying.

Legibility

Lynch's notion of legibility was described in his classic work, *The Image of the City* (1960). It explores the meaning of a space to its users by breaking down that space into a semiotic lexicon of codes. The analysis reveals five basic elements, namely nodes, landmarks, districts, edges and paths. Their nature, position and relative importance provide the codes from which meaning can be inferred.

Nodes and landmarks are the most useful elements as regards design and symbolism. Lynch defines nodes as 'strategic spots in a city into which an observer can enter', and the 'focus and epitome of a district, over which their influence radiates and of which they stand as symbol'. Landmarks are similar to nodes, except 'in this case the observer does not enter within them, they are external. They are usually a rather simply defined physical object: building sign, store, or mountain. Their use involves the singling out of one element from a host of possibilities' (Lynch 1960: 47–8). A legibility analysis of Times Square sought to understand how its spaces are assembled and perceived.

Using this analysis, the central bow-tie spaces within Times Square are each experienced as separate nodes, some of which contain landmarks. In 2002, landmarks contained in the south bow-tie spaces were the Times Tower, NYPD station, and the armed forces recruiting station. Although these could be entered by all, few people apart from those that worked there did; the important element was their landmark and symbolic value. Travelling north the next landmark in the bow-tie spaces was the Faces Fence (see below), then the statue of George M. Cohan, followed by the statue of Father Duffy (Figure 9.4) and the ticket booth TKTS. The most northern bow-tie space landmark was the advertisements opposite the Times Tower on the Renaissance Hotel. Surrounding landmarks from

9.3 The Times Square canyon

south to north included the Condé-Nast Building, Reuters Building, Toys 'R' Us shop, Marriott Marquis Hotel, and the Morgan Stanley Building.

The other legibility elements (districts, edges and paths) offered other insights. The roads are wide and noisy, and form definite edges, particularly Broadway and 7th Avenue. The pavements offer busy paths for pedestrians, as do the cycle lanes on Broadway and 7th Avenue. Finally, the study area itself represents a definite district as the ambience and design of the surrounding areas are quite different (excluding 42nd Street to the south-west). This is deliberate and a result of the zoning regulations, creating a distinct and valuable identity for place marketing purposes.

Land uses

A land-use survey allows for an assessment of the kinds of functions performed by that space, and how they relate to the space itself. This is important as it is the buildings and their occupants that frame the public space and create much of its symbolism. Times Square BID's own map (Figure 9.1) illustrated the key land uses in the area, at least those considered significant for place marketing purposes. In fact the BID's jurisdiction contains 25 per cent of the total hotel beds in Manhattan, putting tourists in the heart of theatreland and close to transport links.

Concentrating on the square itself, the stretch of six blocks containing the bow-tie spaces and all façades and corners that front onto the blocks

9.4 Father Duffy

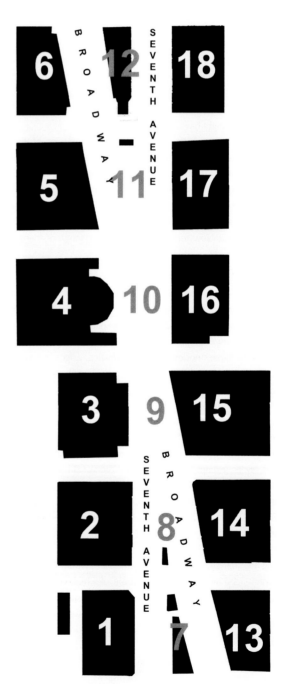

on Broadway and 7th Avenue, it is possible to distinguish 18 structural elements that make up the area: six central bow-tie sub-spaces, with the Times Tower to the south and the Renaissance Hotel to the north; six blocks on the west, from 7th Avenue to Broadway, and six blocks on the east, from Broadway to 7th Avenue. The bow-tie spaces form Times Square itself, while the blocks on either side act as framing elements for the space. For ease of reference each of the 18 elements has been numbered in Figure 9.5.

The occupants of Times Square are distinguished by ground- and upper-floor uses, but are united in that all are predominately large multinational companies. At ground level this tends to be internationally branded eateries such as McDonald's, Starbucks, Haagen-Dazs, Pizza Hut, TGI Friday's, or Planet Hollywood; or retailers such as Virgin or Sunglasses Hut. Many independent businesses still exist in the study area, yet are increasingly displaced by the bigger and brasher global brands. The location of these multinational outlets in Times Square reinforces the claim that there is a global process of homogenisation of public space, with those same businesses increasingly present in most high-profile public spaces around the world (Table 9.1). Until the late 1990s in Times Square only two of these brand names were represented: Pizza Hut and McDonald's.

9.5 Times Square structural elements

9.6 Times Square signs and lights

Upper-level land uses are dominated by office space, particularly for media companies and hotels. In 2002 seven out of twelve blocks had very large office buildings, with an average of 40 storeys. The main tenants were large multinational companies, for example Morgan Stanley (investment banking), Bertelsmann (media conglomerate), Reuters (news agency), and Condé-Nast (publishing). Three of the other blocks had tall hotel buildings, with the Marriott Hotel taking up a whole block.

Both the large multinationals and the smaller independent businesses use their location as a selling point. This creates a duality between multinational companies contributing to a homogenisation of the space through their presence, yet at the same time trying to give some individuality to their branch by reinforcing their links with the place.

Signage

The symbolic implications of the signage and commercial advertising in the space and how this shaped its character and meaning to users was explored. The sheer amount of signage in Times Square is spectacular, and can be analysed in two ways: in terms of the product/company/brand that is being advertised; and with regard to the physical form of the sign, for example, billboard or video screen.

Signage in Times Square is largely concerned with advertising, and thus strongly associated with consumption and commerce. Yet, taken holistically it also has a civic and communitarian function, as users of the space observe and enjoy the spectacle, feeling part of a greater whole (Berman 1997: 77). Most of the signage is on billboards or posters and is varied in terms of the products it advertises, with clothing, movies, financial services, and Broadway shows all featuring prominently. Due to the high price of advertising in this location, it is the large multinational companies that dominate (Figure 9.6). However, this was not always the case, and, as Sagalyn (2001) notes, up to the 1970s many smaller New York advertisers could rent space.

By 2002, the only locally based advertising was on the blocks between 46th and 47th Streets, where there were several billboards for different Broadway shows, some imported from London, and replicated in many parts of the world. At this time a minority of billboards diverged from purely commercial concerns to address civic issues reflecting the events of 9/11. One billboard had been leased by Yoko Ono to display a quote from John Lennon's ode to peace 'Imagine all the people living life in peace', a peace plea with a twin in London's Piccadilly Circus. A large Virgin Atlantic billboard above the Virgin Megastore shop displayed the message 'United We Stand' with a Stars and Stripes flag behind. Chase Bank sponsored the same message on a small rotating sign on the side of a building.

A look back at the way signage has changed on the square suggests that recently the signage has become more prosaic. In particular, Sagalyn (2001: 322–36) cites the art critic Dan Bischoff who laments the replacement of automated or interactive signage with corporate logos and video screens. This has made Times Square's signage more homogenised in form and less original and unique than it once was. Historically the advertising in Times Square was epitomised by the Camel cigarette advertisement which featured a cowboy blowing smoke rings with the caption 'I'd walk a mile for a camel'. The advert rather than being a standard billboard was individual and place-specific, and thus contributed to a unique sense of place. Another example that reinforced the special character of Times Square was the flowing waterfall flanked by two giant sized male and female mannequins on the roof of the Bond Clothing shop. Hiss (1991: 81–3) notes that the sound of the water helped drown out the traffic noise.

This 'interactive' 3-D signage style has almost died out and only a few examples remained at the time of the survey. The Cup Noodle advertisement on the Times Tower was 3-D and steamed as if warm. The Discover Card ATM sign on the corner of the 46th Street side of the Marriott Hotel was 3-D, featuring a mock-up ATM with a cash card and $20 bill that moved in and out of the machine. Planters Peanuts had a tipping nuts tin and the Coke bottle on the Renaissance Hotel had a moving straw that extended in and out of the bottle. The Motorola sign on the Condé-Nast

Table 9.1 The global credentials of brands located in Times Square

Business	Number of worldwide branches	Number of countries business operates in
Planet Hollywood	40	22
TGI Friday's	695	55
Haagen-Dazs	700	54
Starbucks	4,700	24
Pizza Hut	12,000	88
McDonald's	30,000	121

Building featured two mobile phones telling each other jokes and HSBC had a video screen that displayed the face of visitors to the Times Square Visitors Centre, showing them as part of a comic stereotypical scene from a number of countries such as India, Japan, and Spain (an unwitting digital reference to the long history of caricaturing in public space). These interactive signs are to be weighed against 13 different video screens advertising sponsors, and over 100 billboards, many of which could be in any city, and which contribute little to the individuality that once existed in the district.

A similar story of commercialisation applied to the 16 different electronic message reader boards, or 'zippers', that emulate the original 1920s system introduced by the New York Times around the Times Tower as a way of spreading the latest and most important news. Eight of the 16 zippers now have exclusively commercial functions promoting products or businesses, such as the Wrigley's or ESPN zippers. In this form, a vehicle once associated with civility has been re-appropriated by commercial interests in a homogenised form.

Signage is also linked to the image that the various building occupiers around Times Square wish to project of themselves. So, rather than associate themselves with 'kitsch' or 'tacky' signage on their buildings, the large multinational companies that dominate the above ground levels have negotiated the signage regulations by using video technology and 'zippers' to advertise their corporate services. Examples of this are the seven video screens on the Reuters Building; Nasdaq, who reputedly have the most expensive video screen in the world; and Morgan Stanley who provide three huge stock market zippers mixing news and self-promotion.

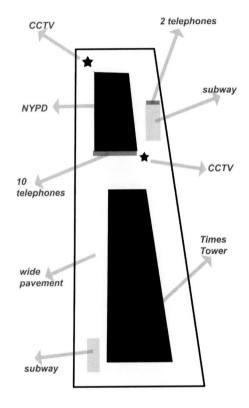

9.7 Diagrammatic plan of element 7

Microanalysis: the space close up

The sections above have tried to deconstruct the signs and codes that structure Times Square. They provide the visual and sensorial background to the space. However, the real experience of the place from the users' point of view happens at a much smaller scale. This section describes the key structural elements that make up Times Square from the perspective of a user trying to negotiate them at ground level. The area managed by Times Square Alliance covers several blocks, but Times Square proper, particularly as visitors are concerned, covers the 18 structural elements which include and surround the central bow-tie spaces (7–12 in Figure 9.5) that make up for the non-existent square.

ELEMENTS 7 AND 8

The southernmost element (7) is framed by the Times Tower and contains the one-storey NYPD police station (Figure 9.7). The tower functions as perhaps the prime advertising frontage to Times Square with ground floor retail space below. Counterpointing the commercial role of the tower was a reflection of civility in the form of the news zipper and the NYPD station; a strong state symbol, whose windows (at the time of survey) were plastered with messages of thanks from the public after 9/11. This strong state presence was enhanced by two CCTV cameras, one seemingly tracking the traffic interchange, and one trained on 10 telephones on the back of the police station. The wide pavements have ensured that this is the only block the BID has not widened. The space formed also contains two subway entrances/exits.

Moving north, element 8 is the most popular 'photo opportunity' bow-tie space, as it contains only one small structure – the Armed Forces

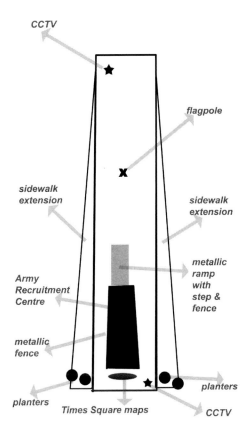

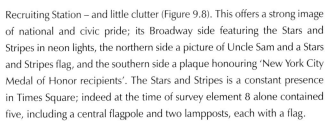

9.8 Diagrammatic plan of element 8

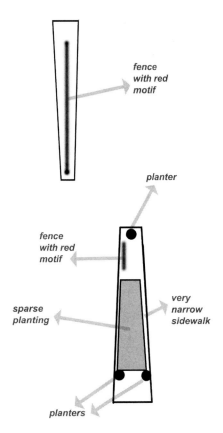

9.9 Diagrammatic plan of element 9

Recruiting Station – and little clutter (Figure 9.8). This offers a strong image of national and civic pride; its Broadway side featuring the Stars and Stripes in neon lights, the northern side a picture of Uncle Sam and a Stars and Stripes flag, and the southern side a plaque honouring 'New York City Medal of Honor recipients'. The Stars and Stripes is a constant presence in Times Square; indeed at the time of survey element 8 alone contained five, including a central flagpole and two lampposts, each with a flag.

The management and control presence in this space took the form of a CCTV camera at each end of the space. It also stemmed from the BID's interventions to create something more akin to a civic space: first in the form of pavement widening to both sides of the space, and second in the form of the Times Square information board, displaying street and subway maps. The recruiting station had a ramp for disabled visitors at the northern side, offering one of the few places to sit down in the bow-tie; a small step that is often in use.

ELEMENTS 9 AND 10

If Times Square is the 'crossroads of the world' then element 9 is physically where this occurs as it is geographically where Broadway crosses 7th Avenue. Thus the space contains two traffic islands, only intended for use by motorists to keep in lane or at its northern and southern extremities by pedestrians (Figure 9.9).

The southern island contained a low wall enclosing planting and several small round planters used by some as a place to sit down, although this was somewhat uncomfortable and only really possible on the southern side. To traverse the side of the island is dangerous due to its narrow width. The northern end of the island contained a lamppost with the Stars and Stripes and a black fence with a red motif. Despite the repelling nature of this space, and the danger involved in traversing the side of the island given its narrow width, users regularly went to the northern tip in order to sit down. The northern traffic island was just a fence, identical to that on

9.10 The Faces Fence

the southern island. The dominance of the traffic use over the pedestrian environment hampers any civic function here, while the poor condition of the contents of the planters' did little to dissipate the impression.

Element 10 contains the Faces Fence by Monica Banks (Figure 9.10). This is a curving decorative fence, similar to the one in element 9, but using a continuous red motif to create characterful Broadway 'faces'. Both sides of the space have been widened by the BID, with planters added (Figure 9.11). There were also two flagpoles both displaying Stars and Stripes.

ELEMENTS 11 AND 12

Element 11 is perhaps the most complex in design, containing numerous structures and fences (Figure 9.12). At the southern end is a statue of George M. Cohan, a famous Broadway impresario and patriotic songwriter, and names of Cohan's songs such as 'You're a Grand Old Flag' are inscribed on the plinth. A small fenced planting bed existed behind Cohan, protected by one small and one tall fence. The perimeter wall of the bed was angled to make sitting on it very uncomfortable. Whether or not this was the intention, the design and management of the space seem to aim at repelling users and thus reduce any potential civic qualities it might have.

In a formally organised fenced-off space at the northern end of element 11 is Father Duffy acting as a 'permanent reminder of the local hero who ministered in the decidedly secular world of Times Square' (caption on the fence below the statue). The Duffy statue has the only reference to religion in Times Square, backed as it is by a large Celtic cross. The area around Father Duffy contains formal planting and a bench, all of which are inaccessible. The steps around the base of the area provided the only place within the bow-tie where several people can sit down. However, the area inside the fence was, at the time of survey, poorly managed, whilst the fence itself is intimidating with its pointed metal spikes. Here again, design and management seemed to repel users rather than invite them in.

At the northern end of element 11 is the TKTS ticket booth, selling cut-price theatre tickets at set times of day. The booth is a semi-permanent structure in the regulation red of Times Square. The queue for tickets transforms the space into a crowd, particularly on Wednesdays and the

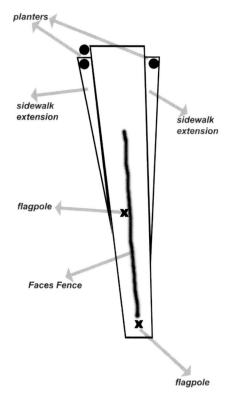

9.11 Diagrammatic plan of element 10

weekend when there are matinées. In doing so the booth fosters the use of the public space, but like much of Times Square this is mainly for tourists. At the time of survey the booth was about to be rebuilt, with stadium style seats on the roof.

Cameras overlooked the north and south of the space, and, like the other bow-tie spaces, were always trained on the road or pavements opposite. Two central flagpoles displayed the Stars and Stripes. Several lampposts had been fitted with additional lighting for safety and both sides of the space had been widened. There were also phone boxes on the northern side of the space which double as billboards.

Element 12 was more typical of the areas surrounding the bow-tie rather than of the bow-tie proper, with most of the site taken up by the Renaissance Hotel building (Figure 9.13). The hotel acts as the northern framing façade, with an advertising tower on its southern side facing Times Square. The element contained a subway entrance, and a range of vending machines/distribution boxes.

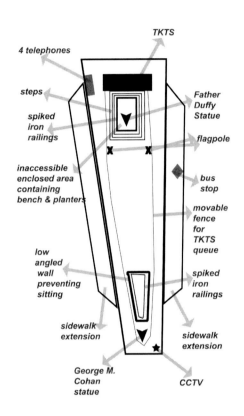

9.12 Diagrammatic plan of element 11

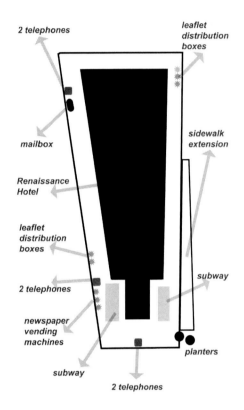

9.13 Diagrammatic plan of element 12

ELEMENTS 1–6 AND 13–18 (AROUND THE BOW TIE)

A number of characteristics shape the public space in the area surrounding the six bow-tie elements described above. Street furniture was one of these, and in this area consisted of: free publication distributing boxes, newspaper vending machines, mailboxes, and public telephones. Street furniture was often clustered together on the street, often near the corners of the block. Total numbers are shown in Table 9.2.

Bus stops, public telephones, litterbins and mailboxes are part of the essential infrastructure of modern civic life. The quantity, arrangement and condition of these pieces of furniture (unlike elsewhere in Manhattan) suggested a concerted effort to bring an element of civility and identity to the square, although with ever-present advertising. The hotdog vendors were the centre of activity on almost every street corner in Times Square, and while clearly a commercial activity, they create a focus of life in the public space and give it character. Another key characteristic was the lampposts, nearly all of which (immediately post-9/11 displayed one or

more Stars and Stripes flags. The vending machines/distribution boxes contained a multitude of publications and leaflets. While many of the publications were 'free guides' (mainly consisting of advertising), there was also a strong civic role reflected in their content and by the numbers of distribution boxes in a six block area (Table 9.3). Another noticeable feature of the area was the lack of greenery, although in the summer the planters in the bow-tie spaces contain some planting. Beyond this, there were several small trees outside the 45th Street side of the Marriott Hotel and some placed evenly along Broadway.

Regarding the position and occurrence of sitting spaces, if Whyte's (1988) analysis is accepted that a lack of sitting space is an explicit code for eschewing civil space, then Times Square fares poorly. Loitering space could equally be described as hanging-out or relaxing space, and is defined here as space that one can stand still in. This could be a space to meet someone, observe activity going on around, or to stand quietly. Loitering space is also vital for the civil and social nature of public space.

Table 9.2 Street furniture in Times Square

Street furniture/element	Number
Public telephones	53
Mailboxes	12
News huts	5
Bus stops	4
Hotdog vendors	15
Newspaper vending/leaflet distribution boxes	67

Table 9.3 Publications and leaflets in Times Square vending machines and distribution boxes

Publication	Form	Number of outlets
New York Times	Daily newspaper	5
USA Today	Weekly newspaper	7
Village Voice	Weekly local newspaper	4
New York Press	Free local alternative newspaper	9
Gotham Writers' Workshop	Free creative writing school guide	18
The Learning Annexe	Free local course/learning guide	10
The Seminar Centre	Free local course/learning guide	4
Employment guide	Free local employment guide	3
Employment source	Free local employment guide	4
New York visitors guide	Free guide	1
New York Resident	Local news and listings	1
New York Family	Local information	1
Total		67

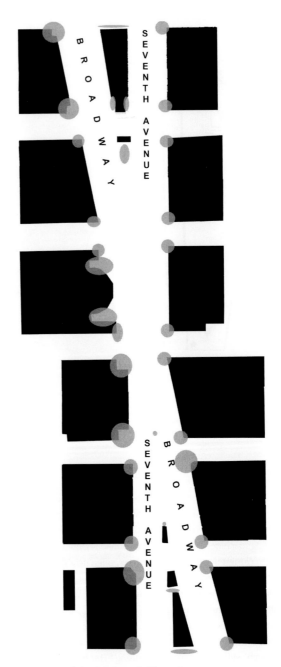

9.14 Sitting/loitering space in Times Square

The sitting and loitering spaces most favoured by the users of Times Square are illustrated in Figure 9.14, the use of which was particularly pronounced at night and at weekends. However, despite the widening of some of the pavements in the square, there was little chance to stop and relax. This is further hampered by shop fronts that contain no steps or walls to sit on or lean against, and by the absence of alcoves to stand in; with all facades flush to the pavement. The exceptions were the two external areas leading through to the drive through/drop off of the Marriott Marquis Hotel in element 4. The spaces in front of the hotel, particularly in the two alcoves, proved very popular with those who wanted to stop and rest. Buskers and hawkers (licensed and unlicensed) recognised this

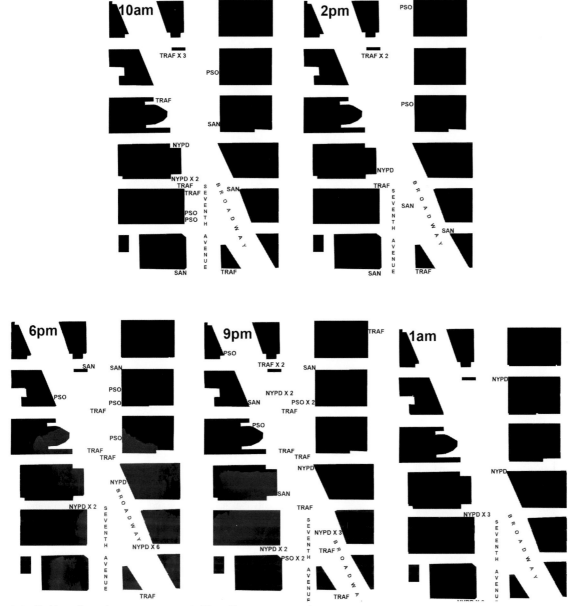

9.15 Position of prominent management in Times Square over one day

by performing and selling in the space, whilst the regularity of element 16 (opposite) effectively prevented such activity.

The corners of buildings, particularly where the corners were indented (elements 1 and 3), almost always had activity. This had a lot to do with the presence of hotdog and other vendors, human and mechanical, at these locations. These indented corners were used as places to meet others, smoking or break areas for the local office workers, and as impromptu arenas for buskers and street performers. Described as spaces to 'loiter', these spaces became centres of activity, essential in keeping the kinetic nature of Times Square alive. They provided an important contrast to the sleek commercial homogeneity of parts of the square.

Control and surveillance

The location and activities of management figures in the form of BID public safety officers (PSO), sanitation workers (SAN), NYPD, and NYPD traffic division (TRAF) were noted over a 15-hour period on Saturday 23 March 2002 (Figure 9.15). A Saturday was chosen as the time when the area is at its most lively and busy. The number of management figures fluctuated throughout the day, and dropped substantially at lunchtime. In the evening there was a large presence of NYPD and public safety officers, reflecting the large crowds flocking to the Broadway shows in the area. By 1.00am the square was still busy but only NYPD were left in the study area.

The observations revealed that there were relatively few public safety officers at certain times of day. These officers work in shifts from 9.30am–6.00pm, 11.30am–8.00pm, and 3.30pm–12.00am, although with many more on duty at night. This meant that during the day illegal peddlers and buskers (mainly breakdancing kids) could sell and perform at certain times without interruption. Despite this, there was a constant to-ing and fro-ing between performers and safety officers (sometimes involving the NYPD), although the BID management confirmed that it is very rare for performers to be arrested. By contrast it is policy for public safety officers to move pan-handlers on immediately, although in practice this is done with a stated warning of a couple of minutes.

Acoustic buskers can perform without a licence, but amplified buskers have to apply for a licence from the NYPD, which involves an audition. Street vendors selling food must have a permit from the Consumer Affairs and the Food Departments of the city government, while non-food vendors must have a permit from the former only. Permits are often site- or district-specific.

The positions of CCTV cameras in the bow-tie spaces are shown in Figures 9.7–9.9 and 9.11–9.13. Officially, the BID had only one camera, placed outside the visitors' centre with a panoramic view of most of the study area. Most of the other CCTV cameras were owned by the traffic division of the NYPD. There were also a few private cameras, one on the 47th Street side of the Morgan Stanley Building and a few others well-hidden from view in private buildings.

User behaviour

Detailed observation of the users of Times Square revealed, particularly up to lunchtime, that most were alone. The majority of these people in the week were white-collar office workers on their way to work, typically between 8.00–9.00am, and then at noon on their lunch break. Apart from smoking in corners outside the lobbies of their office blocks, these people did not engage with other users of space, and were always moving through Times Square at speed.

The second main user group, tourists, built up in numbers from lunchtime through the afternoon, until just before curtains went up on Broadway. At that time all the pavements and bow-tie spaces were packed. This pattern is repeated in reverse after the curtain comes down as people spill out on to the streets and gradually disperse. From 11.00pm activity slowly subsides until it is relatively quiet by 3.00am. At all times tourists stayed in groups and only interact with other groups when street activity took place, particularly busking.

The observation revealed that busking played a critical role in preventing Times Square from degenerating into a dystemic space;[1] creating instead the necessary activity and space for engagement between the diverse range of space users to occur. The example revealed how implicit codes in

the fabric of the case study area collectively produced an ambience, and distortions to that ambience, that are understood and interpreted by users. A sample of activities is included in Table 9.4.

Conclusions

Times Square conforms to some of the characteristics ascribed in the literature on contemporary public space (see Chapter 3). Three major negative aspects were identified during the study: the displacement of 'the other', a restriction of impromptu activity, and an increasingly artificial and dystemic environment. All three are associated by critics with a reduction of civility and community, as well as with an overall homogenisation of public space. They reflect a general trend towards the increasingly commercialised public realm epitomised by Times Square.

In Times Square the analysis of place, activity, and signage suggests a space that plays its role on a global stage but is heavily commodified in its use and marketing. Here to some extent 'the other' has been displaced through commercial gentrification, and the lack of places to sit, stand still, or loiter is part of that process. In discouraging 'the other' from the public space, elements that foster civility and community are also removed. Moreover, the absence of sections of society has led to a homogenisation of ambience and function.

The attempts at restricting impromptu activity, largely through surveillance and control mechanisms, have a similar effect. Through the displacement of 'the other' and the removal of spaces to loiter, impromptu activity is discouraged. Because surveillance and control prevent this on a day-to-day basis, Times Square has become a more predictable experience.

A third major negative aspect is the dystemic and artificial environment produced by the factors already discussed and by the commodification of the space. Rather than simply being a space in which to be, public space becomes a space in which to consume. The consumption emphasis of Times Square and the dominance of global entertainment outlets, with standardised designs as well as products, introduce a corporate, impersonal and delocalised feel to the space.

However, while the dominance of office headquarters of multinational companies and financial institutions and of global fast-food and restaurant chains and retailers creates a space much like other central districts in other world cities, Times Square still retains several elements that make it unique. These include the interior and exterior design of some of the multinational chains, the presence of some small business, the concentration of entertainment functions, the interactive and automated signage, and of course the urban form of the space itself.

Similarly, the BID management system reinforces a duality of character in the public space. While it provides many elements that reinforce a sense of civility and community, it is financed for, and by, large corporate interests, which actively promote their commercial ends. Nevertheless, the consumption itself has also created scope for fostering civility and community, through the existence of cafes, restaurants, and shops in the area that provide the type of third-place environments for civility and community for those that can afford to consume.

Notwithstanding all the pressures for homogenisation, the activity behaviour table (Table 9.4) depicts a rich variety of activity occurring in Times Square. This partly reflected the management of the area which was not 'zero tolerance' (whether or not intentionally), and which instead gave some leeway to some who might be defined as 'the other' or strangers as well as vendors and performers.

Performers and vendors in particular managed to survive and even thrive because of the indented corners of some blocks in the framing spaces (aided by the widening of elements 8 and 11 – see Figure 9.5) around the bow-tie, and particularly in front of the Marriott Hotel (element 4). Performers were also helped by the lack of CCTV used by the BID and police, often allowing time to perform before being spotted. The fact that many spontaneous/irregular activities happen in spite of the restricted space, overt control and direct competition with established high-profile entertainment attractions is testament to the vigour and positive nature of Times Square as a public space. Many of the decried characteristics of contemporary public space were certainly in the ascendant, but the essential elements of what gives Times Square its vibrancy and ambience were still also in evidence.

Finally, the social, racial, and cultural mix of users remained very diverse, though it did seem biased in favour of the middle classes and tourists, both characteristics which were to be expected. Many of the vendors and performers, both legal and illegal, were ethnic minorities and once would have been defined as 'the other' in Times Square. The sanitising of ethnic culture through such activities as breakdancing for tourists might have been appropriated in a way that place-markets New York as the centre for African-American street culture, but this can arguably benefit all involved: the performers, the businesses, and the tourists. By making the place more up-market the BID has contributed to the displacement of the ethnic minorities who socialised and hung-out in the area in previous decades. This would suggest that the 'other' has not so much been removed but rather sanitised into something more controllable and less threatening.

Note

1   An environment for purely impersonal and abstract relationships: Wallin 1998.

Table 9.4 Study area sample activity/behaviour table

Event/behaviour	When/frequency	Place/position (see Figure 9.1)	Connection to design	Connection to management	Connection to 'the other' and dystemic place	Connection to place marketing	Photograph
BID public safety officer	All week between 9.00–12.00am	Tend to not stand in bow-tie spaces	NA	Main management figures in Times Square. No powers to arrest, radio to NYPD	Mitigate presence by not employing zero tolerance	Red motif on uniform	
BID sanitation workers	All week between 9.00am–9.00pm	All over	NA	Do not involve themselves in management issues	Sanitation workers are 'the other'	Uniform of bright red with BID logo	
Percussion performance on boxes, crowd gathered	Observed once only	Space 4	Widest pavement in study area, room for performers, dancing, and crowd	Legal, acoustic	All performers Black, crowd mixture of locals and tourists	Ethnic and musical	
Black market/pirate handbag and DVD street salesmen	Present all weekend, but not in week	Space 2	Near corner with 44th Street, activity tends to occur on the corners of streets	Presume no permits from the abuse given while taking photographs. No BID employees in sight	All salesmen Black, no customers seemed to be local or ethnic	Association with making a 'quick buck'	
Kids practising break-dancing with loud stereo	Present all weekend, but not in week	Space 7	Wide pavement on side of Times Tower	Illegal, amplified. No public security or BID employees in sight	All Black	Association with Black/hip-hop culture and New York	
Professional-looking break-dancers, MC, and loud music, large crowd	Present all weekend, day and night, but not in week	Space 8	Widening and open nature of space allows a crowd to form	Illegal, amplified. No public security or BID employees in sight, lookout is ready to tip off if they need to move	All performers are ethnic, crowd is a mixture of cultural backgrounds, but all seem to be tourists	Association with Black/hip-hop culture and New York	
People meeting, vendors talking	Constant throughout fieldwork	Space 1	Corner of building (Reuters) set back and not leased, creating space to stand and talk	All legal	None	Association with hot dog vendors and New York	
Man eating lunch on floor in doorway	Observed once	Space 13	No where else to sit and eat lunch outside in Times Square	No one has moved him	NA	NA	
Vendor selling 'I love NY' t-shirts	Constant throughout fieldwork	Space 15	Corner of 44th Street, space for stall	Legal	NA	Strong association with 1970s financial crisis and 9/11	
Naked Cowboy posing for photographs with tourists	Constant throughout fieldwork	Space 10	Lack of enclosure on the bow-tie spaces, particularly in the centre, increases exposure to others	Legal, acoustic	Only White busker/performer. Particularly popular with groups of white teenage girls	Association of American cowboy identity, and nudity and Times Square	

Event/behaviour	When/frequency	Place/position (see Figure 9.1)	Connection to design	Connection to management	Connection to 'the other' and dysemic place	Connection to place marketing	Photograph
Ticket tout offering Broadway tickets to TKTS queue	Observed during weekend only	Space 11	NA	The tout manages to look as if he is himself queuing	NA	Black market of place marketing through Broadway shows	
Users sitting on steps around Duffy Statue/ area. Rest of space 11 defensible space with spiked railings	Observed throughout fieldwork at all times	Space 11	Only place where a group of several users can sit together outside in study area	Legal	Mainly inhabited by tourists resting or in TKTS queue	NA	
Man drinking beer from a paper bag	Observed once only on a Friday night	Space 11	NA	Illegal, but finishes drink and is not moved on/arrested	Man is Black	NA	
Tourists taking photographs	Present at all times of day and night	Space 8	Most open and easily accessible section of bow-tie	NA	NA	Promoted as a 'photo opportunity' in the BID map	
Two men sitting talking and eating amongst traffic on corner of planter	Observed once only	Space 9	No place to sit outside in Times Square	NA	Na	NA	
Steel pan busker. Played when matinee TKTS queue formed	Wednesday, and all weekend during morning/afternoon	Space 11	Open space in middle of snaking queue around space 11	Illegal	Black	Link with Times Square and music	
Xylophone busker/ performer	Most of weekend	Times Square subway station	Near main entrance/ exit	Legal, official permit	Ethnic	Link with Times Square and music	
Professional photographer offering to take photos of tourists in Times Square	Present every evening	Space 4	The spaces outside the Marriott Hotel are a set back and a good place to stand still for photographer and users alike	Presume legal	One of many photographers who were Black	Link with photography and Times Square	Did not want photo taken
Homeless-looking man begging with cup and notice	Witnessed begging only once	Space 15	Begged outside non-leased part of block	Illegal, moved on within 10 minutes	Man was 'the other' and was Caucasian	NA	Did not want photo taken
Jewish leafleters (approx. 6)	All day Wednesday and Thursday	Spaces 1, 2, 7, and 8	On street corners, near subway station	Legal	Not dysemic, talking to those who wanted to ask questions or debate religion	Link with immigrants and Times Square	Did not want photo taken
Portrait artists clustered around the 42nd Street side of space 1	Every evening, many more at weekend	Space 1	On indented street corner that had not been leased	Presume legal	All artists were oriental	Link with immigrants and Times Square	Did not want photo taken
Man selling sex positions booklets for $1	All weekend, moved around	Spaces 2 and 3	Tended to stay near street corners	Presume illegal	Black	Link with sex and Times Square	Did not want photo taken

10.1 Leicester Square today

10.2 Piccadilly Circus today

Chapter 10

Two linked iconic civic spaces

Managing Leicester Square and Piccadilly Circus, London

This chapter takes a detailed look at two neighbouring iconic civic spaces in London, Leicester Square and Piccadilly Circus. As in Chapter 9, the objective is to explore how the introduction of a new public space management regime based on greater private-sector involvement has shaped users' experience of these spaces. A first part describes recent management initiatives for Leicester Square and Piccadilly Circus, how they coalesced into a business improvement district (BID), and what this has meant for the way private and public interests interact. The second part comprises a micro-analysis of the public spaces and their components, as well as the uses and activities they foster. The main purpose of this section is to understand how management and control affects how visitors' experience the spaces and the way they use it. The discussion links the particular arrangement of public and private actors found in Leicester Square/Piccadilly Circus, with a prominent role for the local council, to the kind of transformation that the spaces are undergoing. This greater partnership of interests means that the transformation is less dramatic and less univocal than that experienced in Times Square and similar places elsewhere.

The move to BIDs

This chapter continues the detailed analysis of public space carried out in Chapter 9, but the focus now is on Leicester Square and Piccadilly Circus (Figures 10.1 and 10.2), neighbouring iconic civic spaces in London. As with Times Square, Leicester Square and Piccadilly Circus have come to exemplify the commodified, privatised and homogenised character of much contemporary public space. They have become areas of hedonism and consumption, and it is the consequence of this that management processes are now attempting to control.

However, the programmed interventions in Leicester Square and Piccadilly Circus do not encompass the vast regeneration programme and change in uses that have occurred in Times Square. No major changes in land use have been proposed, nor has there been an overt campaign to increase the number of visitors and the appeal of the space to corporate users and investors. Instead, the issue has been about how to regulate and control the use of the public space and its intensity. As a result, the management approaches adopted in Leicester Square and Piccadilly Circus have been less intrusive than in Times Square.

BIDs in England

The UK, like the US, witnessed dramatic falls in public-sector funding for public space throughout the 1980s and 1990s (see Chapter 5). As a consequence, 'new additions to urban space are often developed and managed by private investors, as the public authorities find themselves unable or unwilling to bear the costs of developing and maintaining public places' (Madanipour 1999: 888).

In this context, BIDs along the lines of their North-American counterparts (see pp. 160–1) were suggested in the mid-1990s as a way of drawing private resources into public space management in a more consistent and formalised way than through voluntary arrangements such as the ubiquitous town centre management schemes. In April 2001, the Prime Minister announced that BIDs in England would be funded by an additional levy on the business rate – the local tax paid by occupiers of commercial property – if agreed by local business and councils. This extra funding would help to pay for new projects, including initiatives designed to make streets and other public open spaces safer and cleaner. Regulations allowing the formation of BIDs were enacted on 17 September 2004:

the Business Improvement Districts (England) Regulations 2004 (Statutory Instrument 2004 No. 2443).

English regulations for BIDs are in large part similar to the US model. The legislation now stipulates that where a majority of non-domestic ratepayers agrees to it, a BID can be set up. Once the BID is formally approved, all non-domestic ratepayers in that area will be charged a levy that will fund the BID and its activities. The main difference with the US model is that in the US the levy falls on property owners, whereas in the UK they fall on business occupiers. The level of the levy, the chargeable period (not exceeding five years), the non-domestic ratepayers who are liable to contribute, and the date of commencement are to be specified in 'BID Arrangements', prepared in partnership by the local authority and local businesses. BID proposals are then only regarded as approved if the majority of non-domestic ratepayers in the proposed BID approve these arrangements in a ballot (TSO 2003; 2004). By April 2006 there had been 27 positive ballots to establish BIDs in England and five negative ones.

The London BIDs

BIDs in London include: Kingston Town Centre; Bankside; Holborn; Paddington; New West End (Oxford Street, Regent Street and Bond Street); London Bridge; Camden Town; Waterloo; Ealing; Hammersmith; and Heart of London (Piccadilly Circus, Coventry Street (which joins them) and Leicester Square). In the capital, the idea of BIDs was first raised in a report commissioned by the Corporation of London (the local authority for the Square Mile of the City of London) in 1996. While cautious of the role BIDs could take in the UK, its authors concluded that 'an experiment with BIDs in Britain could bring considerable advantages ... the time has come to test BIDs in Britain' (Travers and Weimer 1996: 29). This led to the formation in 2000 of The Circle Initiative, which created five partnerships in different areas of central London intended as pilot studies for potential BIDs. The initiative was organised though the Central London Partnership, a public–private partnership established in 1995 and comprising eight central London local authorities, other public organisations, and a number of private-sector partners. Most of its funds came from a successful bid to the Single Regeneration Budget, the main source of UK government funding for urban regeneration in the late 1990s.

In a Circle Initiative brochure, the foreword by the chairman suggests that aspirations for the London BIDs were very much along the lines of the Times Square model:

> Specific goals have been defined, centred around raising standards
> of management and maintenance of the public realm, reducing
> crime and the fear of crime, improved public transport and

its usage, building commercial activity and local jobs, and the creation of local community groups to have an input into each BID. Unlocking private sector investment is crucial, in order to pave the way for quality streetscape enhancements and on-street wardens, improved security, extensive marketing and promotion and enhanced employment opportunities.

> (The Circle Initiative 2000: unpaginated)

The production and management of public space in Leicester Square and Piccadilly Circus

Piccadilly Circus Partnership

One of the five pilot BIDs of the Circle Initiative was the Piccadilly Circus Partnership (PCP), which encompassed an area covering part of Piccadilly Circus, Coventry Street, and a small corner of Leicester Square (Figure 10.3). Although the areas around Leicester Square and Piccadilly Circus have not faced anything similar to the actual and perceived levels of decline, stigmatisation and abandonment of Times Square, they had been sliding down towards a degree of shabbiness and seediness often found in heavily visited and used public spaces in the central area of large European cities. The key symptoms were declining standards of maintenance, increasing levels of crime, anti-social behaviour and illegal activities in general, downgrading of occupiers of commercial property, and increasing vehicular and pedestrian congestion. At the root of this process were the sheer number of visitors, the nature of the activities concentrated in the area and the inability of normal management systems to cope with the pressures of intense and continuous use. The pressures were compounded by the declining levels of public funding for public space services.

PCP was formed in September 2001 as a company limited by guarantee, to:

> address the key problems of the cluttered and dirty street
> environment, illegal street trading, high levels of crime and anti-
> social behaviour and constant congestion ... [and to] make sure that
> Piccadilly Circus, Coventry Street and Leicester Square continue
> as a vital and cosmopolitan business, office, and retail centre and
> also remain the heart of London's film, theatre, entertainment and
> leisure scene.

> (The Circle Initiative 2001: 8–9)

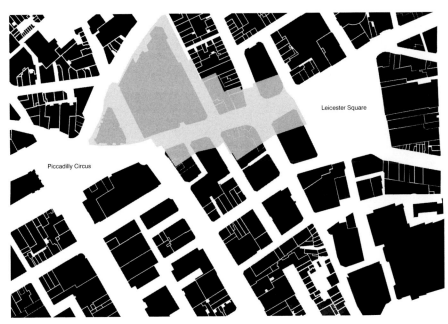

10.3 Piccadilly Circus Partnership Business Improvement District jurisdiction

It included not only business occupiers but many of the main property owners as well, all of which were charged a voluntary contribution until the partnership's formal constitution as a BID in 2005.

In spite of its ambitious aims, its programme of activities was actually quite modest. Given the limited funds obtained from voluntary contributions, there were no plans for extensive re-design or infrastructure works and emphasis was put on improving on-going management systems through small-scale initiatives, especially those addressing safety and cleanliness (reflecting broader government priorities).

Leicester Square Action Plan

However, in a clear demonstration of the fragmented nature of public space management in London, the area covered by PCP excluded most of Leicester Square. Instead, an action plan for changes in the design and management of the square was prepared by Westminster City Council concurrently with, but unrelated to, that of the neighbouring PCP.

The Leicester Square Action Plan was part of an urban renewal programme which stemmed from two comprehensive studies of London's West End: the 'West End Entertainment Impact Study' and the 'West End Public Spaces Study', both published in October 2001. The plan itself was adopted by Westminster Council in April 2002, with the hope that it would make the run-down but crowded Leicester Square:

> once again the jewel in the crown of a truly world class city, a place characterised by its strong business base, vibrant local community, supporting infrastructure and its cultural attractiveness for the rest of the world.
>
> (City of Westminster 2002a: 1)

The perceived problems of Leicester Square were in many ways similar to those of Times Square. Over the last few decades Leicester Square and neighbouring Soho had become globally renowned late-night hedonistic areas, with a large concentration of night-life leisure establishments. In the late 1990s, in an attempt to create a 24-hour economy, Westminster Council granted licences to many large drinking establishments, unwittingly contributing to an increase in episodes of alcohol-fuelled anti-social behaviour. However, in a marked difference to Times Square, the increased seediness of the area did not lead to its abandonment by middle-class and corporate users and its takeover by minority groups, but caused instead an increase in conflicts and of pressure upon the physical infrastructure and social fabric of the space. As stated in the introduction of one of the studies referred to above:

> The informal as well as the formal economy has boomed, creating ever growing problems of unlicensed street traders, buskers, beggars, squeegee merchants, fly posters and carders competing for limited pavement space whilst unlicensed clubs and tables and chairs add to the impression that there is quick money to be made, anything goes.
>
> (EDAW 2001: 3)

Consequently, the measures put in place by the action plan tended to focus on the control of uses and users rather than on their replacement, and on the management of conflict among those same uses and users. The vision contained in the plan stressed 'a family atmosphere in Leicester Square, where at least one PG or U certificate film is being shown on any evening, al fresco dining is encouraged around the gardens, and an events programme is put in place (City of Westminster 2002b: 6). The measures adopted to bring about this vision relied on a more careful policy of licensing activities in and around the square and a more effective enforcement of existing regulations related to the use of public space.

As part of the action plan, overall management of the square was contracted out to a private security firm who employed a number of

10.4 Leicester Square wardens

fulltime wardens (Figure 10.4). These wardens did not have police powers, much like their Times Square Public Safety Officer counterparts, but were used to 'meet and greet' lost tourists, direct delivery vehicles, or liaise with the police or the enforcement agents of the council. To support the work of the wardens and other enforcement agents, the council implemented a network of fixed and mobile CCTV cameras to survey the square. These cameras were put there to allow enforcement agents and wardens to spot issues that might need their attention, but also to provide evidence to be used in court if necessary. In addition, nearly all of the businesses around the square had their own private CCTV, which could be used by the council if necessary.

Heart of London BID

The separate arrangements for Leicester Square and Piccadilly Circus, under Westminster Council and the PCP respectively, were unified when the pilot Piccadilly Circus BID became a fully-fledged BID in February 2005. By then, the common issues faced by both areas were evident, as were the interests of businesses in both places and the desirability of a wider tax-base for the BID to be economically viable. Consequently, the new 'Heart of London' BID came to life with a much larger jurisdiction than the pilot, encompassing Piccadilly Circus, Coventry Street, the whole of Leicester Square and the southern end of Shaftesbury Avenue (Figure 10.5). In April 2007 this BID became the first in the UK to secure a second term of operation.

The objectives of this new BID are similar to those of Times Square, and range from a more targeted delivery of public services, especially those concerning safety and cleanliness, to advocacy for, and involvement in, re-design projects, to the branding and marketing of the location to users and investors. Underlying those is the perceived benefit of involving business in the management of the area.

To achieve these objectives, the new BID relied from the start on extending the management processes set up under the council's area action plan for Leicester Square to the whole BID area. This reflected both the relative success of the action plan in tacking the problems of Leicester Square, and the important role the council had in the public–

private partnership which created and now manages the BID. Although the Heart of the London Business Alliance is an independent management company meant to provide additional and complementary services to those provided by the local authority, it derives much of its ability to act and its effectiveness from the local authority that it partners. This close relationship between BIDs and local councils seems to be a characteristic of nearly all English BIDs (DCLG 2007).

In Central London, the BID is currently funded by both a compulsory levy and voluntary contributions from landlords in roughly equal proportions. The resulting budget of around £1 million per year has allowed the BID to fund on its own, or in partnership with others (including Westminster City Council), several initiatives related to their clean and safe agenda (Heart of London: 2007).

SAFETY AND CLEANLINESS

To address issues of safety, the BID has created a dedicated crime reduction manager. In partnership with Westminster City Council, it operates a Business Watch Scheme to deal with shoplifting, and a team of wardens – the City Guardian Team – whose job it is to help with '... deterring crime, reporting incidents and acting as ambassadors for the thousands of visitors to our area every day. They provide a uniformed reassurance to the public at a time when London is on a state of high alert' (Heart of London 2004a). Moreover, in conjunction with the Metropolitan Police, the BID has launched the 'Heart of London Pavilion', a permanently staffed police station located on the junction of Shaftesbury Avenue and Coventry Street, with jurisdiction over the BID area. This serves as the base for a dedicated police unit for London's West End.

To tackle issues of cleanliness, the BID has created a 'Clean Team' that operates in addition to the council's regular street-cleaning services, and which is charged with a daytime, evening and night-time janitor service, street sweeping and cleansing, additional litter picking, and removal of graffiti and fly posting from both the public realm and private property.

PROMOTION AND INFRASTRUCTURE

The BID has also been active in place promotion, and reflecting the experience in Times Square, although not as intensely, there has been an effort to promote and market the Heart of London area through advertising campaigns, dedicated websites, and street events. These initiatives have been aimed jointly at attracting customers to the businesses located there and increasing the profile of the area among potential incoming businesses and investors.

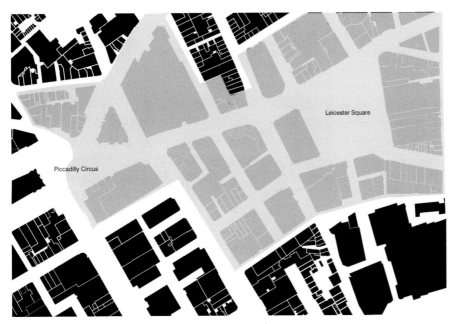

Leicester Square

Piccadilly Circus

10.5 Heart of London Business Improvement District jurisdiction

However, the main emphasis of the BID has been on the delivery of enhanced services to the Leicester Square/Piccadilly Circus area and its businesses. This reflects the nature of the BID mechanism in the UK, with its focus on additional levels of 'public' services and its levy base of business occupiers rather than property owners. Effectively it denies the power and resources for the more ambitious infrastructure and re-design initiatives that have characterised some of the larger US BIDs. When, therefore, the Heart of London BID became involved in a £1m pedestrian improvement scheme on Coventry Street, it did so as a minor player, with Westminster City Council in the driving seat; as they were in the much larger refurbishment of Leicester Square.

Recent developments in the latter case include a £15m scheme that includes new lighting and signage, new street furniture, a new theatre ticket booth, a new performance area and the replacement of the existing fence around the central gardens with a retractable fence (City of Westminster 2006). So far there has been no financial involvement of the BID in the project and private sector involvement has happened through sponsorship arrangements coordinated directly by Westminster City Council.

JUDGING THE BID'S EFFECTIVENESS

As, at the time of writing, the BID had only been in place for two years, it was not yet clear how successful it had been in changing the profile of the area. The available evidence from the BID's own system of targets and indicators suggests it is succeeding in making the area feel safer and cleaner (Heart of London 2006). Whether this will eventually change the nature of what goes on in Leicester Square and Piccadilly Circus, and how these spaces are used, it is too early to say.

The emphasis on fighting anti-social behaviour in particular has the potential to function as a mechanism to exclude from the public space sections of the community whose norms of behaviour diverge from accepted standards; changing in the process the mix of uses and users that give these public spaces their vibrancy. Similarly, the emphasis on the commercial success of public spaces reinforces a consumerist view of what these spaces are for, with equally exclusionary consequences. As argued in the previous chapter, these issues can only be investigated through a detailed analysis of the space and the way it has been used. This is the purpose of the next section.

The spaces and their components

As in the previous chapter, the in-depth reading of Leicester Square/ Piccadilly Circus begins with a look at the area as a whole. Analysis then turns to the scale of the individual user and examines how the area and its elements relate to user behaviour.

Experience of place

As in the previous chapter, Hiss' (1991) concept of 'experience of place' was used to capture the sensory perception that users will have of the spaces. Visual and non-visual impressions make up the ambience of the space which users experience. In the case of Leicester Square/Piccadilly Circus, the public space is made up of sections of quite different character, each providing a different experience.

The ambience of Leicester Square is very different to that of Times Square, mainly because of its central green garden space and absence of vehicular traffic. This creates a space that is tranquil until the afternoon, when, like Times Square, visitors slowly increase in numbers. Local office workers create a rush through the square at 9.00am and 5.00pm, and are also prominent in and around the gardens at lunchtime. Users in the area are socially, culturally and racially very diverse.

The north terrace of the square is a heavily used pedestrian thoroughfare. During the day the pervasive presence of street performers adds to the liveliness of the space (Figure 10.6). As the day turns to afternoon and to evening and night the ambience becomes far more alcohol-based as drinks can be seen and smelt all over the square. By late night the area has built up to a raucous feel, which, on occasions, becomes unpleasant and intimidating.

The gardens in the centre of the square are typical of London's Georgian squares today, in that the layout of the square is community- and civil-minded with many benches and low walls for sitting on. Further symbolism is added through the two statues and four busts in the gardens. The Victorian and typically British elements are the grass, tall trees, flowerbeds and serpentine paths, all enclosed by iron railings.

The architecture of the surrounding buildings speaks little of the rich history of Leicester Square for, like the gardens, it dates only as far back as the late nineteenth century. Cinema replaced theatre as the main form of entertainment in the 1930s, and still features prominently around the square. The square is dominated by the 1930s art deco Odeon Leicester Square on the east side, and more recently the Alhambra Theatre. On the north side there is the Empire Theatre, which was one of the first cinemas in London in 1896. On the south side is the Odeon West End, dating from the 1920s. On the west side there is a low-rise office block, the Communications Building. The building heights are much lower than Times Square with no building rising beyond 10 storeys, the tallest structure being the tower of the Odeon Leicester Square.

Coventry Street leads from Leicester Square in the east, through to Piccadilly Circus in the west, in what feels like a dynamic pedestrian thoroughfare (Figure 10.7). Walking from Leicester Square the first space is Swiss Court with the Swiss Centre to the north. The court is cobbled and at the time of survey contained several unlicensed henna tattooists. The Swiss Centre contains offices above ground floor, with a small cinema and nightclub. The ground floor use of the centre is dominated by low-quality tourist shops. On the side of the centre is a clock and several times a day a mechanical procession with music moves along the face of the building. Further along Coventry Street is the Trocadero Centre, an entertainment mall with shops, cafés, cinema, games arcades, and a funfair. The overall feeling of the street is of movement and entertainment, the latter as much from the colour and diversity of the street life, as from the various global-type destinations that line its edges.

Piccadilly Circus is a shock to the senses after Leicester Square and Coventry Street due to the huge flow of traffic. The wide pavement around the famous Eros statue and the Criterion Theatre building – now mainly occupied by Lillywhites sports shop – are obvious magnets for tourists and visitors, whilst the pedestrianised space and the Eros statue still

10.6 Portrait artists in Leicester Square

10.7 Coventry Street

10.8 Piccadilly Circus: a place of movement

10.9 The study spaces

possess a sense of a meeting and stopping place set amongst the constant flow of the busy vehicular and pedestrian traffic all around (Figure 10.8). The space, the statue and the properties surrounding it are public space, but in a form that is different from most other pavements and streets in London. They are not owned by the local authority, but instead belong to the Crown Estate, a corporate body that manages a portfolio of former royal properties on behalf of the state, including the famous Regents Street which extends from one corner of Piccadilly Circus up to Regent Park to the north.

Legibility

Using Lynch's (1960) analysis of urban elements, the gardens in Leicester Square represent a dominant node in the area and the epitome of the district with four entrances and exits. The gardens contain the landmarks of the central statue of William Shakespeare, and the smaller busts of famous residents of the square at each corner. The TKTS ticket booth (mirroring that in Times square) and the Odeon and Empire cinemas create strong visual landmarks around the edge of the square.

The route between Leicester Square and Piccadilly Circus contains no nodes, instead having the characteristics of a dynamic path. Landmarks include the Swiss Centre, and particularly the clock on the Leicester Square corner and the 'Swiss Cantonal Tree' on the west side of Swiss Court. Other landmarks are McDonald's, TGI Friday's, Planet Hollywood, and the entrance to the Trocadero Centre, itself an internal node.

Piccadilly Circus is an important node on a London-wide scale, as the convergence of five major and two minor streets. Although spatially poorly defined, it contains the Eros statue (Figure 10.2) as a visually and perceptually important landmark. Set amongst a wide surrounding pavement, this can be interacted with by standing and sitting around its raised steps. Other key visual landmarks include advertising on the

north-east corner, the London Pavilion, Tower Records, and Lillywhites. Collectively the images of Leicester Square and Piccadilly Circus are amongst London's most memorable and legible.

Land uses

The analysis of the land uses around the Leicester Square/Piccadilly Circus study area, as at Times Square, concentrated only on the framing buildings – those that face Leicester Square, Coventry Street, or Piccadilly Circus. For the purpose of this analysis the space was divided into three main spaces. Leicester Square is space 1; travelling east is space 2, which starts at Swiss Court on the easterly side, and leads down Coventry Street until Haymarket; space 3 is the most westerly and covers all the buildings and spaces around Piccadilly Circus. (Figure 10.9).

In general, patterns of land use in the study area are similar to those in Times Square with tourist and entertainment uses predominating. The main difference is the absence of office skyscrapers and therefore of a large tier of upper-floor uses. Another significant difference is in the large number of venues (bars, public houses and night clubs) that serve alcohol. Three large bars and two night clubs front onto Leicester Square, for example, with many more in the side streets leading off the square. The small size and low quality of the tourist souvenir shops in the study area also contrasts with those in Times Square.

The dominance of multinational companies and chain stores was evident from the land use survey. Not surprisingly, all the multiplex cinemas are linked to international operators: a 12-screen Odeon, a 7-screen UCI, and a 9-screen Warner cinema. The dominance of chain operators was also evident in the public houses and bars in the area, including: Yates, Moon Under Water, and All Bar One, all of which front on to Leicester Square. Global branded shops, cafés and restaurants such as Starbucks, McDonald's, Pizza Hut, TGI Friday's, Planet Hollywood, Haagen-Dazs,

10.10 Film symbolism in Leicester Square

Virgin Megastore, The Body Shop, HMV, Tower Records, and Sock Shop were all also present at the time of survey, sometimes with more than one branch, reinforcing a sense of similarity to equivalent public spaces elsewhere. Indeed, many of these also featured in Times Square. In Leicester Square and Piccadilly Circus this homogenisation of land uses is more obvious because most of the chains have done very little to tailor the design of their outlets (interior or exterior) to the location.

Other land uses include the two theatres, the Prince of Wales and the Criterion, the latter, at the time of survey, staging 'The Full Monty' which was playing simultaneously on Broadway in New York. However, the Leicester Square/Piccadilly Circus study area contained only two hotels, the Radisson Edwardian Hampshire and the Thistle, both of modest size when compared to those in Times Square. Small, low-quality tourist souvenir shops are also present.

Upper-floor land uses in the area are mainly offices, in buildings between five and seven storeys tall. Few are high profile in the sense that the upper-floor occupiers of Times Square are, although London's largest commercial radio station – Capital Radio – occupies space on Leicester Square. A small residential presence remains in the area.

Signage

The advertising signage in the study area hardly compares to the quantity and scale of signage in Times Square. Nevertheless, after Eros, Piccadilly Circus is mostly associated with electric advertising, and indeed it had an electric sign in 1890, a year before Broadway and Times Square (GLC 1980: 11). The signs also have a more immediate impact than those in Times Square, reflecting the relatively small size of the space and their relative dominance over it. Today, the electric adverts in space 3 of the study area are all in the northeast corner of Piccadilly Circus. They advertise a range of international brands, and make no reference to their location and context.

At the time of the survey, prosaic electric signs advertised the Japanese and Korean electronics firms of Sanyo, TDK, and Samsung, while two video screens advertised the twin American giants of McDonald's and Coca-Cola. There was also a small neon advert for the Danish beer Carlsberg. The electric signs all had moving neon lights, as opposed to the dominance of billboards in Times Square. There was also an electronic message reader, as in Times Square, repeating a set pattern of sports news, weather, the time, and advertising for Samsung. The civil/community function was largely negligible.

Finally, in early 2002 Piccadilly Circus also featured an exact double of the billboard message Yoko Ono posted in Times Square. The fact that Ono chose Piccadilly Circus as the only other place to display the message

emphasises the equivalence of the two spaces as perceived international hubs. This was the only large-scale signage element in Piccadilly Circus at the time that was not directly linked to commerce or corporate publicity.

Space 2 contained very little signage above fascia level. At the time of survey a basic video screen on the corner of Coventry Street and Haymarket advertised Vodaphone whilst zippers could be found above the entrance to the Trocadero Centre and Swiss Court. Like most of those in Times Square these zippers had no other function than to advertise the shops and attractions contained in the two centres. The only other signage was temporary in the form of a police sign on the corner of Coventry Street and Whitcomb Street asking for witnesses to a 'serious assault'.

Space 1, Leicester Square, contained advertising mainly through one huge billboard on the façade of the Odeon Leicester Square, advertising the films that were showing. This is part of a richer symbolism of cinema represented by film premieres that take place in the cinemas around the square, as well as actors' handprints on the pavement (Figure 10.10). The handprints extend signage to the ground just outside the railings of the gardens, with around 45 handprints from British and American actors and studio logos, in a similar display to that on Hollywood Boulevard in Los Angeles.

Other signage was limited but included small billboards advertising films, a video screen on the corner of Leicester Street and Leicester Square owned by the concert and event promoters, Mean Fiddler Group, and zippers on each side of the TKTS booth. Temporary signage was also found here, including two large boards on the north terrace of the square displaying messages warning criminals about the use of CCTV and undercover police in the area, and users to keep their valuables safe. In the centre of the gardens on the edge of the circular central path around Shakespeare's statue was a circle of brass plates giving the distances to 50 different cities in 50 Commonwealth countries. The display reflects the symbolism of the space as the historic 'hub of Empire', and compares with the 'crossroads of the world' symbolism of Times Square.

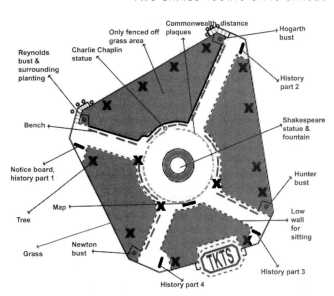

10.11 Plan of Leicester Square Gardens

Microanalysis: the area close up

Like the previous chapter, the following sections move from a general overview of the area and its components and seek instead a more detailed dissection of the three spaces that make it up. The spaces are examined to assess how ideas of civility, community and inclusion – historically present in public space – interact with the pressures for control, exclusion, homogenisation and commodification that would be expected in the core of a global metropolis.

SPACE 1

Space 1 is Leicester Square proper (Figures 10.11 and 10.12). In its centre, the gardens have historically been intended to foster behaviour associated with civility and community (see Chapter 2). As well as containing 30 benches, with each bench designed to take four people, there are low walls and grass for sitting on and trees to shelter under in summer. Each of the four busts is placed in a carefully tended flowerbed. The statues of Shakespeare and Chaplin will appeal to tourists, but neither exist to facilitate commerce, whilst the byelaws listed opposite the busts at each entrance to the gardens provide the major visible projection of state authority in the design of the gardens. The Westminster City Council logo also appears on the 24 litterbins in the gardens, projecting the presence of the state into an element of civility. The gardens are open to everyone during daylight hours and there is no visible exclusionary element at play, related either to social status, age or ethnicity. Nonetheless, the closure of the gardens at dusk is a reminder of the limits on public use set up by the controlling organisation, in this case, the elected local authority.

The street furniture around the square contains little reference to the presence of state or BID management and control. Indeed, the most visible displays of some form of management are the 16 red telephone boxes around the square, and the one letterbox at the southern tip of Leicester Place. However, even more than the litterbins in the gardens, the telephone and the letterboxes have become integral elements of London streets, and therefore are much more strongly associated with notions of civility and community. Other strong symbols of civility and community are the large public toilets on the north terrace – clean, free of charge and with a 24-hour attendant – and the bike racks on the south side of the square (Figure 10.13).

Reflecting what has arguably become the most surveilled city in the world (thisislondon.co.uk/news_2007) there were 19 CCTV cameras around Leicester Square alone, a fact that might suggest an Orwellian 'Big Brother' presence surveying and controlling the square and its users. In

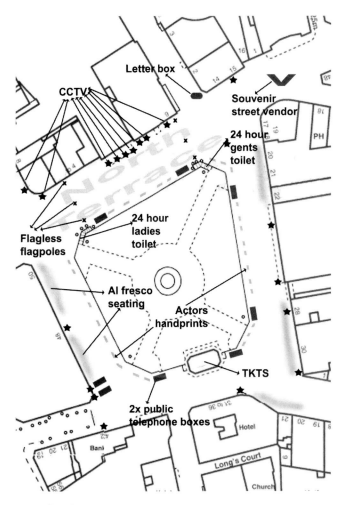

10.12 Plan of Leicester Square

10.13 Signs of civility: public toilets and telephone boxes in Leicester Square

fact, as is common in many English town centres, most of the cameras are owned and operated by private businesses around the square, and used for the surveillance of their own premises.

SPACE 2

Space 2 consists of Swiss Court and Coventry Street (Figure 10.14) and contained 16 more CCTV cameras. There are few elements to describe in space 2, apart from the more general features of the Swiss Centre and Court mentioned earlier. Of note, at the time of survey, were four licensed vendors grouped together, selling newspapers, souvenirs and ragga jungle mix cassettes, and through their position creating a focus for activities in that space. Street furniture was restricted to four telephone boxes and one letterbox.

SPACE 3

Space 3 is Piccadilly Circus (Figure 10.15). At its centre is the statue of Eros whose base has always been a place to sit and meet because it has steps that users can sit on and survey the misshapen circus and surrounding activity. In addition to Eros, another strong visual feature of the space is a large equestrian statue and fountain set into a recess on the corner of Piccadilly and Haymarket.

Piccadilly Circus is a very complex and busy traffic interchange as well as a major tube station with two lines of the London underground. Its role as a nodal public space – and the notions of civility and community this conveys – is represented by the five entrances to the underground station around the circus. The station also constitutes another layer of public space below ground, with shops, vendors and public toilets.

Surprisingly, there were fewer CCTV cameras in this space, eleven at the time of survey. Of the five cameras, two were directly linked to the control and management functions of the BID. These survey the statue and the steps, both monitored by the Leicester Square wardens from the

upstairs room of the TKTS booth. Apart from these cameras, there are no strong reminders of state or BID control in Piccadilly Circus.

A large number of licensed vendors trade in this area, and in the process gave the space some of its life, some selling souvenirs, others ice cream, and several selling newspapers. The newspaper vendors also displayed selected headlines on their stalls, providing an element of civic life in the form of news. There were only two telephone boxes in the area, both full of the ubiquitous prostitutes' cards that characterise public telephones in central London.

Managing the area

As with Times Square, the analysis attempted to determine how the expected increase in control and surveillance that tends to come with emerging forms of public space management was felt in the case study area. The fieldwork in the area took place in April 2002, well before the Heart of London BID was formed. At that time, only Leicester Square was under a special management regime, and wardens were beginning to appear as the visible embodiment of a new form of public space management.

In 2007, wardens patrol the whole of the BID area, and there are uniformed cleaners and police officers and the police station dedicated to the BID area. The existence of a system of active control of the space and its use is therefore far more evident, although still far from the idea of all-encompassing total control that is often conjured up by critics.

SPACE FOR SITTING AND LOITERING

In comparison to Times Square the London study area contains an abundance of spaces for sitting and loitering (Figures 10.16–10.18). Leicester Square Gardens has 120 bench spaces, and at least room for several hundred more people on the grass and the low walls. Unlike the Father Duffy statue at Times Square, Eros has not been railed off, and at least 50–100 people can and do sit on the steps around it.

The pedestrianisation of Leicester Square and the southern side of Piccadilly Circus in the 1990s also created far more activity and loitering space. The north side of Leicester Square is a favourite site for buskers and street performers, while the west side contains licensed pitches for street artists. This is in stark contrast to Times Square with its dominance by vehicular traffic. The wide pavements in Coventry Street and Piccadilly Circus allow for licensed and unlicensed vendors to bring activity to the space, although the space to loiter is countered by the dominance of CCTV cameras throughout the area.

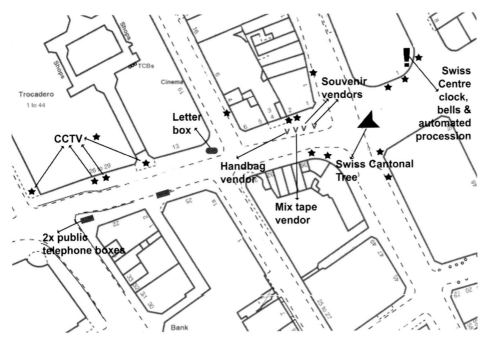

10.14 Plan of Coventry Street

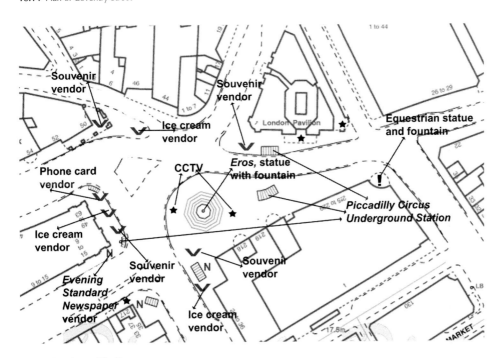

10.15 Plan of Piccadilly Circus

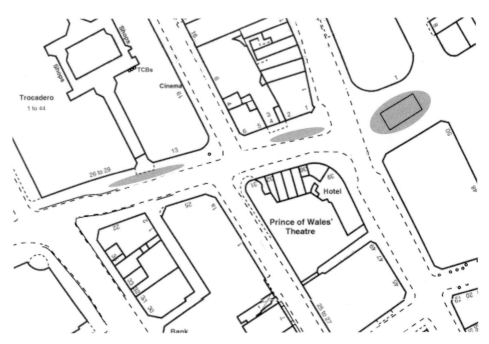

10.16 Sitting/loitering/activity space in Coventry Street

10.17 Sitting/loitering/activity space in Piccadilly Circus

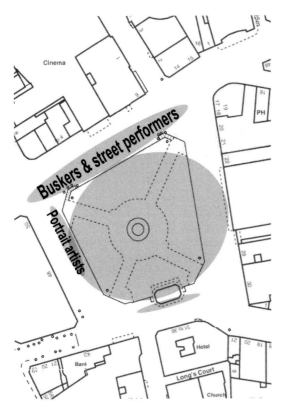

10.17 Sitting/loitering/activity space in Leicester Square

Control and surveillance

At the time of the survey, the main signs of control and surveillance were those traditionally found in urban public spaces elsewhere – council signs and services – together with the typically British phenomenon of CCTV monitoring. A total of 39 CCTV cameras were found in the study area, compared to just six in Times Square. Leicester Square alone contained at least 19 cameras, some of which were movable, some of which could cover 360 degrees.

As many users of the space report that the CCTV cameras make them feel safe and therefore more inclined to use the square and its businesses more often (EDAW 2001), it could be argued that these cameras are less an instrument of control and exclusion than facilitators of commerce, of interaction and of public space use, and thus of civility. The counter argument is that in spaces such as Leicester Square, with their strong appeal to consumption-orientated leisure, CCTV cameras, increasing surveillance and the potential regulation of impromptu activity are in fact constraining notions of civility and community by tying them to ideas of safe consumption. This display of civility and consumption combined can also be seen in the use of the public space by many eating and drinking establishments.

A survey of the position of Leicester Square wardens and police officers was completed on the same day and times as the survey in Times Square, a Saturday at 10.00am, 2.00pm, 6.00pm, 9.00pm and 1.00am. Significantly, very few management figures were in evidence: two wardens in or around the gardens at 2.00pm and 6.00pm, a single policeman

walking through Swiss Court, also at 2.00pm, and a pair of policemen on the south side of Leicester Square at 6.00pm. At Piccadilly Circus, a police van was parked on the pedestrianised area near Eros, and another was present from 6.00pm through to 1.00am in the junction between Coventry Street and Swiss Court. Compared to Times Square, there was comparatively little presence of control and management agents, even though weekends are generally a busy period. This absence of visible security has since been acknowledged as a problem by both Westminster City Council and the Metropolitan Police, and with the advent of the BID, the visible presence of management agents has greatly increased (Heart of London 2004b, 2007).

But, just as many of the CCTV cameras contributing to the surveillance of public space were privately owned, so too was some of the human security infrastructure. For example, many of the restaurants and bars in the area had bouncers and security of their own, including three security staff on a Saturday night in the Swiss Court McDonald's.

User behaviour

Finally an observation study of human movement, activity and behaviour in the London study area also sought to explore users' perception and interpretation of the public space and its ambience. Activity and behaviour was recorded, as for Times Square, over one week, to produce an activity/behaviour table, a sample from which is given in Table 10.1.

The most obvious difference between behaviour in the London study area and Times Square related to the consumption of alcohol. In London, alcohol was being consumed at all times of day, and the nightlife was nearly all alcohol-based, with pubs, bars and clubs serving until late. The law in the UK allows drinking in the street, unlike in New York where it is illegal and the ban is strictly enforced. A large variety of people were engaged in drinking, from the local homeless community, to tourists, to groups of office workers. On the Friday and Saturday nights, when this was far more intense, the result was an area that felt raucous, reeked of alcohol, and featured significant amounts of drink-related rubbish.

Due to pedestrianisation, street life in terms of performers and musicians was more prolific than in Times Square. Musicians on the north terrace were generally of a high quality, with traditional Chinese music, rock ballads, jazz and Latin music. Performers varied from circus-style tricks to mime artists. Others included a regular three-card monte man, and henna tattooists. All this gave Leicester Square a hustle and bustle that seems to have been removed from Times Square.

Conclusions

The Leicester Square/Piccadilly Circus area partially conforms to the characteristics attributed to contemporary public space in the literature. Homogenisation of the experience of the place under the pressure from a globalised economy and culture is manifest in the nature of the shops, restaurants and bars that dominate its space. In that regard, the London study area is not and does not feel dissimilar to the centre of many world cities.

Even though dominated by commercial uses, the area does not feel as commodified as might be expected for such a prominent public space. The existence of abundant space for sitting and loitering and the ease with which this exists under the management regime certainly contributes to that. Equally, the lack of any intense effort to market the place (such as that mounted by the Times Square BID) might also have contributed to a weak association between the public space of Leicester Square/Piccadilly Circus and consumption of that space as a brand. This might change as the Heart of London BID gets more established.

The indications of a monitored and controlled space were as strong here as they were in Times Square, albeit in different ways. Whereas a highly visible presence of management figures characterised control and monitoring of public space in Times Square, here it was the ubiquitousness of electronic surveillance through CCTV cameras. So far, the increasing control and monitoring through CCTV and through a growing presence of wardens and other authority figures has not lead to exclusion of the 'other' from the public space. The Leicester Square/Piccadilly Circus area still retains a degree of spontaneity and social mix which is stressed by the continuing presence of street performers, vendors and musicians and space users from a broad social background. In this sense, the London study area still retains a strong sense of vibrancy, civility and community (broadly defined), and the feeling of a sanitised space is not as strong as in Times Square.

As a consequence, in spite of similar management regimes, the public space of the Leicester Square/Piccadilly Circus area appeared more socially inclusive than that of Times Square, as the contrasting sample activity/behaviour tables illustrate (Tables 9.4 and 10.1). Whether this is related to different degrees of corporate control in both cases, or to different levels of tolerance for 'otherness' in the two societies, or to differing histories of spatial segregation by race and class, it is difficult to say.

However, it is also the case that direct private involvement in public space management in the Leicester Square/Piccadilly Circus area is still in its infancy and so far the trends are less stark than in Times Square. Indeed, one characteristic has been a more explicit partnership between public and private interests, with Westminster City Council taking a particularly prominent role in both the day to day management of the area, and in planning for the new management regime envisaged in the future. The area might never achieve the same degree of power and independence from elected local government that has been obtained in New York, but it will certainly evolve and consolidate. As it does, tensions between inclusion and exclusion, spontaneity/vibrancy and control/safety, private and community interests, are likely to grow and lead to the reshaping of management priorities and methods, and consequently to changes in the character of the public space.

Table 10.1 Study area sample activity/behaviour table

Event/behaviour	When/frequency	Place/position (see Figure 10.9)	Connection to design	Connection to management	Connection to 'the other' and dystemic space	Connection to place marketing	Photograph
People eating lunch and talking on grass. Indian group party with large picnic. 'Others' passed out on grass. Variety of users and 'others' drinking alcohol	During every day. Much busier when good weather	Space 1, Gardens	Three out of the four grass areas of the gardens can be used to sit on with no fence or attempt at defensible space	Gardens kept clean by private firm. Drinking is not allowed in the gardens, but wardens have no powers to remove alcohol from users	'The other' increases in early morning and just before closing at dusk. 'The other' is from a variety of cultural backgrounds	NA	
Users sitting on all 30 benches in gardens	Users even when weather bad	Space 1, Gardens	Users will sit on benches if provided	NA	Mixture of groups and individuals. Variety of activity	NA	
Users sitting on low wall	Depended on weather	Space 1, Gardens	Low wall acts as extra sitting space, not turned into defensible space	NA	Mixture of groups and individuals	NA	
'Human statue' standing on Union Jack, dispensing Covent Garden promotional magazine. One of three regular human statues	Appeared regularly during the week in the daytime	Space 1, north terrace	Large pedestrianised thoroughfare	Legal, no objection by wardens	Face obscured by balaclava and skiing mask	Promoting tourism through flag and Covent Garden magazine	
Homeless man busking by singing ad lib melodies through a traffic cone, and collecting change in a McDonald's cup	Various times of day and night by a variety of homeless people	Space 1, north terrace. Space 2, Coventry Street and Swiss Court	Large pedestrianised thoroughfare	Legal as not amplified. Police and wardens not bothered. Would be illegal if just begging with cup	'The other' is homeless. Visitors amused	Association of high profile public space and music, but unrelated to specific place marketing of Leicester Square or London	
Chinese musician playing traditional bowed instrument. Instrument case for donations/tips and CDs of his own for sale	Every afternoon/ evening. Same position on north terrace	Space 1, north terrace	Large pedestrianised thoroughfare	Amplified accompaniment. Illegal, but police do not object between 5-11pm through 'gentleman's agreement' with buskers. Wardens object as local businesses complain about noise and increased crime	Busker ethnic minority. Users stop and listen to music, non dystemic	Association of high profile public space and music, but unrelated to specific place marketing of Leicester Square or London	

Event/behaviour	When/frequency	Place/position (see Figure 10.9)	Connection to design	Connection to management	Connection to 'the other' and dystemic space	Connection to place marketing	Photograph
Man in traditional African dress performing circus style tricks to amplified music	Observed three times during one week	Space 1, north terrace	Large pedestrianised thoroughfare	Illegal and against agreement with police as before 5pm, police walk by and do nothing. Wardens threaten to ring Westminster enforcement team to boos from crowd. Performance continues for 1 hour with no sign of enforcement team	Busker Black, audience enjoy and partake with applause	Association of high profile public space and music, but unrelated to specific place marketing of Leicester Square or London	
Latin jazz guitar trio. Instrument case for donations/tips and CDs of trio for sale	Performed every evening	Space 1, north terrace	Large pedestrianised thoroughfare	Illegal, but within agreement with police as after 5pm	All performers ethnic minorities. Large number of users watching	Association of high profile public space and music, but unrelated to specific place marketing of Leicester Square or London	
Timotei shampoo advert and promotion	Observed once	Space 1, north terrace	Large pedestrianised thoroughfare	Legal	No users interested	Possible Leicester Square as centre of London theme for advert	
Users adapting movable fence to sit on when lent against garden railings. Also users sitting against railings	Various times	Space 1, north terrace	Large pedestrianised thoroughfare. No attempt to turn into defensible space	NA	Variety of users using the garden side of the north terrace to watch life going by, read, drink, or chat	NA	
Man juggling with flaming clubs	Observed once during the afternoon	Space 1, north terrace	Large pedestrianised thoroughfare	Legal, no amplified music	No users interested	Association of high profile public space and performing, but unrelated to specific place marketing of Leicester Square or London	
Three card monte man	Observed once during the afternoon	Space 1, north terrace	Large pedestrianised thoroughfare	Illegal, Gaming Act	Card man from Southern Italy. Variety of users watching and partaking. Would seem some of those partaking are 'plants'	Association of high profile public space and quick money	

Event/behaviour	When/frequency	Place/position (see Figure 10.9)	Connection to design	Connection to management	Connection to 'the other' and dystemic space	Connection to place marketing	Photograph
Jesus Army conversion team and car dispensing leaflets	Observed once during a Friday night	Space 1, north terrace	Large pedestrianised thoroughfare	Illegal to drive car onto square, otherwise legal	Variety of users stopping to talk with Jesus Army reps. All White	Association with public space and debate. Historic connection with religion	
Muslim conversion group dispensing leaflets	Observed once during Saturday night	Space 1, north terrace	Large pedestrianised thoroughfare	Legal	All Black	Association with public space and debate. Historic connection with religion	Did not want photo taken
Crowds of people, many drinking and in large groups	Every night, much worse at weekends	Space 1, north terrace	Large pedestrianised thoroughfare	Legal to drink in public space. Police and council thinking of implementing scheme of no alcohol bye law and giving the police powers to seize glasses or bottles	Huge variety of users	Leicester Square seen as area for drunk debauchery. Supported by council through number of late licensed premises	
The 'other' passed out	Observed several times in different places in the study area	Space 1, next to Bella Pasta	Space to sleep, though unsheltered	Illegal, Vagrancy Act. Police and wardens usually move unconscious 'others' on in day time	Drinkers from a variety of cultural backgrounds	Leicester Square and Piccadilly Circus are a magnet for alcoholics/ homeless/ beggars	
Al fresco dining and drinking	From lunchtime through to midnight	Space 1, west side of square	Separated from highway by bollards	Police claim al fresco dining is a focus of crime and oppose more. Vision of Westminster Council for new plan	The 'other' approaches tables to beg or commit crime?	Recreate Leicester Square as centre for eating and not drinking	
Portrait artists	Present from morning to dusk	Space 1, west terrace	Pedestrianised area	Legal. All artists have to pay for license	All portrait artists are foreign	Association of high profile public space and portrait artists, but unrelated to specific place marketing of Leicester Square or London	

Event/behaviour	When/frequency	Place/position (see Figure 10.9)	Connection to design	Connection to management	Connection to 'the other' and dystemic space	Connection to place marketing	Photograph
Unlicensed henna tattoo artist	Observed several times	Space 2, Swiss Court	Large pedestrianised thoroughfare	Illegal, just been fined £6,000 by magistrate. Made £70,000 last summer. Observed wardens trying to move man on, and him ignoring them	Italian	Association of London and punk tattoos	Did not want photo taken
One of two security guards. Also note CCTV	Security guards at weekends	Space 2, McDonald's doorway	-	Open 24 hours at weekend	Both security guards are Black. Huge variety of users in McDonald's	No attempt to match interior of restaurant to location	
Unlicensed hotdog salesman	Observed only on Saturday night	Space 2, outside KFC	Wide pavement	Police opposite did nothing, out of warden jurisdiction	Ethnic origin	Only association through Americanisation	
Old man playing harmonica with hat held out for change. Not playing any discernible melody	Observed only on Saturday night	Space 2, outside HMV	Man beneath canopy and in playing in area that is slightly indented	Legal	White man. No users interested	-	
Users left rubbish, mainly alcohol related, in fountain	Observed on Saturday night	Space 3, interior of Eros fountain	Place to drink (on steps) and dump rubbish	Not illegal to drink on and around statue	Large community of local drinkers who meet up daily and drink on statue steps. Always disappeared at night, presumed to be because they have to get into a hostel by a certain time	Association of Piccadilly Circus and hanging out	
Impromptu game of football with a ball stolen from Lillywhites	Observed during the afternoon	Space 3, Pedestrianised area outside Lillywhites	Large open space through pedestrianisation	Game continues unstopped until ball is lost in traffic	All players part of local community of Eros loiterers	Association of Piccadilly Circus and hanging out	
Users sitting on Eros statue steps	All times of day, very few people after 2am	Space 3, Eros steps	Comfortable area to survey local area, passing activity, meet people, and loiter	2 CCTV cameras surveying each side of statue. Cameras linked to Leicester Square wardens room	Tourist in foreground, local Eros community in background	Eros and London magnet for many	

Event/behaviour	When/frequency	Place/position (see Figure 10.9)	Connection to design	Connection to management	Connection to 'the other' and dystemic space	Connection to place marketing	Photograph
Large party of young language students sitting on Eros steps	Observed many times during the day	Space 3, Eros steps	Comfortable area for tourists to rest	NA	Tourists seem oblivious to local Eros community	Eros and London magnet for tourists	
Bagpipe busker	Every afternoon at 4pm	Space 3, outside Lillywhites	Large pedestrianised area	Legal	Appalling musician, but no one seems to notice	Association with bagpipes and the UK	
Urine in street	Evening and night	Various	Often in side streets just off main thoroughfares	Illegal. Council do provide toilets in Leicester Square, and mobile pissoirs for men during the weekend in Leicester Place	Most offenders seem to be drunk visitors rather than the 'other'	Association with London and alcohol	

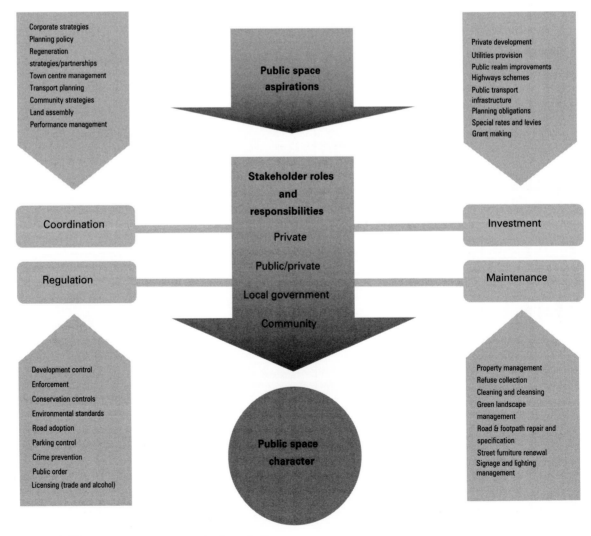

11.1 Public space management: a matrix of contributions

Chapter 11

Theory, practice and real people

This final chapter brings together the various strands of research reported in the book as a means to draw out conclusions about the management of public space in the post-industrial world. In doing so, gaps are identified between the increasingly dominant and accepted theory of public space decline exacerbated by management processes, and the realities of actual management practice on the ground. From the research a set of 13 key lessons are offered for the development of public space management practice in the future. In a postscript to the book, and in order to focus attention where it is really needed, the results of a fourth and final empirical study are briefly presented that traces what people really want from public space, and what is important to them in making these judgments. The study reveals that academic preoccupations are not always directly reflected in the lived experience of public space. The chapter and the book therefore draws to a close by briefly considering whether practice as it is developing, is meeting the challenges being laid down by the people that really matter, the everyday users of public space – the public.

The theory and practice of public space management

A matrix of contributions

In Chapter 1, the case was made that public space represents a hugely complex stage – physically, functionally, socially – and, as a consequence, managerially. A failure to understand this complexity, and to appropriately value the benefits that flow from high-quality public space seem to be amongst the key reasons for a widespread deterioration in the quality of public space around the world.

The complexity of the management task can be characterised as a 'matrix of contributions' that input into the overall process of public space management, and that impact either positively or negatively on the greater whole. This matrix of contributions is shaped by the stakeholders' combined objectives; operationalised through a wide range of discrete delivery processes; and is finally felt on the ground as collective outcomes that shape the character of public space. It encompasses:

1 Stakeholders' roles and responsibilities: the sixteen key stakeholder groups identified in Table 1.3; each encompassing a complex range of stakeholders with different roles, interests and influences.
2 Public space aspirations: which should, but it seems too often do not, inform public space provision and management. Ten key qualities of public space were defined in Table 1.2 that can be recast as aspirations for better quality public space.
3 Public space character: which is determined by the inherent quality of the 'kit of parts' that constitute the space (i.e. the uses and physical components of public space) and the socio-economic and physical/spatial context in which public space sits (see Figures 1.6, 1.7 and 1.8). This character is also decisively influenced by how space is managed.
4 Delivery processes: organised into the four key means through which stakeholders contribute to the management of public space (see Figure 4.1):
   • coordinating the actions of themselves and others;
   • direct investment in the public realm (either themselves or by levering-in resources from others;
   • regulation utilising statutory powers;
   • the ongoing processes of public space maintenance.

5   These four types of process each break down into a wide range of actions that those with public space management responsibilities engage in, and which are distributed across the range of public services.

The matrix is represented in Figure 11.1. The aspirational objective should be for all stakeholders to play their part in instigating processes and delivering outcomes that continue to change the character of public space for the better. This, however, will need to start by understanding the complexity itself, and the web of interconnected aspirations, processes, services and stakeholders that collectively manage (or not) public space. Mapping this web of connections for an indicative English local authority (see Figures 11.2–11.5), it is easy to understand why the connections so often are not made, and why the quality of public space continues to suffer.

Some lessons from history

This complexity is nothing new. As the brief review of public space through history contained in Chapter 2 revealed, a rich variety of functions, themes and meanings have always characterised public space. Indeed, a powerful lesson from history was that in providing for the multifarious needs of urban populations at large, public spaces are bound to contain a certain element of disorder and tension, and that this is part of the rich mix that makes public space eternally varied and fascinating. Conversely, a recurring struggle against disorder in public space has also long been a feature of urban management strategies, a struggle that takes many different forms, including, at its most extreme, pressures for privatisation, conformity, and exclusion. Positively, the history also suggests it can encourage civility, a sense of pride, aesthetic fulfilment, and help to facilitate economic, social and political exchange.

In retrospect, historic public space has often been idealised, depicting a much greater inclusiveness and participation than actually existed, something that has influenced criticism of the management of contemporary public space. This tendency to look back with rose-tinted spectacles had been exacerbated in recent years by the undoubted impact of mass consumption and globalisation that characterises post-industrial economies. These find expression in perceived pressures for a more actively managed public realm, and a homogenisation in the character of public space.

Throughout history, the dominant issue dictating management strategies has been (and remains) the balance between public and private power and responsibilities, with private interests often seeking to mould or even remove public space to meet their own commercial and social objectives. Equally, even the most perfunctory analysis of public space

through the ages would reveal that this use of power to favour the interests of some groups over others has not always been one way, and that the state and its organisations has often been the instigator of practices designed to both control and exclude. It would also reveal the massive impact of management, as opposed to original design, on how public space is used and perceived, and, as a result, on how the quality and users of space can change – often dramatically – over time.

The changing forms of public space

Another relatively recent lesson from history relates to the form that public space takes, and to the explosion in the twentieth century of forms of public space that go against the centuries of producing 'positive' spaces that act as places for exchange, as well for communication. Instead, the spread of modernism, and more recently the all-pervasive impact of private transportation, has generated a range of new public space types, many of which are entirely 'negative' (as far as the experience they offer to people on the ground), and which throw up a diversity of new management challenges.

In Chapter 3, a new typology of public space was offered that demonstrates this complexity (see Table 3.1). Twenty distinct types of urban public space were classified, four of which are 'negative' forms of space, ten are ambiguous in terms of their role and ownership, and three are entirely private. Increasingly the negative and ambiguous forms of space have come to dominate the contemporary urban landscape, breaking down the sharp traditional divisions between public and private, and in so doing blurring boundaries between management responsibilities. Today, cities are made up of a patchwork of management responsibilities, reflecting in turn the patchwork of public space types, and requiring as a result a far more integrated, negotiated and nuanced approach to public space management than has been the case in the recent past.

Critiquing public space

It is perhaps the absence of such approaches, however, that is leading to the overwhelmingly negative critiques of public space seen in the literature and explored in Chapter 3. In summary, those responsible for the design and management of contemporary public space have been criticised for:

- neglecting public space, both physically and in the face of market forces;
- sacrificing public space to the needs of the car, effectively allowing movement needs to usurp social ones;

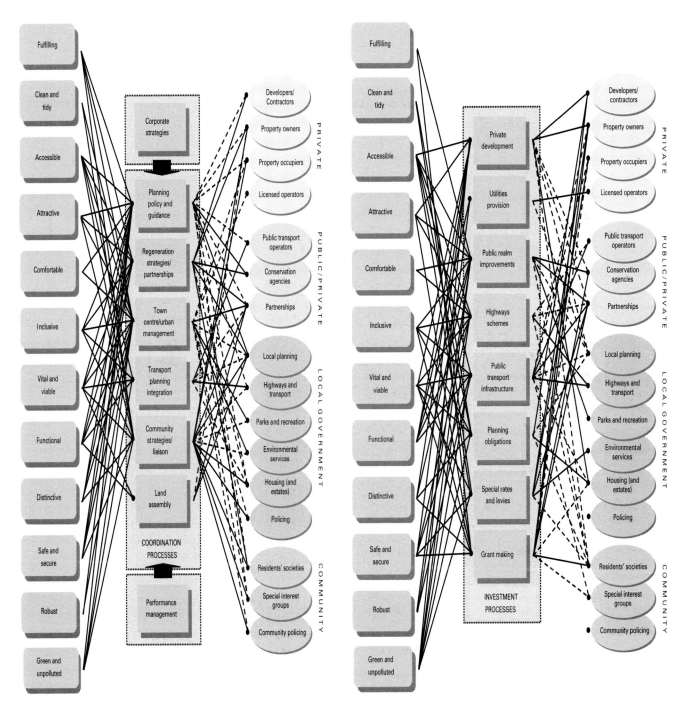

11.2 An indicative coordination web

11.3 An indicative investment web

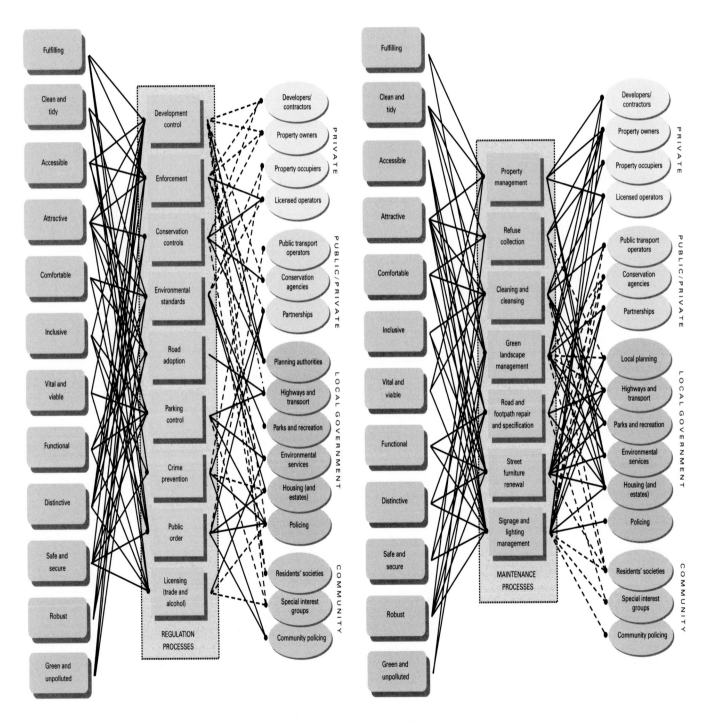

11.4 An indicative regulation web

11.5 An indicative maintenance web

- allowing fear of crime and the stranger to dominate public space design and management strategies;
- failing to address the needs of the least mobile and most vulnerable in society;
- allowing public space to be commercialised and privatised, with a knock-on impact on political debate and social exclusion;
- failing to halt a retreat from public space into domestic and virtual space
- condoning the spread of a placeless formulae-driven entertainment space;
- generally presiding over a homogenisation of the public built environment in the face of the relentless forces of globalisation.

Given that the critiques (particularly the first three) are so widely and consistently made, it is hardly surprising that private interests are choosing to turn their back on public space in favour of the more controlled, specialised and ambiguous forms of space that are increasingly seen across the Western world. But whether these critiques are any more pertinent today than in the past remains a moot point, with arguments also made that the reported decline in the quality of public space has been much exaggerated. Instead, counter arguments go that we are simply seeing new forms of contemporary space that, although different, are not necessarily any less worthy or valuable than those they displace, and which reflect (as public space has always done) the fragmentation and complexities of society.

Whether this is accepted of not, the challenge for the managers of public space is to work with the new forms of space and the increasing diversity of stakeholders to achieve the best outcomes within any given context. The best outcomes may very well be those that use management strategies to overcome the critiques that are prevalent in the literature, and to do that, the critiques themselves must be aired, debated and understood.

A new governance context

Notwithstanding the relevance of the key critiques, the debate often tends to assume a somewhat dichotomous view of the complex history and dynamics of urban government, and overlooks the complex processes through which rights, roles and attributions regarding public spaces are continually defined and re-defined. In fact the meaning and function of public spaces, their forms of management, and the distribution of power and responsibilities over both are all contingent on the historical context of places and their governing practices.

Therefore, as argued in Chapter 4, recent trends in the management of public spaces need to be seen as part of the context-specific process whereby 'government' is being replaced by 'governance'. Changes to public space management can therefore also be understood as the re-shaping of the specific sets of institutional arrangements structuring this field of policy, and that this is happening in a context of more general changes to urban governance more widely. Today, no one single social actor can claim to have all the solutions for the policy problems at hand, and the management of public space can no longer be seen as the exclusive, or even necessarily the natural, province of the public sector.

Changes in public space management are therefore a reflection of wider changes in the relationship between government, especially local government, and society, including both community and private interests. This embedding of public space management changes into wider changes in urban governance provides a framework for understanding the positive and negative potential implications. It raises the possibility that governing routines and coordinating mechanisms that have served public space needs well in one particular context and time might have become a problem under different conditions.

The management response

Despite the changes and debates over who legitimates it, and how, the purpose of public space management has not changed, and was defined in the book as

> the set of processes and practices that attempt to ensure that public space can fulfil all its legitimate roles, whilst managing the interaction between, and impacts of, those multiple functions in a way that is acceptable to its users.

Within this broad definition, public space management was subdivided into four key interlinked processes – coordination, investment, regulation and maintenance – reflected in Figure 11.1 and discussed in Chapter 4.

These processes apply whichever organisation(s) or sector(s) is/are delivering the actual services. But increasingly, the evidence suggests that the highly specialised and fragmented models of local government that had grown up from the middle of the twentieth century were increasingly serving the management of public space poorly in the late twentieth and early twenty-first centuries. Thus although the activities that make up the various functions of managing public space largely existed in local government, public space itself was rarely an explicit policy focus. More often than not it was simply the context within which a range of disparate management activities occurred, and public space as a concept was often limited to the parks and iconic civic spaces that make up only a tiny portion of the public realm.

The switch from government to governance was nevertheless decisive in encouraging an increased focus on the complex, crosscutting and seemingly intractable policy problems, of which recent trends in the management of public space were a part. It has meant a redefinition of how such services are funded and delivered, including a new emphasis on the potential of private and community stakeholders to play a role across each of the four process of public space management. Today, in different places, different management approaches are being adopted, each representing variations or combinations to different degrees of three key management models:

- the state-centred model, the traditional approach to public service delivery offering clear accountability in the public interest but often at greater cost, unresponsiveness and bureaucracy;
- the market-centred model, allowing a greater flow of resources from a wider constituency, and often greater efficiency and responsiveness, but at the cost of fragmentation and the potential for exclusion and commodification;
- the community-centred model, offering a greater sensitivity and commitment to local user needs, but with the danger of fragmentation and inequality in the provision of services.

Management on the ground: everyday practice

The choice involves an increasingly complex trade-off between different dimensions of service quality and control that also underpin many of the critiques of contemporary public space and its management. It is predicated on the idea of an enabling public sector, rather than an all-delivering one, and in places like the UK this has led to a multiplicity of delivery agencies, actors and organisations; although not necessarily to better services. In England, for example, the New Public Management reforms of the 1980s and 1990s led to a switch from state to market provision, although, the weight of evidence presented in Chapter 5 suggests, in pursuit of a cost-cutting agenda, rather than one of service improvement. The quality of public space was a prime casualty, a process that continued well into the second New Labour administration (post-2001) when the political significance of what had been seen until then as the rather prosaic concern for the quality of the local environment, began to dawn.

Surveying the state of public space management in England in the immediate aftermath of this realisation, the evidence suggested that although this area of public sector responsibility was in need of significant reform, the new emphasis from national government was beginning to inspire a burgeoning range of bottom-up initiatives from local government below. Elsewhere, services were fragmented, partial, and lacking any real vision about how they might be improved in the future, indeed the range of stakeholder groups interviewed concluded that problems across coordination, investment, regulatory and maintenance dimensions of the public space management process were endemic. In such places, the concept of public space as a complex yet unified single entity was completely lacking, and instead, the focus remained on the delivery of discrete tasks, that may, but usually did not, add-up to an integrated management strategy.

Where evidence was found of more sophisticated practice, often this has been accompanied by a redefinition and redistribution of roles and responsibilities within the state sector, and between the state and the private and community sectors. Indeed, amongst many stakeholder groups, the limitations of the state-centred model were obvious, and greater use of market and community-centred models were seen as desirable. This, however, needs to be achieved on the basis of a mutually supportive three-way partnership, with an end to the 'us' and 'them' exploitative philosophy. Overwhelmingly, however, the balance of power, and therefore responsibility, for this area of policy still remains with local government, and moves away from state-centred provision have been tentative.

Management on the ground: innovative practice

The political realisation that the quality of public space represented a significant local political factor in England, led, from 2001 onwards, to an increasing range of policy pronouncements, reports, and initiatives designed to shake up the sector. The outcome was an endorsement of local government's role as the central provider of public space services, whilst extending the rights of private interests to play a decisive role through the creation of business improvement districts (BIDs). Although the results of the national survey tended to endorse the introduction of BIDs (no doubt reflecting the poor performance of the public sector across large swathes of the country), the range of innovative local authorities explored in Chapter 6 tended to confirm the British government's faith that local government could – given the right level of resourcing, support and know-how – substantially improve public space services.

This phase of the work largely confirmed the complex range of barriers to the better coordination, regulation, investment and everyday maintenance of public spaces, but also a willingness to work with community and private sector partners to overcome these. The former focused primarily on involvement rather than the devolution of power, and amounted to a community-oriented, rather than a community-centred view. The latter also shied clear of a full-scale transfer of power, and instead amounted to a pragmatic willingness to work with the private sector as and when market involvement was seen to deliver benefits. Both the community and private sector where therefore viewed as partners, rather than drivers of the public space management agenda. By contrast, these more proactive local authorities increasingly saw their role as the guardians of a more controlled environment, in what they saw as the interests of the majority, rather than of any particular groups.

The reality seems to be more complex than much of the literature would have us believe, and indeed that in a minority of places, the type of more integrated, negotiated and nuanced approaches to public space management that contemporary public space seem to require are beginning to be delivered. Rather than a battle between private- and public-sector interests, with the community squeezed out altogether, the reality is more often a limited transfer of powers to a range of stakeholders, with the public sector still in the driving seat. In such places, authorities are more concerned to establish a seamless network of public space that is all subject to the same high standards of management, rather than, necessarily, continuous public ownership or management responsibility.

To some degree these new collaborative arrangements seem to be emerging strongly (if inconsistently) in this field of public sector management because public space management is a new area of policy. It suggests a new acceptance of the need to structure services and policies around an emerging view of public space as a holistic entity, and the focus for policy. In turn, this more defined policy focus on public spaces is placing these services in a better position to argue for greater attention and resources, whilst the emergence of multi-sector public space management mechanisms such as town centre management, area management partnerships, neighbourhood management schemes, and (most controversially) BIDs, are the clearest examples of this.

Management on the ground: exceptional practice

In Chapters 9 and 10, the opportunity was taken to explore the implications of one of these models – BIDs – through an analysis of Times Square and Leicester Square/Piccadilly Circus. On both sides of the Atlantic, this new market-centred delivery vehicle is being used to manage these iconic civic spaces; spaces that at one time or another have faced all the pressures (and critiques) explored in the public space literature. The analysis suggested that to a greater or lesser extent the spaces conform to the characteristics so often ascribed to contemporary public space – displacement of some groups, homogenisation and commodification, surveillance and control, and a reduction in elements that foster civility and community. Moreover, that these characteristics are both fostered and legitimised by the management models being adopted.

Yet, despite these characteristics, several positive aspects could also be identified. First, the cited historic functions of urban public space

were all still evident, even though actors and symbols had changed. Second, while there is a clear trend towards the facilitation of commerce via consumption in these spaces – encouraged by the management practices – consumption itself fosters scope for civility and community, for example, through cafés, restaurants and shops (for those that can afford to consume). Indeed, it can be argued that consumption of globalised brands and entertainment remain a popular choice for city users, giving in the process new meaning to public space in a context where elsewhere it is being eschewed. Third, arguably, the majority also have a preference for omnipresent management and surveillance in public space, whilst, to some degree the contemporary characteristics of these spaces also give order to an otherwise often fragmented public realm.

Fourth, in each of the spaces, the characteristics that gave them their status and reputation in the first place had largely been preserved, and continue to include a rich variety of activities and obvious tolerance for difference and diversity. Finally, in each case the management regimes have charted a deliberate path 'up-market', which has contributed to the displacement of some dispossessed groups (particularly in Times Square). However, with this, have come benefits for tourists, performers, businesses and the other everyday users of these public spaces.

The analysis revealed that, although still a legitimate cause for concern, the extension of private involvement to the management of public space does not automatically lead to high levels of intolerance and control and to an irreconcilable shift in the balance of power in public space. Instead, it confirmed that even in the most high-profile examples of where this is happening, the characteristics that make public space 'public' are typically robust. It confirmed that the shift to a more market-led approach can even reduce the control required on public space through pursuit of the sorts of liveable qualities sought in Chapter 1 (see Table 1.2). In this respect, the more balanced public/private approach so far seen in Leicester Square and Piccadilly Circus, where the public sector has been able to reclaim large areas of space from private traffic, may be better able to lever the advantages of both sectors, to the benefit of both, and the wider community.

The international context: adding value

Moving from this explicitly localised view on the practices and paradoxes of public space management, to a broader world-view, and at the same time from a focus on internationally iconic public space, to everyday public space, it is possible to draw out recommendations for future practice with broader relevance. In England, the interplay of national initiatives and local responses is undoubtedly shaping a new policy field that is more effective, integrated, responsive to local circumstances, and confident about its role and significance. Looking beyond the UK, it is equally clear that the field is developing equally quickly in a wide range of cities globally. Focusing on public open (green) space, Chapters 7 and 8 first examined the context for public space management in 11 cities around the world, before focusing on the delivery of management via the four key management processes.

Although none of the cities examined would regard their public space management practice as beyond criticism, all reported considerable benefits from their emphasis on green space quality. These benefits were not accidental, but resulted from the emphasis on, and investment in, their public open spaces and their management, and the associated efforts by municipal agencies to promote these benefits to a wider audience. They included:

- the enhanced reputation of the cities for their high quality living environments;
- their enhanced reputation for sound urban governance;
- city marketing benefits in the light of the increasingly competitive economic environment;
- raised environmental awareness and citizen involvement;
- social benefits through better health, accessibility, recreational opportunities and quality of life.

Lessons from an international stage

Each of the benefits goes well beyond the immediate policy objectives of public space management, and therefore potentially have wider local political advantages for the responsible city administrations. The common experiences from the international cases, combined with those from the English cases explored in Chapter 6 give rise to thirteen key lessons:

1. START WITH POLITICAL COMMITMENT

The need for strong political commitment to deliver public space quality was reinforced throughout the research. Success in the management of public spaces seems to result from a mix of political will by successive administrations, reinforced by the technical skills of public space managers. It is also self-perpetuating, with a positive perception of public spaces leading to greater political commitment and so on. Thus political and administrative commitment needs to exist side by side if a strong organisation to manage public space – both strategically and operationally – is to be built. This is likely to require support for public space issues at all levels of the administration and across the political spectrum. The inclusion

in the local government structures of at least one cabinet-level politician with direct responsibility for public spaces would seem a minimum starting point to build a greater political commitment to public space.

2. MAKE A LONG-TERM STATUTORY COMMITMENT

A long-term commitment went hand-in-hand with a political commitment, as a prerequisite for not only delivering high-quality public space, but for ensuring that quality remains high thereafter. This commitment requires foresight, long-range planning and the fostering of a wider civic commitment to urban public spaces. The direct public benefits from sustaining high-quality public space in cities that have managed to do so over a long timeframe are significant. In such cities, in different ways, the management of public space is invariably a statutory responsibility of the city authorities, and the need to invest in the management of public space is therefore non-negotiable. A carefully constructed set of statutory public space roles and responsibilities might therefore create the necessary incentive to raise the quality of public space management (to at least a minimum acceptable level) in places where practice is currently poor.

3. TAKE A STRATEGIC (POLICY) VIEW

A clear public space strategy can help to ensure that public space priorities infuse other key policy areas, including spatial planning, giving public space management a welcome continuity regardless of political changes and helping to consolidate the importance of public space management in relation to other services and priorities. A statutory provision for local authorities to create public space strategies as an element of their spatial planning framework might offer the necessary incentive to deliver a more strategic and community-centred view of public space. Such strategies and plans should include a clear spatial vision for public (and private) space, as well as policies for the provision, design and long-term management of public spaces. They should provide the basis for more detailed public space maintenance plans to structure, coordinate, and resource the day-to-day delivery of public space maintenance.

4. ADAPT TO LOCAL CIRCUMSTANCES

The research illustrated the importance of taking a coherent local view on public space management that adequately reflects the priorities of local populations. The aspiration should be that public spaces remain a matter of social and cultural interest in order that citizens are convinced that public spaces are necessary elements for the life and identity of their environment. The continual engagement of residents as users and customers was the favoured approach, hand-in-hand with improvement in communications between city administrations and their citizens. Cultural issues seem to play an important role in determining the nature and extent of participation, and the public attitude generally to public space. Apathy of some local populations towards public spaces might be addressed through a far greater emphasis on proactively educating local citizens about the benefits of public space, and by involving them more directly in public space decision-making processes.

5. DELIVER ADEQUATE AND RELIABLE RESOURCES

Not all of the successful case study cities were generously funded, but all were funded to a level that allowed them to at least meet their ongoing management responsibilities. A key lesson was therefore that there is not only a need for adequate funding, but also for reliable sources of funding over the long term. Long-term rolling funding plans, for example, allow administrations to commit themselves to projects spanning several years. This means that the constraints of annualised budgetary rounds need to be overcome in order to ensure longer-term planning for public spaces, whilst the capital and revenue funding available for public space management should be clearly published at both local and national levels to allow adequate local scrutiny of available resources. As the most successful case studies suggested, the need to protect revenue funding streams is paramount in order that maintenance can be prioritised across existing public space networks. The value of exploiting all potential supplementary income streams was also demonstrated. However, these funds should be collected and spent directly by public space management departments and need to be viewed strictly as additional funding over and above core income streams.

6. MAKE THE CASE INTERNALLY

Winning resources against other competing claims represents a key and increasing skill amongst public space managers. This requires strong leadership and the strength of conviction and ability to present public space issues to key political and organisational audiences. Publicising public space successes to both internal and external audiences may be an important part of this process in order to secure political support and a willingness to spend. Public space managers therefore have to be advocates for the benefits of high-quality public space, not least of their soft economic benefits. Public space managers need also to understand that half the battle lies in repeatedly demonstrating the value they add

through their work, in so doing garnering cross-political and public support. Indeed, the most successful international examples are founded on this ability to continually make the case for resources to a wide range of audiences.

7. INVEST IN THE SKILLS BASE

The key to success in some of the case studies was a well-trained and engaged staff that knew how to combine political, economic, organisational and design skills and how to take advantage of the variety of opportunities available to them. This requires a stable staff in order to build up detailed knowledge and expertise of the diversity of public spaces. This also requires a continual renewal and investment in skills, not just at management levels, but also at the operational end of public space management. In this regard, departments staffed with marginalised, low-status staff were never found in the successful cities. Elsewhere, the transformation of public space management services from the Cinderella service of local government to a first division service will require a similar and continual investment in staff. The creation of dedicated degree programmes and continual professional development opportunities in the sector may offer a valuable starting point. The aim should also be to create long-term stability in organisational structures so as to nurture staff stability and commitment and the building of internal links.

8. CONSISTENTLY FOCUS ON QUALITY

The case studies illustrated both the dangers of an emphasis on quantity over quality, resulting in the provision of standardised public spaces with little regard to the needs of surrounding communities, but also the benefits of a long-term commitment to high-quality public space for generating lasting economic, social and environmental value. The latter requires high-quality, robust design solutions designed to reflect positive aspects of the original character and context. Public space managers need to be involved from the start in the design and planning of new public spaces, as do skilled landscape designers in the ongoing management of existing spaces; particularly as and when new interventions are planned. A key lesson is that designing high-quality, resilient public spaces can not only save on public space resources through the proper consideration of lifetime costs, but can ensure that local communities engage more fully in their ongoing management through the provision of the spaces they want, rather than simply the spaces that policy says they need.

9. EMPHASISE EFFICIENCY (BY DEVOLVING RESPONSIBILITY)

A considerable emphasis is placed on efficient management by the most successful cities. The reduced cost that flowed from such approaches seems to require, first, an investment in modern management methods, learning where appropriate from private-sector practices, second, the introduction of clear and direct decision-making structures, and third, investment in a skilled and specialised workforce. On the issue of localised versus centralised management and operations, benefits were apparent in both models, and the balance needs to struck in the light of local circumstances. The key aim should be to establish the optimum cost/quality ratio by distinguishing those elements of the service that are best devolved to the neighbourhood level, from those that require a more strategic organisation. This can be achieved through a clear typologically-driven view of public space, with management strategies for particular types of space defined by their function, ownership, and perception (see Table 3.1) and by local and national aspirations.

10. INVOLVE OTHERS

Different cases reported success with both heavily privatised, and largely public models, and all combinations in between. Most cities saw this relationship as a partnership that needed nurturing and careful management over the long-term. Thus the aim should be good collaborative relationships that aim to increase expertise and responsibility for quality on the part of contractors and the creation of a transparent but competitive environment for the authority. Dogmatic approaches to service delivery should be rejected in favour of carefully considering which aspects of public space management can be more efficiently and effectively delivered by the private sector, and which are best left to the public sector. The former are likely to be the more routine and easily specified maintenance tasks, whilst tasks requiring a greater degree of creative interpretation and adaptability in the field might be retained in-house. Other key stakeholders may also beneficially have a direct role in the management of public space. Examples in the case studies included voluntary and community groups, users in all their guises, educators, health professionals, private-sector operators, and other relevant local government departments.

11. INTEGRATE RESPONSIBILITIES (BY COORDINATING ACTIONS)

The imperative to coordinate local government public space responsibilities with the public space activities of other organisations was clear. This can be achieved by devolving responsibilities to a lower level to better integrate service delivery at the coalface. Equally, integration at a more strategic level is valuable to secure broader buy-in to public space management objectives. In both models, the benefits of having one strong central organisation with responsibility for all or the majority of public space management functions was evident. The proviso remains, however, that it is more important that aspirations and actions are coordinated, than that ownership and responsibilities for public space reside in one place. This requires a simple commitment to work in an integrated manner within and between all organisations and stakeholders. Fully integrating responsibilities for public space management in one organisation nevertheless remains a laudable aspiration. A step on the road may be the more frequent dissociation between the ownership of public space and its management.

A related issue concerns enforcement powers, which need to be taken more seriously, properly resourced, and coordinated with other public space management activities. The need for proper feedback loops between enforcement work and policy, design and maintenance activities was a key finding from the UK and international case studies. Without joining up these roles to other public space management activities, the quality of public space can be quickly undermined.

12. CONSIDER A DEDICATED MANAGEMENT MODEL

Dedicated and semi-independent agencies seem to have been particularly effective at achieving their ends, in part because of the absence of competing calls on expenditure. Unfortunately, the conditions that have made such models successful are not always easy to replicate as the political and financial independence required and the narrow focus on public spaces stands outside of normal local government structures. As a consequence, it is highly unlikely that local governments today would relinquish tax-raising powers and political accountability to, for example, an independent parks agency, except in exceptional cases. The latter might include relatively rare but nevertheless important circumstances where new settlements or other major developments are being planned, and where it is important to capture the rising land value to pay for long-term management needs. In such circumstances, a hypothecated funding model might offer the appropriate tool. More common, however, will be the establishment of dedicated agencies within the purview of local government of along the lines of the BIDs model. For both, care should be taken to ensure that clear public interest goals are reflected from the start in their management practices and priorities.

13. MONITOR INVESTMENTS AND OUTCOMES

Monitoring activities ranged from regular assessments of management performance, to more fundamental systems designed to both record existing, and play a part in delivering new, quality. The benefits included both more efficient and effective maintenance processes and more outcome-focused management. The need to accurately record the state of public space, and thereafter to monitor the delivery of public space management goals should therefore not be underestimated as a means to ensure that other policy and management goals are being delivered. The most sophisticated systems might track depreciation of assets over time so that the condition of new investments can be monitored and lessons leant, and so that costs can be factored into ongoing work programmes as part of a continuum of replacement and maintenance activities. In localities where these systems are largely absent, they can bring significant benefits in a context where continual improvement is dependent on adequate feedback to inform decision-making.

So to conclude

The five years during which the research in this book was conducted have represented a time of change during which public space policy has continued to develop. However, as argued above and in more detail in Chapter 6, this is still a new policy field (internationally), and its further development will be dependent on further research to understand the appropriate scope, nature and limitations of public space management.

To this process the research reported over the previous 10 chapters can contribute the following insights:

- The 'rose-tinted' view of the nature of public space through history, accompanied by the 'doom-laden' view of the nature of public space today are actually two sides of the same coin; neither are correct.
- Public space is in fact the site of massive economic and social potential; potential that can be either suppressed or released by management practice.
- Fears over exclusion and commodification are real and significant, but despite the impression given in the literature, privatised space remains only a tiny proportion of the total in most Western countries.
- For the majority of public space, the public sector will remain the dominant provider of public space management services.

- Nevertheless, there is a great opportunity to supplement public services by tapping into the real economic gains that the private sector derives from better quality public space.
- A system that involves all stakeholders in caring for the quality of public space should be the aspiration.
- This is particularly the case in a context where the public sector have often done such a poor job when left to their own devises, in particular on those issues that really matter to people – the provision of clean, safe and fulfilling streets (see postscript below).
- There is no moral or practical superiority of one model (state-, market- or community-centred) over another, each, and different combinations of them, can provide the right solutions in particular contexts.
- The key is to recognise the advantages and disadvantages of each and from there decide where and how they should be appropriately used.
- The aim should always be to deliver the 'public good', whilst avoiding any unintended consequences, perhaps through the safeguards offered by tight legal agreements, planning conditions, strong enforcement, coordinated partnerships, and the checks and balances provided by an overview role for the public sector for all public space.

All this suggests that it is time to stop being dogmatic about the management of public space, and instead to embrace pragmatic solutions, using whatever balance of approaches and responsibilities is appropriate locally, and that delivers the most effective public space management service. The empirical research explored in Part Two of the book illustrated a range of approaches to provision on a continuum from fully devolved to entirely public in provision, and each can be made to work given the right resources, commitment and vision. Management organisations and strategies should be put in place to achieve this. The thirteen lessons set out above give some indication of what will be required.

Postscript: but what do people really want?

Notwithstanding the conclusions above, in this final postscript to the book, the discussion finishes where it began, in the continuing search for an understanding of public space and in particular of what constitutes 'high-quality' public space. In this instance the analysis is based on user perceptions of public space and on what that means for attempts to develop appropriate management responses. The lessons above derive from a broad cross-section of experiences from around the world, but wherever they are implemented, it will be important to ask: have we understood properly what people really want?

A final empirical research study, conducted as the book was being compiled, concluded that actually it is the run-of-the-mill issues that dominate the concerns of everyday users, across social scales, and that largely these pertain to the run-of-the-mill, everyday types of public space that exist all around us (Carmona and de Magalhães 2007). The project focused on the measurement of local environmental quality, including, but going beyond, public space, and building on the foundation provided by the research reported in this book. It asked:

- What are people's aspirations for the quality of their local environment?
- Which aspects are important and which are less so?
- Does this vary from context to context and community to community?

Research methodology

To address this part of the work, a qualitative survey of attitudes and aspirations involving 12 focus groups distributed geographically around the English regions was undertaken. Locations were chosen to take in communities from a range of socio-economic and physical contexts (inner city, suburban, rural), whilst groups were selected to reflect a balanced distribution of age, ethnicity, family circumstances (children or no children) and household type and tenure. Groups consisted of around eight residents, and were focused on establishing, first, the basic parameters by which people judge their local environment, and, second, what are realistic, meaningful and consistent definitions of acceptable standards.

A second stage of the work brought together key stakeholders (professional and political leaders) from the various communities involved in the research to discuss the perceptions emerging from the qualitative survey. Two workshops of this nature were undertaken, each comprising a half-day session with around 20 key people covering local councillors, local government officers, private contractors and representatives from community groups and interested NGOs. Both the focus groups and workshops used the 'universal positive qualities' for public space identified in Chapter 1 as the basis for discussion and analysis (see Table 1.2). These collectively summarise a broad range of inter-connected and inter-dependent dimensions of 'quality' as identified in the literature. They were used as a tool to 'drill down' beneath the surface of headline environmental qualities, and to understand in some depth how the quality of public space is perceived.

What users really want

The focus groups revealed that people generally find it difficult to discuss qualities of their local public space in an abstract way, and found some qualities more difficult to understand than others, e.g. 'functional' (described for the purposes of the focus groups as 'can be used harmoniously for a variety of purposes'). Participants in the focus groups generally felt that many of the qualities overlapped, and often cross-referenced between the different qualities e.g. 'clean and tidy' and 'robust' (the latter described for the groups as 'well-maintained'). The professionals had a similar reaction, with some concern that terms would be difficult for their user communities to comprehend.

With prompting, however, both sets of participants (public and professionals) were able to grasp each of the 12 qualities and understand their importance. Although they sometimes had a different take on the qualities, they were nevertheless able to identify and articulate a range of sub-qualities or issues that each encompassed. As such, there was no quality that the participants regarded as unimportant, all qualities have significant merit, and all contribute to how public space is perceived. All were also seen as inter-related in complex and mutually reinforcing ways.

Despite this, some qualities were regarded as particularly significant in helping to improve or undermine the quality of people's lives. 'Clean and tidy', 'safe and secure' and fulfilling (understood by many in the focus groups as engendering a sense of 'community and belonging' were of this type. At the other end of the scale, qualities such as 'attractive', distinctive' and 'functional'[1] tended to be cited.

Partly explaining the priorities was a belief that some of the qualities related more to the initial design of an environment than to its subsequent management, and therefore that aspects of these concerns were fixed and not open to influence (at least in the short-term). The aesthetic quality and distinctiveness of buildings fell into this category (confirming the discussion of the 'kit of parts' in Chapter 1). Although it was recognised that such aspects contributed strongly to the quality of space, and residents either liked them or not, they did not feel able to change them, and therefore such concerns were not generally prioritised.

Focusing on the qualities singled out in the focus groups as either more or less important, with other qualities sitting somewhere in between, a hierarchy of qualities can be constructed (Figure 11.6). Seen in this way, some qualities might be regarded as more fundamental than others, although:

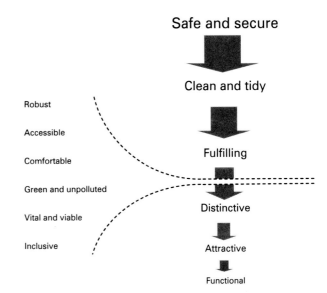

11.6 A hierarchy of universal positive qualities for public space

- it was clear that lower order concerns were not considered unimportant, simply lower priorities;
- each of the lower order concerns were, in different ways, understood to be intimately related to the higher order ones;
- the more satisfied local communities are with their local environment, the more they focus on, and are critical of, the lower order concerns.

The true test of high-quality public space, therefore, will be one in which success in each of the qualities is achieved.

What standards of quality are expected

Answers to the more tricky question of what are or are not acceptable levels of quality were difficult to address during the research, with both the professional and non-professional audiences finding it hard to articulate what is or is not 'acceptable' in any given context. For professionals, levels of acceptability are dictated by public expectations which differ between contexts, and which are dictated by levels of resource, consultation, and, in some (more affluent) areas, by levels of complaints. For them, receipt of complaints means that levels of unacceptability are being reached; conversely, a lack of complaints signifies satisfaction.

However, the analysis of public aspirations and attitudes revealed the problematic nature of such assumptions. For many communities the research confirmed that levels of quality are not satisfactory, but are not so unacceptable that residents and users are driven to complain. In other words, they are resigned to accepting the level of quality that they are used to. Instead of articulating what is an acceptable quality for a particular dimension of the public space agenda, they tend to simply prioritise one quality over another; prioritisation that varies between individuals.

Whether residents should be able to drive levels of quality was open to debate, with some concerned that such activity inevitably shifts resources to more affluent places; a finding supported by research reported in Chapter 1. (Hastings et al. 2005: viii–ix). Others argue that services should be more responsive to resident demands and perceptions. Despite the debate and inherent difficulties, public consultation was seen by the professionals to be an essential tool for gauging levels of satisfaction with the local environment and with the provision of public space management services.

The difficulties experienced by professionals and the public alike in articulating how they judge levels of acceptability in the quality of the local environment meant that it was not possible to clearly identify commonly held perceptions of what exactly is 'quality'. Nevertheless, most of the non-professional participants were able to indicate the kind of factors that influenced whether they felt positively or negatively about

their neighbourhood. By contrast, the professional audiences found this difficult to do, seemingly often preferring to discuss definitions of the dimensions of quality, rather than levels or quality, and preferring to rely on user complaints rather than professional judgements to identify negative factors.

Table 11.1 summarises and compares views on acceptability across these two constituencies, summarising the range of positive and negative factors that were identified as being important in determining perceptions of local acceptability. The analysis revealed that a range of factors are noticeable by the regularity with which they feature in different categories, particularly the visual signals of how well a place is looked after:

- anti-social behaviour
- state of repair e.g. roads, street furniture, etc.
- general cleanliness
- levels of lighting
- availability of facilities for young people
- perceptions of security
- parking/traffic problems
- visual quality/greenery
- walkability/ease of movement
- feeling of community cohesion.

Other factors were noticeable by their absence, particularly any concerns for commercialisation, privatisation or homogenisation that so dominate much of the academic literature discussed in Chapter 3.

Moving practice forward

This final piece of research confirms one important conclusion from the research reported above, and adds three additional findings. It confirms, first, that many professionals continue to think in silos, and find holistic, cross-cutting concepts of public space quality difficult to engage with.

Table 11.1 Perceptions of ascceptability: positive and negative factors

	Public positive	Professional positive	Public negative	Professional negative
Clean and tidy	Small quantities of litter are to be expected (acceptable)	Absence of litter, flytipping, graffiti Well-mown verges	Traffic fumes leading to poor air quality and dirty walls Litter (of all types) Dog foul Needles Graffiti Chewing gum Rubbish from shops / takeaways Rubbish bags piled up	Commercial rubbish
Accessible	Accessible for less mobile Good signposting Access by foot Adequate parking Adequate public transport	Good signposting and access to information Barrier free environments Good lighting Disabled access Perceptions of safety	Traffic congestion Cracks and holes in the pavement Lack of pedestrian crossings Problems caused by bad parking Children playing in the street Lack of parking provision Problems caused by deliveries	
Attractive	Trees, greenery, planting, flowers Maintained green areas Christmas lights Architectural quality Building maintenance Clean and tidy Murals Coordinated signage/street furniture Good street lighting	Architecture and heritage Clean and well-maintained Public art Coordinated street furniture	Vandalism Graffiti New housing estates	
Comfortable	Better-maintained benches, shelters, public toilets Green, well kept and attractive Confident and safe Walkable space Good street lighting Police on the street Adequate parking and signage Traffic calming	Ease of living in an area Feeling at home Continuity of care e.g. dedicated police, street cleaners, etc. Transport access Parking convenience Sustainability	Poor quality benches, shelters, public toilets Graffiti problems Broken glass Traffic congestion and noise Litter and cleanliness Lack of parking Potholes	
Inclusive	Adequate facilities for teenagers Tolerant of minority groups Welcoming to all users	Accessible for disabled Mixed communities Mixed age profile	Teenagers hanging around Poor integration of different groups Late night noise	Racism and ageism
Vital and viable	Variety of shops and services Availability of cash points High occupancy levels Building renovations Community spirit/interaction Events and activities Essential shopping available locally Healthy housing market Feeling of safety and community	Community satisfaction	Derelict buildings Litter, vandalism and fights Inundated streets	Level of dereliction

	Public positive	**Professional positive**	**Public negative**	**Professional negative**
Functional	Use without intimidation or danger Separate facilities for young people Suitably calmed traffic Controlled parking, balancing different users	Easy parking Free flowing	Congestion Parking problems Lack of play space Illegal activities e.g. drugs dealing	
Distinctive	Socially distinctive e.g. friendly, ethnic mix, relative affluence Physically distinctive e.g. features, history, buildings Availability of facilities and amenities Well maintained historic fabric	Distinctive features, history, buildings Visitor satisfaction Distinctive landscape	Possessing a bad 'reputation' Physically bland	Levels of deprivation
Safe and secure	Perception of personal security Child physical safety Freedom from intimidation A well cared for place – looks safe Low level disorder is acceptable (e.g. drunkenness) Visible police presence	Feeling safe and secure	Discomfort and fear at night Poor lighting Obvious drugs paraphernalia Threatening groups Frequency/quality of road crossings Obvious illegal activities	High perceptions of crime Signs of anti-social behaviour Poor lighting Unkempt environment Speeding and traffic problems
Robust	Parks in good condition Pavement and road condition (some wear and tear acceptable – small potholes/slightly uneven paving) Good lighting Tree and shrub maintenance Accessible paths Flower displays Condition of community facilities Safe, accessible and well signed Buildings well-maintained	Road and pavement quality Longevity of surfaces Resilience of street furniture Street lighting Building maintenance	Potholes and uneven paving (when it can cause an accident)	Roads being dug up Chewing gum Graffiti and vandalism
Green and unpolluted	Level of greenness Flowers and colour Fresh air Low traffic and congestion Open space in walking distance (e.g. 10 minutes) Well-maintained play areas Cycle lane provision Recycling facilities Well tended (but not like private gardens)	Keeping healthy Air quality Well-kept flowers and plants	Poor quality green space Rubbish, litter and dog foul Poor surveillance Poor lighting Visible air pollution Poor quality seating Anti-social behaviour Failure to replace trees/ planting Overgrown foliage	Noise pollution
Fulfilling	Interaction with neighbours Feeling comfortable (at home) Community spirit Levels of involvement Events and activities Facilities for young people	Community engagement (all sections of society) Sense of belonging / satisfaction Information in different languages	Intimidation leading to alienation Transient communities e.g. students, bedsit tenants Rapid in-migration Feelings of insecurity	Increased personal mobility

Instead they focus on the limited objectives of particular services, and not on what each service contributes to the whole, to what the public actually experience.

Everyday public space users, by contrast, find it difficult to break their view of the local environment down into its constituent parts, because they do not think in that way. Instead:

- They take a holistic view of public space, and equate the quality of their local environment directly to broad socio-physical constructs such as their sense of community.
- Certain factors repeatedly emerge as key priorities for individuals using public space – safety and security, cleanliness and tidiness, and a sense of belonging – as do a wide range of other interrelated factors that they do not immediately associate with this agenda, for example, how attractive an area is, the levels of pollution, or whether retail units are in active use.
- Levels of deprivation influence these priorities and perceptions of local environmental quality, with some (particularly lower income) communities more accepting of the levels of quality they are provided with than others.

The key challenge will be to cut through the complexity whilst raising the game through extending the notion of holistic public space quality across all services with a role to play in its delivery. In essence this typically means dealing with the unglamorous everyday stewardship of public space that impacts so disproportionately on users' sense of well-being, pride and belonging. This means actively managing streets to keep them safe, secure, clean and tidy.

The evidence presented earlier in this book suggests that even some of the most advanced societies have a long way to go to meet the basic aspirations for high quality public space demanded by the people that really matter; the public. There is nevertheless significant cause to be optimistic, with an increasing number of cities and communities – worldwide – recognising the importance of such concerns, and putting in place the necessary policy, resources, regulatory and management frameworks to deliver on this agenda.

Notes

1   As described in the focus groups, 'functional' was the least understood term, perhaps explaining its lowly rating;

Bibliography

Ackroyd, P. (2000) *London: The Biography*, London, Chatto and Windus.

Andersen, H. and R. van Kempen (eds) (2001) *Governing European Cities: Social Fragmentation, Social Exclusion and Urban Governance*, Aldershot, Ashgate.

Appleyard, D. (1981) *Livable Streets*, Berkeley, CA, University of California Press.

Atkinson, R. (2003) 'Domestication by cappuccino or a revenge on urban space? Control and empowerment in the management of public spaces', *Urban Studies*, 40(9): 1829–43.

Audit Commission (2002a) *Street Scene*, London, Audit Commission.

Audit Commission (2002b) *Community Safety Partnerships*, London, Audit Commission.

Aurigi, A. (2005) *Making the Digital City: The Early Shaping of Urban Internet Space*, London, Ashgate.

Bailey, N. (1995) (with A. Barker and K. MacDonald) *Partnership Agencies in British Urban Policy*, London, UCL Press.

Banerjee, T. (2001) 'The future of public space: beyond invented streets and reinvented places', *APA Journal*, 67(1): 9–24.

Baulkwill, A. (2002) 'Lots of conviviality', *The Garden*, September 2002: 693–7.

Beck, U. (1992) *Risk Society: Towards a New Modernity*, London, Sage.

Bell, K. (2000) *Urban Amenity Indicators: The Liveability of Our Urban Environments*, Technical Paper No.63 – Urban Amenity, Wellington, New Zealand Ministry for the Environment.

Ben-Joseph, E. (2005) *The Code of the City: Standards and the Hidden Language of Place Making*, Cambridge, MA, MIT Press.

Ben-Joseph, E. and T. Szold (2005) *Regulating Place: Standards and the Shaping of Urban America*, London, Routledge.

Bentley, I. (1998) 'Urban design as an anti-profession', *Urban Design Quarterly*, 65: 15.

Bentley, I. (1999) *Urban Transformations: Power, People and Urban Design*, London, Routledge.

Bentley, I., A. Alcock, P. Murrain, S. McGlynn and G. Smith (1985), *Responsive Environments: A Manual for Designers*, London, Architectural Press.

Berman, M. (1997) 'Signs of the Times', *Dissent*, 44(Fall): 76–83.

Berman, M. (1999) 'Too much is not enough: metamorphosis of Times Square', in L. Finch and C. McConville (eds), *Gritty Cities: Images of the Urban*, Annandale, NSW, Pluto.

Blakely E.J. and M.G. Snyder (1997) *Fortress America: Gated Communities in the United States*, Washington DC, Brookings Institution Press, and Cambridge, MA, Lincoln Institute of Land policy.

Boyer, C. (1994) *The City of Collective Memory: Its Historical Imagery and Architectural Entertainments*, Cambridge, MA, MIT Press.

Boyer, M. (1993) 'The city of illusion: New York's public places' in P. Knox (ed.) *The Restless Urban Landscape*, Englewood Cliffs, NJ, Prentice Hall.

Brill, M. (1989) 'Transformation, nostalgia and illusion in public life and public space', in I. Altman and E. Zube (eds) *Public Places and Spaces*, New York, Plenum Press.

Broadbent, G. (1990), *Emerging Concepts of Urban Space Design*, New York, Van Nostrand Reinhold.

Brook Lyndhurst (2004a) *Liveability and Sustainable Development: Bad Habits and Hard Choices*, London, ODPM.

Brook Lyndhurst (2004b) *Research Report 11: Environment Exclusion Review*, London, ODPM.

Buchanan, P. (1988) 'What city? A plea for place in the public realm', *Architectural Review*, 1101: 31–41.

Burgers, J. (ed.) (1999) *De Uistad: Over Stedelijk Vermaak*, Utrecht, Van Arkel.

CABE (Commission for Architecture and the Built Environment) (2002) *Streets of Shame, Summary of Findings from 'Public Attitudes to Architecture and the Built Environment'*, London, CABE.

CABE (Commission for Architecture and the Built Environment) (2005a) *Does Money Grow on Trees?*, London, CABE.

CABE (Commission for Architecture and the Built Environment) (2005b) *Decent Parks? Decent Behaviour?* London, CABE.

CABE (Commission for Architecture and the Built Environment) (2007) *Living with Risk: Promoting Better Public Space Design*, London, CABE.

CABE (Commission for Architecture and the Built Environment) and ODPM (Office for the Deputy Prime Minister) (2002) *Paving the Way, How We Achieve Clean, Safe and Attractive Streets*, London, Thomas Telford.

CABE Space (2004a) *Is the Grass Greener…? Learning from International Innovations in Urban Green Space Management*, London, CABE.

CABE Space (2004b) *The Value of Public Space, How High Quality Parks and Public Spaces Create Economic, Social and Environmental Value*, London, CABE.

Canter, D. (1977) *The Psychology of Place*, London, Architectural Press.

Carmona, M. (2001a) 'Implementing urban renaissance – problems, possibilities and plans in south east England', *Progress in Planning*, 56(4): 169–250.

Carmona, M. (2001b) *Housing Design Quality Through Policy, Guidance and Review*, London, Spon Press.

Carmona, M. and C. de Magalhães (2007) *Local Environmental Quality: A New View on Measurement*, London, DCLG.

Carmona, M. and L. Sieh (2004) *Measuring Quality in Planning, Managing the Performance Process*, London, Spon Press.

Carmona, M., J. Punter and D. Chapman (2002) *From Design Policy to Design Quality: The Treatment of Design in Community Strategies, Local Development Frameworks and Action Plans*, London, Thomas Telford.

Carmona, M., T. Heath, T. Oc and S. Tiesdell (2003) *Public Places, Urban Spaces: The Dimensions of Urban Design*, Oxford, Architectural Press.

Carr, S., M. Francis, L.G. Rivlin and A.M. Stone (1992) *Public Space*, Cambridge, Cambridge University Press.

Castells, M. (1996) *The Rise of the Network Society*, Oxford, Blackwell.

Chatterton, P. and R. Hollands (2002) 'Theorising urban playscapes: producing, regulating and consuming youthful nightlife city spaces', *Urban Studies*, 39(1): 95–116.

Chesluk, B.J. (2000) 'Money jungle: race and real estate in "The New Times Square"', University of California at Santa Cruz, PhD Thesis.

Circle Initiative (2000) *The Circle Initiative*, London, The Circle Initiative.

Circle Initiative (2001) *The Circle Initiative: One Year On*, London, The Circle Initiative.

City of Westminster (2002a) *Consulting on the Future of Leicester Square*, London, Westminster City Council.

City of Westminster (2002b) *Leicester Square Action Plan*, London, Westminster City Council.

City of Westminster (2006) *Future West End: Creating a World Class Location*, London, Westminster City Council.

Civic Trust (2002) *Open All Hours? A Report on Licensing Deregulation by The Open All Hours?* London, The Civic Trust.

Clarke, M. and J .Stewart (1997) *Handling the Wicked Issues: A Challenge for Government*, Birmingham, INLOGOV, University of Birmingham.

Clout, H. (ed.) (1991) *The Times London History Atlas*, London, BCA.

Coleman, A. (1985) *Utopia on Trial: Vision and Reality in Planned Housing*, London, Shipman.

Colin Buchanan and Partners (2007) *Streets Paved With Gold: The Real Value of Good Street Design*, London, CABE.

Congress for the New Urbanism (1993), 'Charter for the New Urbanism', http://www.cnu.org/charter (accessed 5 February 2008).

Conolly, P. (2002) 'The human deterrent', *Regeneration and Renewal*, 4 December: 16–17.

Cornes, R. and T., Sandler (1996) *The Theory of Externalities, Public Goods and Club Goods*, Cambridge, Cambridge University Press.

Crang, M. (1998), *Cultural Geography*, London, Routledge.

Davies, M. (1992) *City of Quartz*, New York, Vantage.

DCLG (Department for Communities and Local Government) (2006) *Strong and Prosperous Communities*, London, DCLG.

DCLG (Department for Communities and Local Government) (2007) *The Contribution of Neighbourhood Management to Cleaner and Safer Neighbourhoods*, London, DCLG.

Deakin, N. (2001) 'Putting narrow-mindedness out of countenance: the UK voluntary sector in the new millennium', in H. Anheier and J. Kendall (eds) *Third Sector Policy at the Crossroads*, London, Routledge.

DEFRA (Department for Environment, Food and Rural Affairs) (2002) *Living Places: Powers, Rights, Responsibilities*, London, DEFRA.

DEMOS (2005) *People Make Places: Growing the Public Life of Cities*, London, DEMOS.

DETR (Department of the Environment, Transport and the Regions), (1998), *Modern Local Government – In Touch with the People*, White Paper, London, DETR.

DETR (Department of the Environment, Transport and the Regions) (2000) *Our Towns and Cities: The Future: Delivering an Urban Renaissance, The White Paper on Urban Policy*, London, DETR.

DETR (Department of the Environment, Transport and Regions) and CABE (Commission for Architecture and the Built Environment) (2000), *By Design: Urban Design in the Planning System: Towards Better Practice*, London, DETR.

Diabetes Prevention Group (2002) 'Reduction in incidence of type 2 diabetes with lifestyle intervention or Metformin' *New England Journal of Medicine*, 346: 393–403

Dines, N. and V. Cattell (2006) *Public Spaces, Social relations and Well-being in East London*, Bristol, The Policy Press.

Ditchfield, P.H. (1925) *London's West End*, London, Jonathan Cape.

DoE (Department of the Environment) and ATCM (Association of Town Centre Managers) (1997) *Managing Urban Spaces in Town Centres: Good Practice Guide*, London, HMSO.

DTLR (Department for Transport, Local Government and the Regions), (2001) *Strong Local Leadership – Quality Public Services*, London, DTLR.

DTLR (Department for Transport, Local Government and the Regions) (2002a) *Green Spaces, Better Places: Final Report of the Urban Green Spaces Taskforce*, London, DTLR.

DTLR (Department for Transport, Local Government and the Regions) (2002b) *Improving Urban Parks, Play Areas and Open Space*, London, DTLR.

Duany, A. and E. Plater-Zyberk with J. Speck (2000) *Suburban Nation: The Rise of Sprawl and the Decline of the American Dream*, New York, North Point Press.

EDAW (2001) *West End Public Spaces Study: A Review of the Management Issues and a Strategy for the Future*, Final Report to Westminster City Council, London, EDAW.

Ellickson, A. (1996) 'Controlling chronic misconduct in city spaces: Of panhandlers, skid rows and public-space zoning', *Yale Law Journal* 105(March): 1172.

Ellin, N. (1996), *Postmodern Urbanism*, Oxford, Blackwell.

Ellin, N. (1999), *Postmodern Urbanism* (revised edn), Oxford, Blackwell.

ENCAMS (2002) *Local Environmental Quality Survey of England (LEQSE)*, Wigan, ENCAMS.

Engwicht, D. (1999), *Street Reclaiming: Creating Liveable Streets and Vibrant Communities*, Gabriola Island, BC, New Society Publishers.

Fabian Society (2001) *Roads for People: Policies for Liveable Streets*, Policy Report 56, London, The Fabian Society.

Fainstein, S. (2001) *The City Builders: Property Development in New York and London, 1980–2000*, 2nd revised edn, Lawrence, KS, University Press of Kansas.

Fainstein, S. and D. Gladstone (1997) 'Tourism and urban transformation: Interpretations of urban tourism', in O. Källtorp, I. Elander, O. Ericsson and M. Franzén (eds), *Cities in Transformation: Transformation in Cities: Social and Symbolic Change of Urban Space*, Aldershot, Ashgate.

Fjortoft, I. (2001) 'The natural environment as a playground for children: The impact of outdoor play activities in pre-primary school children', *Early Childhood Education Journal*, 29(2): 111–17.

Flusty, S. (1997) 'Building paranoia', in N. Ellin (ed.), *Architecture of Fear*, New York, Princeton Architectural Press.

Ford, L. (2000) *The Spaces Between Buildings*, Baltimore, MD, The Johns Hopkins University Press.

Franck, K. and Stevens, O. (2007) *Loose Space: Possibility and Diversity in Urban Life*, London, Routledge.

Franks, M. (1995) 'Covent Garden: The end of an era', *City*, 1(1/2): 113–21.

Frontier Economics (2004) *Quality of Place and Regional Economic Performance*, London, Frontier Economics.

Fyfe, N. (ed.) (1998) *Images of the Street, Planning, Identity and Control in Public Space*, London, Routledge.

Garreau, J. (1991), *Edge City: Life on the New Frontier*, Doubleday, London.

Gehl, J. (1996) *Life Between Buildings: Using Public Space* (3rd edn), Skive, Arkitektens Forlag.

Gehl. J. and L. Gemzøe (1996) *Public Spaces, Public Life*, Copenhagen, The Royal Danish Academy.

Gehl, J, and L, Gemzøe (2000) *New City Spaces*, Copenhagen, The Danish Architectural Press.

Giddens, A. (1999) 'Risk and responsibility', *Modern Law Review*, 62(1): 1–10.

Girouard, M. (1990) *The English Town*, London, Guild.

GLA (Greater London Authority) (2001) *Green Spaces Investigative Committee: Scrutiny of Green Spaces in London*, London, Greater London Authority.

GLC (Greater London Council) (1980) *Piccadilly Circus: From Controversy to Reconstruction*, London, GLC.

Golby, J.M. and A.W. Purdue (1984) *The Civilisation of the Crowd: Popular Culture in England 1750–1900*, London, Batsford.

Goldsmith, M. (1992) 'Local government' *Urban Studies*, 29(3/4): 393–410.

Goldsteen, J.B. and C.D. Elliott (1994) *Designing America: Creating Urban Identity*, New York, Reinhold.

Gospodini, A. (2004) 'Urban morphology and place identity in European cities: Built heritage and innovative design', *Journal of Urban Design*, 9(2): 225–48.

Goss, J. (1996) 'Disquiet on the waterfront: Reflections on nostalgia and Utopia in the urban archetypes of festival marketplaces', *Urban Geography*, 17(3): 221–47.

Goss, S. (2001), *Making Local Governance Work: Networks, Relationships and the Management of Change*, Basingstoke, Palgrave.

Graham, S. (2001), 'The spectre of the splintering metropolis', *Cities*, 18(6): 365–8.

Graham, S. and S. Marvin (1999), 'Planning cybercities? Integrating telecommunications into urban planning?', *Town Planning Review*, 70(1): 89–114.

Graham, S. and S. Marvin (2001), *Splintering Urbanism: Networked Infrastructures, Technological Mobilities and the Urban Condition*, London, Routledge.

Grisso, J.A., K.L. Kelsy and B.L. Stom (1991) 'Risk factors for falls as a cause of hip fracture in women', *New England Journal of Medicine*, 324: 1326–31.

Gulick, J. (1998) 'The "disappearance" of public space: an ecological Marxist and Lefebvrian approach' in A. Light and J. Smith (eds) *The Production of Public Space*, Oxford, Rowman and Littlefield.

Hajer, M. and A. Reijndorp (2001) *In Search of New Public Domain*, Rotterdam, NAI Publishers.

Hajer, M. and H. Wagenaar (eds) (2003*) Deliberative Policy Analysis: Understanding Governance in the Network Society*, Cambridge, Cambridge University Press.

Hakim, A.A. (1999) 'Effects of walking on coronary heart disease in elderly men: the Hondulu Heart Program', *Circulation*, 100: 9–30.

Hall, E.T. (1966) *The Hidden Dimension: Man's Use of Space in Public and Private*, New York, Doubleday and Co.

Hall, P. (1996) *Cities of Tomorrow*, (updated edn), Oxford, Blackwell.

Hall, P. (1998) *Cities in Civilisation*, London, Weidenfeld & Nicolson.

Hall, P. (2000) 'Creative cities and economic development', *Urban Studies*, 37(4): 639–49.

Hall, P. and R. Imrie (1999), 'Architectural practices and disabling design in the built environment', *Environment and Planning B: Planning and Design*, 26(3): 409–25.

Halpern, D. (1995) *Mental Health and the Built Environment*, London, Taylor and Francis.

Hammond, L. (2002) 'The New Urban Public Space', University of London, MPhil Thesis.

Harding, A. (1998) 'Public-private partnerships in the UK', in J. Pierre (ed.) *Partnerships in Urban Governance: European and American Experience*, Basingstoke, Macmillan.

Hartig. T., G.W. Evans, L.D. Jamner, D.S. Davis and T. Garling (2003) 'Tracking restoration in natural and urban field settings', *Journal of Environmental Psychology* 23(2): 109–23.

Harwood, E. and A. Saint (1991) *London*, London, HMSO.

Hass-Klau, C., G. Crampton, C. Dowland, and I. Nodd (1999) *Streets as Living Space: Helping Public Places Play their Proper Role*, London, Landor.

Hastings, A., J. Flint, C. McKenzie and C. Mills (2005) *Cleaning up Neighbourhoods, Environmental Problems and Service Provision in Deprived Areas*, Bristol, The Policy Press.

Heart of London (2004a) Heart of London BID Proposal.

Heart of London (2004b) Heart of London BID supporting documentation.

Heart of London (2006) *Quarterly Review July–September 2006*, London: Heart of London.

Heart of London (2007) Renewal Proposal 2007–2012.

Heckscher, A. (1977) *Open Spaces: The Life of American Cities*, New York, Harper and Row.

Hill, D. (2000) *Urban Policy and Politics in Britain*, Basingstoke, Macmillan.

Hillier, B. (1996) *Space is the Machine*, Cambridge, Cambridge University Press.

Hiss, T. (1991) *The Experience of Place*, New York, Alfred A. Knopf.

House of Commons Environmental Audit Committee (2004) *Environmental Crime: Fly-tipping, Fly-posting, Litter, Graffiti and Noise*, London, The Stationery Office.

Huxtable, A.L. (1991) 'Re-inventing Times Square: 1990' in W.R. Taylor (ed.), *Inventing Times Square: Commerce and Culture at the Crossroads of the World*, New York, Russell Sage Foundation.

ICE (Institution of Civil Engineers) (2002) *The 2002 Designing Streets for People Report*, London, ICE.

Improvement and Development Agency (2003) *A Year of Liveability Challenges 2003–2004*, London, I&DA.

Imrie, R. and P. Hall (2001) *Inclusive Design: Designing and Developing Accessible Environments*, London, Spon Press.

Jackson, P. (1998) 'Domesticating the street: The contested spaces of the high street and the mall', in N.R. Fyfe (ed.) *Images of the Street: Planning, Identity, and Control in Public Space*, London, Routledge.

Jacobs, J. (1961 [1984]) *The Death and Life of Great American Cities: The Failure of Modern Town Planning*, London, Peregrine Books.

Jacobs, A. (1993) *Great Streets*, Cambridge, MA, MIT Press.

Jacobs, A. and D. Appleyard (1987) 'Towards an urban design manifesto: A prologue', *Journal of the American Planning Association*, 53(1): 112–20.

Johns, R. (2001) 'Skateboard City', *Landscape Design*, 303: 42–4.

Jones, P., M. Roberts and L. Morris (2007) *Mixed Use Streets: Enhancing Liveability and Reconciling Conflicting Pressures*, York, The Policy Press.

Kayden, J. (2000) *Privately Owned Public Space: The New York City Experience*, New York, John Wiley.

Kilian, T. (1998) 'Public and private, power and space', in A. Light and J.M. Smith (eds) *Philosophy and Geography II: The Production of Public Space*, Lanham, MD, Rowman & Littlefield.

Kingsford, C.L. (1925) *The Early History of Piccadilly, Leicester Square, and Soho*, Cambridge, Cambridge University Press.

Kitto, H.D.F. (2000 [1951]) 'The polis', in R.T. LeGates and F. Stout (eds) *The City Reader*, 2nd edn, London, Routledge.

Kohn, M. (2004) *Brave New Neighbourhoods: The Privatization of Public Space*, London, Routledge.

Kooiman, J. (ed.) (1993) *Modern Governance: Government–Society Interactions*, London, Sage.

Kooiman, J. (2003) *Governing as Governance*, London, Sage.

Koolhaas, R. (1978) *Delirious New York*, New York, Oxford University Press.

Krieger, A. (1995) 'Reinventing public space', *Architectural Record*, 183(6): 76–7.

Kuo, F.E., W.C .Sullivan, R.L. Coley and L. Brunson (1998) 'Fertile ground for community: Inner-city neighborhood common spaces', *American Journal of Community Psychology*, 26(6): 823–51.

Lang, J. (1994) *Urban Design: The American Experience*, New York, Van Nostrand Reinhold.

Lang, J. (2005) *Urban Design: A Typology of Procedures and Products*, Oxford, Architectural Press.

Leach, R. and J. Percy-Smith (2001) *Local Governance in Britain*, Basingstoke, Palgrave.

Lees, L.H. (1994) 'Urban public space and imagined communities in the 1980s and 90s', *Journal of Urban History*, 20(4): 443–65.

Lefebvre, H. (1991) *The Production of Space*, London, Basil Blackwell.

LeGates, R.T. and F. Stout (eds) (2000) *The City Reader*, 2nd edn, London, Routledge.

Light, A. and J.M. Smith (eds) (1998) *Philosophy and Geography II: The Production of Public Space*, Lanham, MD, Rowman & Littlefield.

Littlefair, P.J., M. Santamouris, S. Alvarez, A. Dupagne, D. Hall, J. Teller, J.F. Coronel and N. Papanikolaou (2000) *Environmental Site Layout Planning: Solar Access, Microclimate and Passive Cooling in Urban Areas*, Watford, BRE.

Living Streets (2001) *Streets are for Living, The Importance of Streets and Public Spaces for Community Life*, London, Living Streets.

Llewelyn Davies (2000), *Urban Design Compendium*, London, English Partnerships/Housing Corporation.

Lloyd, K. and C. Auld (2003) 'Leisure, public space and quality of life in the urban environment', *Urban Policy and Research*, 21(4): 339–56.

Lofland, L. (1998) *The Public Realm, Exploring the City's Quintessential Social Territory*, New York, De Gruyte.

London Assembly (London Assembly Graffiti Investigative Committee) (2002) *Graffiti in London*, London, London Assembly.

Loukaitou-Sideris, A. (1996) 'Cracks in the city: Addressing the constraints and potentials of urban design', *Journal of Urban Design*, 1(1): 91–103.

Loukaitou-Sideris, A. and T. Banerjee (1998) *Urban Design Downtown: Poetics and Politics of Form*, Berkeley, CA, University of California Press.

Loukaitou-Sideris, A., R. Liggett and H. Iseki (2001) 'Measuring the effects of built environment on bus stop crime', *Environment and Planning B: Planning and Design*, 28(2):255–80.

Low, S. and N. Smith (eds) (2006) *The Politics of Public Space*, London, Routledge.

Luther, M. and D. Gruehn (2001) 'Putting a price on urban green spaces', *Landscape Design*, 303: 23–5.

Luttik, J. (2000) 'The value of trees, water and open spaces as reflected by house prices in the Netherlands', *Landscape and Urban Planning*, 48(3/4): 161–7.

Lynch, K. (1960) *The Image of the City*, Cambridge MA, MIT Press.

Lynch, K. (1981) *A Theory of Good City Form*, Cambridge, MA, MIT Press.

Lynch, K. and S. Carr (1991) 'Open space: Freedom and control', in T. Banerjee and M. Southworth (eds), *City Sense and City Design: Writings and Projects of Kevin Lynch*, Cambridge, MA, MIT Press.

Mace, R. (1976) *Trafalgar Square: Emblem of Empire*, London, Lawrence and Wishart.

McKay, T. (1998) 'Empty spaces, dangerous places', *ICA Newsletter*, 1(3): 2–3.

McNeill, D. (n.d.) 'Managing the Micropolis: A Political Geography of Business Improvement Districts in New York City', unpublished.

McNeill, D. (2003) 'Mapping the European urban left: The Barcelona experience', *Antipode*, 35(1): 74–94.

Maconachie, M. and K. Elliston (2002) *Morice Town Home Zone: A Prospective Health Impact Assessment*, Plymouth, Health and Community Research Programme, University of Plymouth and the South West Devon NHS Trust..

Madanipour, A. (1999) 'Why are the design and development of public spaces significant for cities?', *Environment and Planning B: Planning and Design*, 26(6): 879–91.

Madanipour, A. (2003) *Public and Private Spaces of the City*, London, Routledge.

Malone, K. (2002) 'Street life: Youth, culture and competing uses of public space', *Environment and Urbanization*, 14(2): 157–68.

Marsh, D. (ed.) (1998) *Comparing Policy Networks*, Birmingham, Open University Press.

Massam, B. (2002) 'Quality of life: Public planning and private living', *Progress in Planning*, 58(3): 141–227.

Massey, H. (2002) 'Urban farm', *Landscape Design*, 313: 40–1.

Mattson, K. (1999) 'Reclaiming and remaking public space: Towards an architecture for American democracy', *National Civic Renewal*, 88(2): 33–144.

Mean, M. and C. Tims (2005) *People Make Places: Growing the Public Life of Cities*, London, Demos.

Miethe, T. (1995) 'Fear and withdrawal from urban life', *Annals AAPSS*, 539(May): 14–27.

Minton, A. (2006) *What Kind of World are We Building? The Privatisation of Public Space*, London, RICS.

Mitchell, D. (1995) 'The end of public space? People's park, definitions of the public and democracy', *Annals of the Association of American Geographers*, 85(1): 108–33.

Mitchell, W.J. (1996) *City of Bits: Space, Place and the Infobahn*, Cambridge MA, MIT Press.

Mitchell, R.J., and M.D.R. Leys (1963 [1958]) *A History of London Life*, Harmondsworth, Penguin.

Montgomery, J. (1998), 'Making a city: Urbanity, vitality and urban design', *Journal of Urban Design*, 3(1): 93–116.

Moore, J. (2000) 'Placing home in context', *Journal of Environmental Psychology*, 20(3): 207–17.

MORI (2000) *Consumer Focus for Public Services – People's Panel Wave 5*, London, MORI.

MORI (2002) *The Rising Prominence of Liveability*, London, MORI.

MORI (2005) *Physical Capital, Liveability in 2005*, London, MORI.

Moudon, A.V. (1987), *Public Streets for Public Use*, New York, Columbia University Press.

Mumford, L. (1961) *The City in History: Its Origins, its Transformation, and its Prospects*, New York, Harcourt Brace Jovanovich.

Murphy, C. (2001) 'Customised quarantine', *Atlantic Monthly*, July-August: 22–4.

Myers, D. (1987) 'Community-relevant measurement of quality of life', *Urban Affairs Quarterly*, 23(1): 108–25.

New Economics Foundation (2004) *Clone Town Britain, The Loss of Local Identity on the Nation's High Streets*, London, New Economics Foundation.

Newman, O. (1973) *Defensible Space: People and Design in the Violent City*, London, Architectural Press.

Oc, T. and S. Tiesdell (1997), *Safer City Centres: Reviving the Public Realm*, London, Paul Chapman.

ODPM (Office for the Deputy Prime Minister) (2002) *Living Places: Greener, Safer, Cleaner*, London, ODPM.

ODPM (Office for the Deputy Prime Minister) (2003a) *Sustainable Communities: Building for the Future*, London, ODPM.

ODPM (Office for the Deputy Prime Minister) (2003b) *BIDs Guidance: A Working Draft*, London, ODPM.

ODPM (Office for the Deputy Prime Minister) (2004) *Living Places: Caring for Quality*, London, RIBA Enterprises/ODPM.

ODPM (Office for the Deputy Prime Minister) (2006) *State of English Cities: A Research Study*, London, ODPM.

ODPM (Office for the Deputy Prime Minister) and ATCM (Association of Town Centre Management) (2004) *Business Improvement Districts: A Guide from the ATCM*, London, ACTM.

Oldenburg, R. (1999) *The Great Good Place: Cafés, Coffee Shops, Bookstores, Bars, Hair Salons and the other Hangouts at the Heart of a Community* (2nd edn), New York, Marlowe & Company.

Oxford, D. (1995) *Piccadilly Circus*, Stroud, Chalford.

Painter, K. (1996) 'The influence of street lighting improvements on crime, fear and pedestrian street use after dark', *Landscape and Urban Planning*, 35(2–3): 193–201.

Papadakis, A. and H. Watson (eds) (1990) *New Classicism: Omnibus Edition*, London, Academy Editions.

Papayanis, M. (2000) 'Sex and the revanchist city: Zoning out pornography in New York', *Environment and Planning D: Society and Space*, 18(3): 341–53.

Peiser, R.B. and G.M. Schwann (1993) 'The private value of public open space within subdivisions', *Journal of Architectural and Planning Research* 10(2): 91–104.

Pellegrini, A.D. and P. Blatchford (2002) 'Time for a break', *The Psychologist*, 15(Part 2): 61–3.

Phillips, P.L. (2000) *Real Estate Impacts of Urban Parks*, Issue Paper, Washington, DC, Economic Research Associates.

Pierre, J. and B.G. Peters (2000) *Governance, Politics and the State*, London, Macmillan.

Pirenne, H. (2000 [1925]) 'City origins', in R.T. LeGates and F. Stout (eds) *The City Reader*, (2nd edn), London, Routledge.

Planning Officer's Society (2000) *A Guide to Best Value and Planning*, Barnsley, Barnsley Metropolitan Borough Council.

Pollitt, C., X. Girre, J. Lonsdale, R. Mul, H. Summa and M. Waerness (1999) *Performance or Compliance? Performance Audit and Public Management in Five Countries*, Oxford, Oxford University Press.

Project for Public Space (2000) *How to Turn a Place Around: A Handbook for Creating Successful Public Spaces*, New York, PPS.

Punter, J. (1991) 'Participation in the design of urban space', *Landscape Design*, 200: 24–7.

Quayle, M. and T.C. Dreissen van der Lieck (1997) 'Growing community: A case for hybrid landscapes', *Landscape and Urban Planning*, 39(2/3): 99–107.

Rapoport, A. (1990) *History and Precedent in Environmental Design*, New York, Plenum Press.

Rasmussen, S.E. (1934 [1960]) *London: The Unique City*, Harmondsworth, Penguin.

Reichl, A.J. (1999) *Reconstructing Times Square: Politics and Culture in Urban Redevelopment*, Lawrence, KS, University Press of Kansas.

Relph, E. (1976), *Place and Placelessness*, London, Pion.

Rhodes, R. (1994) 'The hollowing out of the state: The changing nature of the public service in Britain', *Political Quarterly*, 65(2): 138–151.

Rhodes, R. (1997) *Understanding Governance: Policy Networks, Governance, Reflexivity and Accountability*, Milton Keynes, Open University Press.

Richards, S., M. Barnes, A. Coulson, L. Gaster, B. Leach and H. Sullivan (1999) *Cross-Cutting Issues in Public Policy and Public Services*, London, DETR.

Roberts, M. and C. Turner (2005) 'Conflicts of liveability in the 24–hour city: Learning from 48 hours in the life of London's Soho', *Journal of Urban Design*, 10(2) 171–93.

Rogers, S. (1999) *Performance Management in Local Government: The Route to Best Value*, (2nd edn), London, Financial Times/Pitman Publishing.

Rogers, R. and M. Fisher (1990) *A New London*, London, Penguin.

Sagalyn, L.B. (2001) *Times Square Roulette: Remaking the City Icon*, Cambridge, MA, MIT Press.

Sassen, S. (1994) *Cities in a World Economy*, Thousand Oaks, CA, Pine Forge Press.

Schuster, J.M. (1995) 'Two urban festivals: La Mercé and First Night', *Planning Practice and Research*, 10(2): 173–87.

Sennett, R. (1977) *The Fall of Public Man*, London, Faber & Faber.

Sennett, R. (1990) *The Conscience of the Eye, The Design and Social Life of Cities*, New York, Alfred Knopf.

Shashua-Bar, L. and M.E. Hoffman (2000) 'Vegetation as a climatic component in the design of an urban street: An empirical model for predicting the cooling effect of urban green areas with trees', *Energy and Buildings*, 31(3): 221–35.

Shields, R. (1991) *Places on the Margin*, London, Routledge.

Shoard, M. (2003) 'The Edgelands', *Town and Country Planning*, May: 122–5.

Shonfield, K. (n.d.) 'At Home With Strangers: Public Space and the New Community' Working Paper 8 The Richness of Cities, Urban Policy in a New Landscape, London, Comedia & Demos.

Sibley, D. (1995) *Geographies of Exclusion*, London, Routledge.

Sircus, J. (2001) 'Invented places', *Prospect,* 81(Sept/Oct): 30–5.

Sitte, C. (1889 [1965]) *City Planning According to Artistic Principles* (translated by G.R. Collins and C.C. Collins), London, Phaidon Press.

Slattery, M.L., Potter, J. and Caan, B. (1997) 'Energy balance and colon cancer – beyond physical activity' *Cancer Research*, 57: 75–80.

Smith, N. (1996) *The New Urban Frontier: Gentrification and the Revanchist City*, London, Routledge.

Smith, N. (2002) 'New globalism, new urbanism: Gentrification as global urban strategy', *Antipode*, 34(3): 427–50.

Smith, T., M. Neiischer and N. Perkins (1997) 'Quality of an urban community: A framework for understanding the relationship between quality and physical form', *Landscape and Urban Planning*, 39(2/3): 229–41.

Smyth, H. (1994) *Marketing the City: The Role of Flagship Developments in Urban Regeneration*, London, E&FN Spon.

Sorkin, M. (ed.) (1992) *Variations on a Theme Park: The New American City and the End of Public Space*, New York, Hill and Wang.

Southworth, M. and E. Ben-Joseph (1997) *Streets and the Shaping of Towns and Cities*, New York, McGraw-Hill.

Sport England, Countryside Agency and English Heritage (2003) *The Use of Public Parks in England*, London, Sport England.

Starr, T. and E. Hayman (1998) *Signs and Wonders: The Spectacular Marketing of America*, New York, Doubleday.

Stationery Office, The (2003) *Local Government Act 2003*, London, TSO.

Stationery Office, The (2004) *Business Improvement Districts (England) Regulations 2004 (Statutory Instrument 2004 No. 2443)*, London, TSO.

Stewart, J. (2001) *Roads for People: Policies for Liveable Streets*, Policy Report 56, London, The Fabian Society.

Stoker, G. (2004) *Transforming Local Governance: From Thatcherism to New Labour*, Basingstoke, Palgrave.

Sullivan, H. and C. Skelcher (2002) *Working Across Boundaries: Collaboration in Public Services*, Basingstoke, Palgrave.

Takano, T., K. Nakamura and M. Watanabe (2002) 'Urban residential environments and senior citizens' longevity in megacity areas: The importance of walkable green spaces'. *Journal of Epidemiology and Community Health*, 56(12): 913–8.

Tames, R. (1994) *Soho Past*, London, Historical Publications.

Taylor, W.R. (ed.) (1991) *Inventing Times Square: Commerce and Culture at the Crossroads of the World*, New York, Russell Sage Foundation.

Taylor, A.F., A. Wiley, F.E. Kuo and W.C. Sullivan (1998) 'Growing up in the inner city: green spaces as places to grow', *Environment and Behaviour*, 30(1): 2–27.

Taylor, A.F., F.E. Kuo and W.C. Sullivan (2001) 'Coping with ADD – the surprising connection to green play settings', *Environment and Behaviour*, 33(1): 54–77.

thisislondon.co.uk/news (2007) 'George Orwell, Big Brother is watching your house', 31 March 2007.

Thrift, N. (2000) 'Commodities', in R. Johnston, D. Gregory, G. Pratt and M. Watts (eds), *The Dictionary of Human Geography*, Oxford, Blackwell.

Tibbalds, F. (1988), 'Ten commandments of urban design', *The Planner*, 74(12): 1.

Tibbalds, F. (2001), *Making People-Friendly Towns: Improving the Public Environment in Towns and Cities*, (2nd edn), London, Spon Press.

Times Square Alliance (2005) Times Square Alliance Quarterly Reports 2003–2005, New York, Times Square Alliance.

Times Square BID (1998) Economic Indicators/Annual Report 1998, New York, Times Square BID.

Times Square BID (2000) Times Square BID Annual Report 2000, New York, Times Square BID.

Times Square BID (2001) Times Square BID Annual Report 2001, New York, Times Square BID.

Trancik, R. (1986) *Finding Lost Space: Theories of Urban Design*, New York, Van Nostrand Reinhold.

Travers, T. and Weimer, J. (1996) *Business Improvement Districts: New York and London*, London, Corporation of London.

University of Newcastle-upon-Tyne (2001) *Improving Green Urban Spaces: Beacon Councils Research – Round 3 Theme Report*, London, DTLR.

University of Sheffield (1994) *Breaking the Downward Spiral: Current and Future Responsibilities of Children to their Town Centre*, Department of Landscape and Geography University of Sheffield, Sheffield.

Upmanis, H. (2000) 'The park has its own climate', *Swedish Building Research* 2: 8–10.

Urban Design Skills Working Group (2001) *Urban Design Skills Working Group*, London, CABE.

Urban Parks Forum (2001) *Public Park Assessment: A Survey of Local Authority Owned Parks Focusing on Parks of Historic Interest*, Caversham, Urban Parks Forum.

Urban Task Force (1999) *Towards an Urban Renaissance*, London, E&FN Spon.

van Kamp, I., K. Leidelmeijer, G. Marsman and A. de Hollander (2003) 'Urban environmental quality and human well-being: Towards a conceptual framework and demarcation of concepts; a literature study', *Landscape and Urban Planning*, 65(1/2): 5–18.

Van Melik, R., I. Van Aalst and J. Van Weesep (2007) 'Fear and fantasy in the public domain: The development of secured and themed urban space', *Journal of Urban Design*, 12(1): 25–42.

Wallin, L. (1998) 'Stranger on the green' in A. Light and J.M. Smith (eds) *Philosophy and Geography II: The Production of Public Space*. Lanham, MD, Rowman & Littlefield.

Watson, S. (2006) *Markets as Spaces for Social Interaction: Spaces of Diversity*, York, Policy Press.

Watts, M. (1999) 'Commodities', in P. Cloke, P. Crang and M. Goodwin (eds) *Introducing Human Geographies*, London, Arnold.

Webb, M. (1990) *The City Square*, London, Thames and Hudson.

Webster, C. (2001) 'Gated cities of tomorrow', *Town Planning Review*, 72(2): 149–70.

Welsh, B. and D. Farrington (2002) *Crime Prevention Effects of Closed Circuit Television: A Systematic Review*, Home Office, London.

Whitford Ennos, A.R. and J.F. Handley (2001) 'City form and natural process: Indicators for the ecological performance of urban areas and their application to Merseyside, UK', *Landscape and Urban Planning*, 57(2): 91–103.

Whyte, W.H. (1980), *The Social Life of Small Urban Spaces*, Washington DC, Conservation Foundation.

Whyte, W.H. (1988) *City: Rediscovering the Center*, New York, Doubleday.

Williams, K. and S. Green (2001) *Literature Review of Public Space and Local Environments for the Cross Cutting Review*, Final Report, London, DTLR.

Wilson, E. (1995) 'The rhetoric of urban space', *New Left Review*, 209: 146–60.

Wilson, J. and G. Kelling (1982) 'Broken windows', *Atlantic Monthly*, March: 29–36.

Woolley, H. (2003) *Urban Open Spaces*, London, Spon Press.

Woolley, H. and R. Johns (2001) 'Skateboarding: The city as playground', *Journal of Urban Design*, 6(2): 211–30.

Woolley, H., S. Rose, M. Carmona and J. Freeman (n.d.) *The Value of Public Space: How High Quality Parks and Public Spaces Create Economic, Social and Environmental Value,* London, CABE Space.

Worpole, K. (1999) 'Open all hours, like it or not', *New Statesman,* 26 April: xxvi-xxvii.

Worpole, K. (2000) *Here Comes the Sun: Architecture and Public Space in Twentieth Century European Culture,* London, Reaktion.

Worpole, K. and K. Knox (2007) *The Social Value of Public Spaces,* York, Joseph Rowntree Foundation.

Zeisel, J. (1984) *Inquiry by Design: Tools for Environment-Behaviour Research,* Cambridge, Cambridge University Press.

Zetter, R. and G. Butina-Watson (eds) (2006) *Designing Sustainable Cities in the Developing World,* London, Ashgate.

Zucker, P. (1959) *Town and Square: From the Agora to Village Green,* New York, Columbia University Press.

Zukin, S. (1991) *Landscapes of Power: From Detroit to Disney World,* Berkeley, CA, University of California Press.

Zukin, S. (1995) *The Cultures of Cities,* Cambridge, MA, Blackwell.

Index